U.S. Space-Launch Vehicle Technology

UNIVERSITY PRESS OF FLORIDA

Florida A&M University, Tallahassee
Florida Atlantic University, Boca Raton
Florida Gulf Coast University, Ft. Myers
Florida International University, Miami
Florida State University, Tallahassee
New College of Florida, Sarasota
University of Central Florida, Orlando
University of Florida, Gainesville
University of North Florida, Jacksonville
University of South Florida, Tampa
University of West Florida, Pensacola

U.S. Space-Launch Vehicle Technology

Viking to Space Shuttle

J. D. Hunley

UNIVERSITY PRESS OF FLORIDA

Gainesville · Tallahassee · Tampa · Boca Raton · Pensacola
Orlando · Miami · Jacksonville · Ft. Myers · Sarasota

Copyright 2008 by J. D. Hunley
Printed in the United States of America on acid-free paper
All rights reserved

13 12 11 10 09 08   6 5 4 3 2 1

Library of Congress Cataloging-in-Publication Data
Hunley, J. D., 1941–
U.S. space-launch vehicle technology: Viking to Space Shuttle/J.D. Hunley.
p. cm.
Includes bibliographical references and index.
ISBN 978-0-8130-3178-1 (alk. paper)
1. Launch vehicles (Astronautics)—United States—History—
20th century. 2. Rocketry—United States—History.
3. Space shuttles—United States—History. I. Title.
TL785.8.L3H87 2008
629.4709739–dc22   2007042544

The University Press of Florida is the scholarly publishing agency for the
State University System of Florida, comprising Florida A&M University,
Florida Atlantic University, Florida Gulf Coast University, Florida
International University, Florida State University, New College of Florida,
University of Central Florida, University of Florida, University of North
Florida, University of South Florida, and University of West Florida.

University Press of Florida.
15 Northwest 15th Street
Gainesville, FL 32611–2079
http://www.upf.com

This book is dedicated to Barnet R. Adelman, Wilbur Andrepont, Charles Bartley, Robert C. Corley, Daniel Dembrow, Ross Felix, Robert L. Geisler, Edward N. Hall, Charles Henderson, Kenneth W. Iliff, Karl Klager, Franklin H. Knemeyer, Grayson Merrill, Ray Miller, Edward W. Price, Milt Rosen, Ed Saltzman, Ronald L. Simmons, Ernst Stuhlinger, H. L. Thackwell, and Robert C. Truax.

Contents

Preface and Acknowledgments

This book and the volume that precedes it, *Preludes to U.S. Space-Launch Vehicle Technology: Goddard Rockets to Minuteman III*,[1] address a significant gap in the literature about access to space. There are numerous and quite excellent volumes covering various aspects of missile and space-launch-vehicle development and some general accounts. But there is no study that traces in a detailed and systematic way how the technology evolved from its beginnings with Robert Goddard and the German V-2 missile to the end of the cold war.

Another problem with the existing literature is the lack of agreement among sources about specifics. From measurements of length and diameter to those of thrust and accuracy, sources differ. These two books cannot claim to resolve the differences, but together they do acknowledge them in endnotes and indicate which sources seem most credible.

I first began working on these histories in 1992 when I undertook a much more modest, monographic study comparing the contributions to U.S. launch-vehicle technology of the Wernher von Braun group that developed the V-2 in Germany and then immigrated to America after World War II with those of the group around Theodore von Kármán and Frank Malina at the Jet Propulsion Laboratory near Pasadena, California. I quickly found that the literature would not permit such a comparison without a much broader assessment of rocket technology, which I now provide.

Because the material to be covered is so broad as well as technical, what I had originally conceived as a single volume had to be divided into two separate books. The first book, *Preludes to U.S. Space-Launch Vehicle Technology*, covers primarily missile development, because many of the launch vehicles borrowed technology and whole rocket stages from missiles, most of which developed before the launch vehicles. In a couple of cases, to provide continuity of coverage, early uses of missiles as components of launch vehicles are discussed in *Preludes to U.S. Space-Launch Vehicle Technology*. Then *Viking to Space Shuttle* steps back in time to pick up development of Viking and Vanguard, which preceded most of early missile production. This second book then continues with the Thor-Delta, Delta, Atlas, Scout, Saturn, and Titan space-launch vehicles, concluding with treatment of the space shuttle. Both books are written in such a way that they can be understood by a general audience, but I hope they will also prove useful to scholars, engineers, and others who already possess an extensive knowledge of at least some of the material covered.

Although the two books constitute a continuous whole, intended to be read as such, some readers primarily interested in either missiles or launch vehicles may want to read just one of the volumes. They will find much that is new in each. But readers primarily interested in missiles should be aware that I have not written a complete history of even the ballistic missiles that *Preludes to U.S. Space-Launch Vehicle Technology* mainly covers, while providing only limited treatment of some tactical missiles. Since my focus is on the technologies that contributed to launch-vehicle development, I spend comparative little time with the business end of missiles, the warheads. Instead, I concentrate on propulsion, structures, and guidance-and-control technologies because these carried over—though often with considerable modification—to launch vehicles.

By the same token, those primarily interested in launch-vehicle technology will find that they will miss out on much of the history by starting with *Viking to Space Shuttle*. It is obviously impossible to understand the Thor-Delta family of launch vehicles without knowledge of the Thor missile or the Atlas launch vehicles without knowing how the Atlas missile developed. Links between the Titan I and II missiles and the Titan launch vehicles are equally important. Even the Scout and Saturn launch vehicles borrowed much technology from missiles. And the solid rocket motors for the Titan III and IV as well as the solid rocket boosters for the space shuttle borrowed much technology from the Polaris and Minuteman missiles.

In researching and writing both volumes, I have incurred many debts of gratitude to an enormous number of people. I acknowledge many of them in endnotes, but unfortunately I can no longer remember everyone who assisted me in a great variety of ways. I owe particular gratitude to Roger Launius, who, as my boss at the NASA History Office, first encouraged me to begin studying the subject of rocket technology and has provided unfailing support in more ways than I can recount. (He also was wise enough to discourage me from attempting to cover the entire gamut of American missile and launch-vehicle technology, but I was not smart enough to follow that advice.) Lee Saegesser and Jane Odom, archivists at the NASA History Office, both provided extensive support, as have the rest of the archival staff there, including Colin Fries and John Hargenraether as well as their predecessors, Jennifer and Bill Skeritt.

After I was fortunate enough to become the Ramsey Fellow at the Smithsonian National Air and Space Museum, Mike Neufeld kindly read several chapters and offered me his criticism. Many of his colleagues at the museum, along with the archival staff, were also extremely helpful. I had valuable discussions with John Anderson, Paul Ceruzzi, Tom Crouch, David DeVorkin, Peter Jakab, Mike Neufeld, and Frank Winter about aspects of my work. Also very helpful were a host of archivists, librarians, curators, docents, and volunteers including Marilyn Graskowiak, Dan Hagedorn, Gregg Herken, Peter Jakab, Mark Kahn, Daniel Lednicer, Brian Nicklas, George Schnitzer, Leah Smith, Paul Silbermann, Larry Wilson, and Howard S. Wolko. I am thus very grateful to NASM for granting me the fellowship and thereby allowing me to complete my research for this book in addition to obtaining the assistance of all of these individuals.

Chapters 3, 5, and 9 of *Preludes to U.S. Space-Launch Vehicle Technology* plus chapters 6 and 7 of this volume contain material I published earlier in chapter 6 of *To Reach the High Frontier: A History of U.S. Launch Vehicles*, edited by Roger D. Launius and Dennis R. Jenkins (2002). The material in the present book incorporates much research done since I wrote that chapter, and it is organized differently. But I am grateful to Mack McCormick, rights manager at the University Press of Kentucky, for confirming my right to reuse the material that appeared in the earlier version.

Many people read earlier versions of this book and provided suggestions for improvement. They include Matt Bille, Roger Bilstein, John Bluth, Trong Bui, Virginia Dawson, David DeVorkin, Ross Felix, Mike Gorn, Pat Johnson, John Lonnquest, Ray Miller, Fred Ordway, Milt Rosen, Frank Winter, and Jim Young. Persons who furnished documents or other source materials that would have been difficult to locate without their assistance include Nadine Andreassen, Liz Babcock, Scott Carlin, Robert Corley, Dwayne Day, Bill Elliott, Robert L. Geisler, Robert Gordon, Edward Hall, Charles Henderson, Dennis Jenkins, Karl Klager, John Lonnquest, Ray Miller, Tom Moore, Jacob Neufeld, Fred Ordway, Ed Price, Ray Puffer, Karen Schaffer, Ronald Simmons, Ernst Stuhlinger, Ernie Sutton, Robert Truax, and P. D. Umholtz.

Librarians, historians, and archivists at many institutions assisted my research in a variety of ways. They include Air Force Historical Research

Agency archivist Archangelo Difante; Air Force Space and Missile Systems Center historian Harry Waldron and archivist Teresa Pleasant; Air Force Flight Test Center historians Jim Young and Ray Puffer; China Lake historian Leroy Doig; Clark University Coordinator of Archives & Special Collections Dorothy E. Mosakowski; Bill Doty at the National Archives, Laguna Niguel, California; Dryden Flight Research Center librarian Barbara Rogers; and JPL archivists John Bluth, Barbara Carter, Julie Cooper, and Margo E. Young. Several reference librarians at the Library of Congress should be added to the list, but I don't have their names.

Thanks are due to everyone who consented to interviews (included in endnote references and lists of sources). They not only provided their time, and often editorial comments on the transcribed interviews, but agreed to permit me to use the information in the interviews. In addition, many people discussed technical issues with me or provided other assistance, such as help with photographs. These include Ranney Adams, Wil Andrepont, Stan Backlund, Rod Bogue, Al Bowers, George Bradley, Mark Cleary (who went to great lengths to provide me with nineteen photographs), Robert Corley, Daniel Dembrow, Robert L. Geisler, Mike Gorn, Mark L. Grills, John Guilmartin, Burrell Hays, Charles Henderson, J. G. Hill, Cheryl Hunley, Michael Hunley, Ken Iliff, Fred Johnsen, Karl Klager, Franklin Knemeyer, Tony Landis, Niilo Lund, Jerry McKee, Ray Miller, Tom Moore, Sarah Parke, Ed Price, Milt Rosen, Jim Ross, Bill Schnare, Carla Thomas, Woodward Waesche, Herman Wayland, and Paul Willoughby. To all of the people above, I offer my thanks for their generous help.

I would like to express my special appreciation to James R. Hansen and an anonymous reviewer of this book for their generous comments and their suggestions for improvement. Both have been extremely helpful. I also want to say thanks to Jim Hansen for his encouragement of and support for my work on both aeronautics and space over the years since we first became acquainted and for his publications, which have greatly influenced my work.

I am deeply indebted to Ann Marlowe for her thoroughness, passion for correctness, kindness, and thoughtfulness throughout the grueling process of copy editing both books. I thank Jacqueline Kinghorn Brown for her patience and her exceptional work as project editor, editor in chief John W. Byram for advice and help on many matters, and managing editor Gillian Hillis and everyone else at the University Press of Florida for their contributions to the two volumes.

Last but not least, I thank my agent and friend Neil Soderstrom for his encouragement, editorial advice, fruitful suggestions, help in finding a publisher for both books, and, above all, friendship through thick and thin in the face of many demands on his time that should have taken precedence over helping me. He has made me look at these books in a different way than would have occurred to me on my own. They owe much to him and his generosity of time, hard work, and spirit.

There are undoubtedly many other individuals who have assisted my research and writing over the years but whose help or names have been erased by the passage of time (and old age) from my active memory bank. I can only apologize for the oversight and say a generic "thank you very much." None of the people acknowledged here bears any responsibility for the details and interpretations that appear in these two books. For them, I alone am responsible. But I hope these generous individuals will approve of the way I have used the suggestions, comments, materials, and information they provided or helped me to find.

Introduction

Note: As these pages reprise the introduction to *Goddard Rockets to Minute-man III: Preludes to U.S. Space-Launch Vehicle Technology*, readers of the previous volume may wish to skip to the concluding two paragraphs.

Although black-powder rockets had been around for centuries, it was not until 1926 that American physicist and rocket developer Robert H. Goddard launched the first known liquid-propellant rocket. Despite this auspicious beginning, not until the mid-1950s did the United States begin to invest significant resources in rocket development. Already by the end of January 1958, the United States had launched its first satellite, and within a generation it had developed a series of missiles and launch vehicles of enormous power and sophistication. The Atlas, Titan, Scout, Delta, Saturn, and space shuttle launched a huge number of satellites and other spacecraft that revolutionized our understanding of the universe, including our own planet, and brought events and reporting from all parts of the world into the American living room with unprecedented speed. How could the United States have advanced so rapidly from the relatively primitive rocket technology available on a small scale in the mid-1950s to the almost routine access to space available by the 1980s?

This book and the preceding volume, *Preludes to U.S. Space-Launch Vehicle Technology: Goddard Rockets to Minuteman III*, attempt to answer that question and to trace the convoluted technological trajectory from Goddard's imaginative but problem-prone early rockets to the huge Saturn V and the complex space shuttle, among other launch vehicles. The history of these vehicles has been punctuated by failures on the path to overall success. But on the whole, the achievements have been remarkable.

Perhaps most remarkable have been the unique features of the space shuttles. As the United States approaches the end of shuttle flights in 2010, it is appropriate to reflect that in some ways this astonishing but troubled launch vehicle and spacecraft was the culmination of the development process discussed in these two volumes. It represented a bold dream of converting previously expendable missile and launch vehicle technologies into a reusable source of routine access to and return from space, analogous to airliners and large cargo aircraft for the nearer skies. In one sense the effort was a failure, since the Air Force continues to rely on expendable launch vehicles and NASA is retreating, under budgetary and safety pressures, from reusability to a concept akin to the Saturn launch vehicles of the Apollo era.

In another sense, however, the crew and cargo launch vehicles Ares I and V (discussed in the final chapter of this book) are themselves a legacy of the shuttles since they will use shuttle experience and technology as part of the basis upon which to build a more affordable and safer way to return to the Moon and even go to Mars.

It is also worthwhile to recognize that many of the achievements of the space shuttles would have been extraordinarily difficult to accomplish without the unique features built into the shuttles. To give but one example, after space shuttle *Discovery* launched the Hubble Space Telescope in April 1990, it quickly became apparent that the enormous promise of this astronomical instrument was marred by a small but critical flaw in its primary mirror. Following a partial correction by computer enhancement, a planned routine repair mission turned into a rescue mission in December 1993 in which the huge telescope was recaptured in the payload bay of space shuttle *Endeavour*, outfitted with a corrective mechanism for the optics of the primary mirror, and serviced in other ways to allow the scientific instrument to continue to function as originally envisioned. Some 1,200 women and men were involved in orchestrating, designing, practicing for, and carrying out this delicate and complex resuscitation effort, which could hardly have been performed by an expendable launch vehicle coupled with any other existing spacecraft. *Endeavour*'s astronauts used five spacewalks to install the device using additional mirrors to correct Hubble's optics as well as to replace failed gyroscopes and the wide field/planetary camera. They then installed equipment to improve the telescope's failing computer memory, among other things. Although Hubble had been providing important new scientific data even before the rescue mission, afterwards it began to live up to and even exceed the performance that astronomers had expected from it, including provision of the first solid evidence for the presence in space of black holes (regions of intense gravitational force) and of spectacular images that graced the pages of newspapers and even appeared on a cover of *Newsweek.*

Literally millions of people had watched as the shuttle astronauts performed their repairs, and the entire team responsible for the mission received the 1993 Robert J. Collier Trophy from the National Aeronautic Association "for outstanding leadership, intrepidity, and the renewal of public faith in America's space program by the successful orbital recovery and repair of the Hubble Space Telescope." This and other almost equally astonishing achievements showed the unique value of the space shuttle as the fruition of a comparatively short but intense period of development of space launch capabilities.[1]

Although the focus of these two books is on technology used by launch vehicles, which permitted space exploration such as that carried out by Hubble, *Preludes to U.S. Space-Launch Vehicle Technology* is mostly about missiles and can be read by itself as a history of missile technology. *Viking to Space Shuttle*, likewise, can serve as a self-standing history of launch vehicle technology, although most readers may also want to read *Preludes to U.S. Space-Launch Vehicle Technology* for the technology upon which that for launch vehicles was significantly based. Missiles follow trajectories aimed at places on Earth instead of the heavens; they carry warheads instead of satellites or spacecraft. But especially in the area of propulsion, they use much the same technology as launch vehicles. In fact, many launch vehicles have been converted missiles. Others have borrowed stages from missiles.

One major irony stands out in this process. While the complexity and sophistication of missiles and launch vehicles gave birth to the expression "rocket science," careful study of the vehicles' development reveals many instances in which the designers and operators encountered problems they did not fully understand. They frequently had to resort to trial-and-error fixes to make their rockets perform as intended. Although data about and understanding of the advancing technologies continually increased, each large jump in scale and performance introduced new difficulties. Rocketry was, and is, as much an art as a science, fitting the description of engineering—as distinguished from science—provided by Edwin Layton, Walter Vincenti, and Eugene Ferguson, among others. (Besides engineering as art, these scholars also emphasized engineering's focus on doing rather than knowing, on design of artifacts rather than understanding the universe, and on making technological decisions in the absence of clear understanding— all features that distinguish engineering from science in their view.)[2]

This is not to say that science and scientists did not contribute to rocket technology. For example, Ronald L. Simmons earned a B.A. in chemistry at the University of Kansas in 1952 and went on to work for thirty-three years as a propulsion and explosives chemist with the Hercules Powder Company, a year with Rocketdyne, and thirteen years with the U.S. Navy at Indian Head, Maryland. Among other projects, he worked on upper stages for Polaris, Minuteman, Poseidon, and Trident.

In 2002 he wrote, "I consider myself to be a chemist . . . even though my work experience has been a lot of engineering. I really believe the titles are arbitrary, though I consider myself a scientist rather than an engineer." He added in relation to the issue of rocket engineering versus rocket science, "'Tis amazing how much we don't know or understand, yet we launch large rockets routinely . . . and successfully . . . that is when we pay attention to details and don't let the schedule be the driving factor. . . . By and large, I believe that we understand enuff to be successful . . . yet may not understand why." Although he spent much of his career working with double-base propellants—primarily those using nitrocellulose (NC) and nitroglycerin—he admitted, "There is much no one understands about nitrocellulose (and black powder for that matter) in spite of the fact that NC has been known since 1846 and black powder since before 1300!"[3]

Chronologically, the two books follow the development of American rocket technology through the end of the cold war in 1989–91. Chapter 2 of *Preludes to U.S. Space-Launch Vehicle Technology* covers the German V-2 because it became one of the foundation stones for U.S. rocket technology. Many of the V-2's developers immigrated to the United States after the end of World War II. They became the nucleus of the later NASA Marshall Space Flight Center. Under the leadership of Wernher von Braun, many of these Germans (along with hundreds of Americans) oversaw the development of the Saturn launch vehicles that lifted twelve astronauts on their journey to the Moon in the Apollo program.

The present book ends about 1990–91 with the close of the cold war because, after that, launch vehicle development began a new chapter. Funding became more spartan, and the United States began borrowing technology from the Russians, who had competed with American missile and launch-vehicle technology during the Soviet era.

Most readers of this book will have watched launches of the space shuttle or other launch vehicles on television. For those less familiar with the fundamentals of rocketry, this may help: Missiles and other rockets lift off from Earth through the thrust created by the burning of propellants (fuel and oxidizer).[4] This combustion creates expanding exhaust products, mostly gaseous, that pass through a nozzle at the back of the rocket. The nozzle contains a narrow throat and an exit cone that cause the gases to accelerate, thereby increasing thrust. The ideal angle for the exit cone depends on the altitude and pressure at which it will operate, with different angles needed at sea level than at higher altitudes where the atmosphere is thinner and the outside (ambient) pressure is lower.

Rockets in the period covered by this book typically used multiple stages to accelerate the vehicle all the way to its designed speed. When the propellants from one stage became exhausted, that stage would drop off the stack, so that as succeeding stages took over, there was less weight to be propelled to higher speeds. Multiple stages also permitted using exit cones of varying angles for optimal acceleration at different altitudes.

Most propellants required an ignition device to begin combustion, but hypergolic fuels and oxidizers ignited upon contact, dispensing with the need for an igniter. These types of propellants typically had less propulsive power than the extremely cold (cryogenic) liquid oxygen and liquid hydrogen, but they required less special handling than their cryogenic counterparts. Liquid oxygen and liquid hydrogen would boil off if not loaded just before launch, so they needed a lot more preparation time before a launch could occur. Hypergolic propellants, by contrast, could be stored in propellant tanks for comparatively long periods, allowing almost instant launch upon command. This was an especial advantage for missiles, and for spacecraft launches that had narrow "windows" of time, when the desired trajectory was lined up with the launch location only for a short period as Earth rotated and circled the Sun.

Solid-propellant missiles and rockets also enjoyed rapid-launch capabilities. They were much simpler and usually less heavy than liquid-propellant rockets because the fuel and oxidizer filled the combustion chamber without a need for propellant tanks, high pressure or pumps to force the propellant into the chamber, extensive plumbing, and other complications. Typically, technicians loaded a solid propellant into a combustion chamber with thin metal or a composite structure as the case, insulation between the case and the propellant, and a cavity in the middle where an igniter started combustion. Engineers designed the internal cavity to provide optimal thrust, with more exposed propellant surface providing more instant thrust and a smaller amount of surface providing less initial thrust. The propellant burned from the inside toward the case, with the insulation protecting the case as the propellant burned outward. The disadvantage of solids was the difficulty of stopping and restarting combustion, which could be done with valves in the case of liquids. Thus, for launch vehicles, solids usually appeared as initial stages, called stage 0, to provide maximum thrust for the initial escape from Earth's gravitational field or as upper stages (although the Scout remained a fully solid-propellant, multistage launch vehicle from 1960 to 1994).

Liquid propellants found more frequent use for the core stages, usually stage 1 and often stage 2, of launch vehicles such as the Atlas, Titan, Delta, and space shuttle. They also served in upper stages that needed to be stopped and restarted in orbit for insertion of satellites and spacecraft into particular orbits or trajectories. But the process of injecting fuels and oxidizers into the combustion chamber proved to be fraught with problems. For reasons that have been difficult for engineers to understand, mixing the two types of propellants in the needed proportions frequently resulted in oscillations that could destroy the combustion chamber. Known as combustion instability, this severe problem only gradually yielded to solutions—each scaling up of a particular type of engine usually causing new problems that required their own specific solutions.

Solid propellants also experienced combustion instability. Problems with solids were somewhat different from those with liquids. But as with liquids, the solutions required much research and, often, trial and error before they could be solved, or at least ameliorated.[5]

Besides propulsion systems, rockets required structures that would withstand the high heats of combustion, intense dynamic pressures as the vehicles accelerated through the atmosphere, shock waves as they passed through the speed of sound (referred to as Mach 1), and aerothermodynamic heating from friction while traveling at high speeds through the atmosphere. Because weight slowed acceleration to orbital speeds and altitudes, structural issues required much research to find lighter materials that would still withstand the rigors of launch. Engineers gradually found new materials that were strong, heat resistant, light, and, if possible, affordable.

Another field of research was aerodynamics. Missiles and launch vehicles needed to have as little drag (friction from the atmosphere that slowed flight and increased temperatures) as possible. They also had to be steerable by means of vanes, canards, moveable fins, vernier (auxiliary) and attitude-control rockets, fluids injected into the exhaust stream, and/or gimballed (rotated) engines or nozzles.

Associated with these types of control devices were various guidance and control systems incorporating computers programmed to adjust steering and keep the missile or launch vehicle on course. Such systems varied greatly in design and weight. They involved increasingly sophisticated programming of the computers. But they were essential to the success of both missiles and launch vehicles.[6]

Chapter 1 of *Preludes to U.S. Space-Launch Vehicle Technology: Viking to Space Shuttle* introduces the two rocket pioneers who had the greatest influence on American missiles and launch vehicles, the American physicist and rocket experimenter Robert H. Goddard and the Romanian-German rocket theorist Hermann Oberth. Both were fascinating characters with highly inventive minds. Although Goddard's innovations foreshadowed many later rocket technologies, his failure to publish many details of his research and development during his lifetime limited his influence. Oberth published his more theoretical conceptions in greater detail and had real influence on Wernher von Braun and other Germans who developed the V-2 missile before and during World War II and then immigrated to the United States. Through them, Oberth arguably had greater influence on U.S. missile and launch-vehicle development than did Goddard. But it can also be argued that they had a synergistic effect, with Goddard providing an example of how to develop rockets, at least to a point, while Oberth provided more theoretical details about rocket development in sources that he published early enough for them to be consulted by early rocket developers.

Although the V-2 was only one of many influences on American rocket technology, it was important, if sometimes overrated. Chapter 2 of *Preludes to U.S. Space-Launch Vehicle Technology* discusses the development of this missile and provides the technical information needed for later analysis of ways in which the V-2 was and *was not* a stepping-stone for American rocketry. Chapter 3 covers rocket development in the United States before, during, and shortly after World War II at what became the Jet Propulsion Laboratory (JPL) near Pasadena, California. Chapter 4 discusses other American rocket efforts from 1930 to 1954, culminating in a joining of German and JPL rocket technologies in the Bumper WAC project. Meanwhile, other efforts during and after World War II yielded important solid-propellant innovations that paralleled those at JPL, which had worked on both liquid and solid propellants. These developments at JPL and elsewhere form the subject of chapter 5.

Chapter 6 of *Preludes to U.S. Space-Launch Vehicle Technology* covers the Redstone missile and its modification into the first stage of the Juno I launch vehicle that placed the first U.S. satellite in orbit on January 31, 1958. The upper stages of the Juno I employed JPL technology, which again blended with that of Wernher von Braun's team in Alabama as they had on the Bumper WAC. The Redstone itself combined German and American contributions to rocketry, including some from the Air Force's Navaho missile. Chapters 7 through 9 cover the Atlas, Thor, Jupiter, Titan I and II, Polaris, and Minuteman missiles that produced still other technologies and separate stages used on later launch vehicles. Without the cold war and the developments it prompted, these contributions to launch vehicle technology would have evolved far more slowly they did.

Chapter 1 of *Viking to Space Shuttle* covers the Viking and Vanguard programs, separate American rocket efforts that contributed importantly to missile and launch vehicle technology, including the use of gimbals for steering. Chapters 2 through 7 of this book then discuss the uses of missile technology in development of the Delta, Atlas, Scout, Saturn, Titan, and space shuttle launch vehicles. Development of these vehicles, and of the missiles that preceded or were contemporaneous with them, was by no means problem-free. Besides cold-war threats and resultant funding, factors in the evolution of rocket technology in the United States included the efforts of both technical rocket engineers and others who were often less intimately involved with technical matters than with engineering the social aspects of missile and launch-vehicle development by promoting that cause in Congress, the Pentagon, and the media (as, previously, the V-2 was advocated to the Nazi regime in Germany). Without the promotional efforts of these so-called heterogeneous engineers, even the cold war might not have been enough to overcome the inertia that stood in the way of complex and expensive development, punctuated by many well-publicized failures in the early years.

Other factors in the rapid development of U.S. rocketry included both rivalry between military services (and the rocket firms that supported them) and, at the same time, a high degree of cooperation and sharing of knowledge by the competitors. A wide range of disciplines was required to design and develop rockets, calling for unselfish collaboration among competitors in solving problems that occurred during tests and operational launches. Universities also played a role in this process, as did a growing technical literature. Relatedly, the movement of personnel between firms (carrying technical knowledge from one project to another), professional networks, and federal intellectual property arrangements all helped promote and transfer innovation, leading to increasingly powerful and sophisticated launch vehicles that placed satellites in orbit and sent spacecraft on missions to explore our solar system and beyond.

A final contributing factor to the rapid and ultimately successful development of launch vehicle technology was a variety of management systems that helped to integrate efforts on the many systems in missiles and rockets, to keep them on schedule, and to promote configuration and cost control. All of these factors and others are discussed in these two books.

Both books are organized essentially by project. As the dates in the chapter titles will suggest, there was a great deal of overlapping in time between projects. Since these projects borrowed technologies from one another, there is an inherent problem with presenting technical materials in such a way that readers not highly familiar with the history of rocket technology can easily follow the story. The problem is compounded by the fact that different systems on a given missile or rocket were developed simultaneously. Thus it is impossible to follow a strictly chronological path in the narrative. Even if that could be done, the result would hardly be comprehensible. To assist the reader, I have included in the present book an appendix of Notable Technological Achievements for both missiles and launch vehicles and a glossary of technical terms and acronyms. The roughly chronological list of achievements may provide background for those who have not read *Preludes to U.S. Space-launch vehicle Technology*, and it can serve as a refresher if earlier events are imperfectly remembered during the reading of this volume.

Let me conclude by saying that although I have spent many years delving in dusty archives and poring over highly technical literature to gather the materials for both books, I am acutely aware that my narrative does not and cannot represent the final word on the subject. Practitioners of rocket design and development are extremely numerous. I could interview or locate interviews of only a small fraction of them, and many sources exist in scattered places I could not locate. My experience suggests that many rocket engineers know only small parts of the story I have told and that their memories do not always coincide with those of colleagues in other firms or military services. I hope that the material I have been able to assemble will stimulate others to build on or correct what I have contributed.

1

Viking and Vanguard, 1945–1959

Although only a sounding rocket, the Navy's Viking made direct contributions to launch vehicle technology. It was also the starting point for America's second launch vehicle, Vanguard. Often regarded as a failure, Vanguard
did launch more than one satellite. Together with Viking, it pioneered use of
gimbals for steering. In addition, its upper stages contributed significantly
to the evolution of launch vehicle technology, since they were converted for
use with the Thor-Delta series of space boosters. Additionally, a variant of
its third stage was modified for use in the Scout program. This stage of Vanguard (in one of its two versions) pioneered the use of fiberglass cases and
itself contributed to upper-stage technology for military missiles.[1]

Viking

The Viking rocket contributed to the Vanguard first stage as well as to the
Vanguard guidance system. Milton W. Rosen, who was responsible for the
development and firing of the Viking rockets, went on to become technical director of Project Vanguard and then director of launch vehicles and
propulsion in the Office of Manned Space Flight Programs for NASA. His
work developing Viking and his experiences with it prepared him for his responsibilities with Vanguard and beyond. Finally, Viking came early enough
in the history of American rocketry that it illustrates a good deal about the
evolution of the technologies used on launch vehicles. For all of these reasons, the history of Viking and Rosen's involvement with it deserve a place
in this history. They constitute an early case study of technology transfer and
the process of rocket development.[2]

   After receiving his B.S. degree in electrical engineering at the University
of Pennsylvania in 1937 and working briefly for Westinghouse and other
firms, Rosen found employment at the Naval Research Laboratory (NRL) in

1940. NRL was an interesting institution set up on July 2, 1923, "to conduct programs in the physical sciences and related fields directed toward new and improved materials, equipment, technology, and systems for the Navy." It was a place where researchers could create areas of research "that were at once technologically important and scientifically interesting." Rosen was later to do this after World War II, a conflict that convinced the Navy its "scientists needed to be concerned with predicting, even defining, and solving problems of the next war rather than the last one."

Meanwhile, during World War II at NRL, Rosen worked under Ernst H. Krause on guidance systems for missiles. At the end of the war, the group under Krause began to plan its future, and Rosen, who had been reading G. Edward Pendray's proposal to use rockets for exploration of the upper atmosphere, suggested that Krause's group do just that. Krause supported the idea, and at the end of 1945 the Rocket-Sonde Research Branch came into existence.[3]

Rosen's background was in electronics, but if the group were to develop the sounding rocket he had proposed, it needed an expert in rocketry. Since the field was in diapers in the United States, it seemed unlikely that Krause would be able to recruit someone with that experience, so he asked Rosen to learn the field. Rosen knew that since he had proposed the idea of a rocket, Krause would not let him off the hook easily. He decided to agree but pose a condition he thought his boss could not meet: permitting him to spend a year at an organization with the most knowledge in the United States about rocketry, JPL. To his surprise, Krause agreed.

This conversation took place in November 1945. Rosen, Krause, and another colleague traveled to JPL in March 1946, and the lab agreed to add Rosen to its staff. Meanwhile, Rosen and his colleague C. H. Smith read "the JPL handbook, which was the primer of rocketry of that time." Rosen and Smith began to plan their own rockets when they learned of the captured V-2s and the plans to fire them at White Sands Proving Ground in New Mexico, a place they had never heard of.[4]

Krause became the first chair of the V-2 Upper Atmosphere Research Panel and thus was involved in the research with the German rockets at White Sands. But for Rosen the immediate effect of this project was to relieve the pressure to produce an upper atmosphere research rocket until the V-2 firings were completed. On the other hand, Krause recognized that research with the V-2s would end sooner or later, so he told Rosen to continue developing NRL's own rocket. Rosen also became aware that once the V-2s reached the upper atmosphere, their stabilizing fins no longer served to keep their noses pointed upward. Some somersaulted in the thin stratosphere and ionosphere, spoiling data scientists had hoped to gather from equipment like solar spectrographs.

Rosen and his colleagues wanted their rocket to be designed specifically for such research, unlike the V-2 missile. But Rosen's specifications for a rocket about a third of the V-2's weight with a thrust of 20,000 pounds (compared with 56,000 for the V-2) would show some influence of the German rocket as well as what little he knew of Goddard's work.[5]

The group under Krause actually came up with two basic designs, one by Rosen that became the Viking, and another by Smith that resembled the Aerobee sounding rocket. They sent specifications for both vehicles to five companies, and received bids from three—General Electric (GE), Douglas, and Martin. By the time the bids arrived, Krause, Smith, and Rosen had decided to go forward with the Viking concept, and the lab selected the Glenn L. Martin Company to build it. They did this because (1) the Baltimore firm was close to the Washington location of NRL, so the engineers under Rosen, who became head of the rocket research branch, could work closely with the prime contractor, and (2) they liked the confidence and optimism of the young Martin engineers, even though they could claim limited experience with rockets.[6]

Rosen also got the engine contract for Viking (originally called Neptune) assigned to Reaction Motors, Inc., again because of the firm's general proximity to Washington and because he and his associates liked the spirit of RMI's engineers. The contract, initiated in September 1946, was for a 20,000–pound rocket engine with the propellants fed by turbine-driven pumps. With the contracts in the works, in August 1946 Rosen left Smith as Viking "caretaker" and headed for JPL, arriving in September.[7]

Rosen spent about eight months at JPL. Still paid by the Naval Research Laboratory, he worked in Martin Summerfield's liquid rocket group, where Richard Canright taught him a bit about rocket testing and turned him loose on a project involving ceramic liners in rocket motors. Rosen selected it for his work because it would require him to design, assemble, and test rockets. Using nitric acid and aniline as propellants, he and his associates conducted several hundred static firings of 300–pound-thrust rockets, each test lasting three to four minutes, a long stretch for that period. The hypergolic propellants were reliable but dirty, requiring Rosen to disassemble and clean the motor after each test. Sometimes he encountered burn-throughs, and he had to analyze the causes in a paper he and Canright wrote. In the process, he learned a great deal about heat transfer in rocket motors.

He also witnessed static firings of the WAC Corporal rocket to test the engine, propellant lines, tanks, valves, and control system. Rosen was sufficiently convinced of the importance of such tests that he decided to have every Viking rocket statically tested on the launchpad immediately before launching.

Besides his own work at JPL and what he witnessed, Rosen took two courses at Caltech. One—on theoretical physics from the Nobel Prize–winning Carl Anderson, who had discovered the positron in 1932 and later worked on Caltech's Eaton Canyon rocket project during the war[8]—consisted entirely of problem solving, which Rosen found useful. From Hans Liepmann, whom Rosen described as "another great teacher at Caltech," he took a course in supersonic aerodynamics, a field that was new to Rosen and one he found highly applicable to his later rocket development.[9]

Rosen and others had also spent several hours with von Braun soon after the charismatic German engineer had come to the United States. Rosen did not remember much of the conversation, but he did recall that von Braun knew Rosen was developing a rocket and asked how it would be built. Rosen said they would construct it of aluminum. Von Braun believed the skin heating would be a problem, but Rosen was confident he and his team could make the aluminum work—as, in fact, they did. Early structural studies for the rocket led to selection of a simple cylindrical shape with a 25–degree cone at the nose. Wind-tunnel tests reduced the size of the tail fins, although they remained large. But mere practicality determined the initial diameter of the fuselage, a slim 32 inches. The widest available sheet of rolled aluminum was 100 inches, which yielded a cylinder almost 32 inches across. Von Braun also, according to Rosen, suggested the idea of a gimballed engine but pointed out (incorrectly) that Goddard had tried to use one and had not succeeded in making it work.

During Rosen's time at JPL, not a great deal of progress was made in developing Viking. It took a long time between letting the contract and cutting/bending metal. But both Krause and Smith were clamoring for his return, which occurred in April 1947 as soon as he had finished his project and courses.[10]

In some respects, Viking followed the design of the V-2, although with important variations. For example, as with the V-2, the American rocket's propellants were alcohol and liquid oxygen, pumped into the combustion chamber by turbines driven by decomposed hydrogen peroxide. But where the V-2 had used alcohol at 75 percent strength and hydrogen peroxide at 82 percent, the Viking used 95–percent ethyl alcohol and 90–percent hydrogen peroxide. More significant, the really important elements of Viking—the nature of the guidance system, the gimballed motor, the post-cutoff stabilization—owed nothing to the V-2 but were original American designs.[11]

John Shesta drew up the basic RMI engine design, basing it on the firm's experience but also on recent information about the V-2. RMI's Edward A. Neu did the detailed design work on the combustion chamber and injector. Tests caused parts to fail and need replacement. Burn-throughs of the steel combustion-chamber liner (inner wall) led to the substitution of pure nickel, the first known use of this metal for such a purpose. Its superior thermal conductivity and higher melting point solved the heating problem in conjunction with the regenerative cooling in the original design. One injector caused an explosion, requiring new designs. Valves were a problem until M. E. "Bud" Parker borrowed valve designs from an MX-774B engine RMI was designing for the Air Force. Even after Rosen began launching Viking rockets from White Sands on May 3, 1949, there were component failures and redesigns. In fact, each of the twelve Viking rockets fired in the program (the last being on February 4, 1955) was different from the previous one as engineers learned from problems on one launch and applied the lessons to the next. In Rosen's view, "this was the most important aspect of the Viking program." The result, among other things, was an engine that developed 20,450 pounds of thrust on the first flight and 21,400 on two others.[12]

Even so, the engine itself (apart from its steering mechanism) made no known contributions to launch vehicle technology other than the experience it afforded to Martin and NRL engineers. But Rosen claimed Viking was the first large rocket to use a gimballed engine for steering, a technology that did find widespread emulation in large missiles and launch vehicles. The validity of Rosen's claim hinges in some degree on the definitions of large and gimballed, because Karel J. "Charlie" Bossart, the "father of the Atlas," has been credited with developing a precursor to gimballing in the MX-774 sounding rocket, which first flew on July 13, 1948, at White Sands, almost a year before the first Viking. This earlier technique was swiveling—rotating a rocket engine in a single axis for steering.[13]

The MX-774 project has been described in the preceding volume, *Preludes to U.S. Space Launch-Vehicle Technology*, in conjunction with Atlas. Here it will suffice to mention that the Army Air Forces (predecessor of the Air Force) awarded the Consolidated Vultee Aircraft Corporation (Convair) a study contract on April 2, 1946, for a missile designated MX-774. Bossart, the project manager, patterned it after the V-2 but with some radical innovations, only one of which was swiveling engines. A drastic funding cutback ordered by President Harry S. Truman in December 1946 curtailed the project, and it was cancelled in July 1947. Meanwhile Bossart's team had built some test vehicles that were 31 feet 7 inches long, had a gross weight of 4,160 pounds, and were powered by a four-cylinder RMI engine rated at 8,000 pounds of thrust but actually operating over a range from 7,600 to 8,400 pounds. Each of the four cylinders could swing back and forth on an axis to provide control in pitch, yaw, and roll. After cancellation, the Air Force permitted Convair to fly three test vehicles. The flights, on July 13, September 27, and December 2, 1948, all had problems but did demonstrate the viability of Bossart's designs, including the swiveling of the engines.[14]

The swiveling engines on the MX-774 test vehicles clearly constituted an important innovation. Jet vanes like those on the V-2 and later on the Redstone were eroded by the missile's exhaust, attenuating control. Moreover, swiveling the rocket engines did not significantly decrease their thrust, whereas jet vanes produced drag that did reduce propulsion efficiency. But swiveling was merely a precursor of gimballing, not the same thing. A gimbal is a more complex device than a swivel, permitting a rocket engine (in this case) to rotate in two directions rather than on a single axis. Thus, in this respect, the Viking was more advanced than Bossart's vehicle, and gimballing became the technique of the future.[15]

In addition, the Viking was a larger rocket than the MX-774. Its dimensions varied, but the early Vikings were almost 43 feet long and had a gross weight of 14,912 pounds. Their thrust more than doubled the MX-774's.[16] So Rosen's claim stands.

The initiative for gimballing came from Martin. The Viking specification required use of movable jet vanes to provide stability, but in 1946 the Martin engineers compared jet vanes with gimballing and recognized that jet vanes not only reduced control and thrust but also imposed a weight penalty of some 275 pounds. So in a report dated January 8, 1947, the engineers recommended gimballing. Rosen as project director concurred. The gimballing system that resulted was not as sensitive as the one later developed for Vanguard. It did not permit doing away with the fins on the Viking, and it required a separate roll control system using two aerodynamic tabs on the fins and two "steam jets" from the hydrogen peroxide turbopumps. The rest of the control system was not innovative, involving an adaptation of a commercially available Sperry A-2 autopilot. But a post-engine-cutoff control system using residual gas from the pressurization system for the propellant tanks was indeed new, and provided a basis for the system used on Vanguard.[17]

Rosen recalled that, starting in about 1949, his team and Bossart's at Convair exchanged visits for several years until the Atlas project began in earnest. On these visits they shared a great deal of information. The gimballed engines that were a significant feature on the Atlas missile and space-launch vehicle owed more to Viking than to MX-774, Rosen felt, pointing to the great difficulty his own team and the Martin engineers had in making gimballing work. It was not until the tenth or eleventh Viking launch that they got the bugs out of the system.

"The real contribution of Viking," he stated, "was in making the gimballed motor work, which could not be done until the development of feedback theory at MIT during World War II." When the gimbals got a signal from the control system, servomotors moved the engine to keep the rocket on the desired trajectory. A potentiometer then fed a signal back to the control system to stop the movement once the course was corrected. Adjusting the system so as not to overcompensate for a course deviation was a delicate process, which required negative feedback.[18] Much of the credit for making the Viking system work efficiently and effectively belongs to Albert C. Hall, who wrote his Ph.D. thesis at MIT on negative feedback. As a consultant to the Martin Company, Hall set up the initial parameters for the system and proposed a method for adjusting it. This proved to be the most difficult problem the Viking team faced, but eventually they eliminated the instabilities in the control system. Others who used gimballed engines then adopted their methods, Rosen believed.[19]

There is no need here for a detailed description of Viking's design. Continual problems arose with each of the twelve Vikings launched during the project, and modifications to solve individual issues or to improve performance were numerous. The biggest change came between Vikings 7 and 8, when a redesign increased the capacity of the propellant tanks and lightened other parts of the rocket so that almost 80 percent of the gross weight consisted of propellant, which permitted reaching higher altitudes. Components were also rearranged to make them more accessible for repair. Whereas Viking 7 had been 47.6 feet tall, Viking 8 stood only 41.4 feet but was 45 inches in diameter rather than just 32 inches.[20] In Viking 9, further improvements included modification of the post-engine-cutoff control system to replace residual gas with hydrogen-peroxide rockets for better control.

Overall, the Vikings were successful. Launched from a ship at sea on May 11, 1950, Viking 4 rose to 105 miles, a record altitude at that time for a sea launch. Vikings 7, 9, and 10 (1951–54) all reached 135 or 136 miles, records for single-stage rockets until May 24, 1954, when Viking 11 rose to 158 miles above Earth's surface. More important, Viking paved the way for the use of gimbals to steer large rockets and missiles, and it helped prepare people like Rosen and the Martin propulsion engineer John Youngquist for Vanguard and, in the case of Youngquist, for Titan. It also gave Martin valuable experience in serving as prime contractor for a large rocket, experience that would carry over to Vanguard, the Titan missiles, and the Titan III and IV launch vehicles.[21]

The Viking team did all of this with a handful of people. As Rosen recalled in 1971:

> There were only twelve men on the Viking launch crew. The Martin Company, which built the rocket, had no more than about two dozen engineers involved in the design. There were never more than about 50 men involved in building the vehicle, and we had to borrow those from other projects because Viking didn't have enough production to hold them permanently. Finally, the government project group consisted of two men at first, and was never more than four. These two to four people wrote the specifications, negotiated the change orders, analyzed flight and test data, and wrote all of the final reports on Viking. That is why you could buy a Viking vehicle for $250,000 and conduct an entire Viking operation for $500,000.
>
> Now, we cannot produce today's vehicles that way. Obviously, they are much more complex; they require more in the way of design and test. But we have been using thousands of people to build and launch these rockets. We have been generating tons of paper. And in many instances we have been using ten people to do the work of one. Can the ten people do a better job than the one? Yes, they can; they can do more calculations, they can try more alternatives, they can catch more errors—but they don't do the job ten times as well.[22]

Vanguard

The proposal that NRL submitted for the International Geophysical Year (IGY) satellite project—from which Vanguard was born—stemmed from an Air Force request in July 1954 to investigate use of a modified Viking to study the reentry nose cone issue that the Army later solved through ablation.[23] This request led to the design of Viking versions M-10 and M-15, so designated for their capability to reenter the atmosphere at Mach numbers 10 and 15 respectively. The Air Force then pursued other avenues to study reentry. But the two designs formed the basis of NRL's proposal on July 5, 1955, of a vehicle to launch the IGY satellite.

The NRL proposal included two possible configurations. One would have the M-10 Viking as the first stage with two additional stages, both using solid propellants. The first stage would use GE's Hermes A-3B engine,[24] modified to increase its thrust from the original 22,600 pounds to some 28,000. The Atlantic Research Corporation had designed the two upper stages, which NRL believed were conservative enough to be developed in time for the IGY.

The other proposed configuration consisted of liquid-propellant first and second stages, collectively called the M-15, plus a third stage using solid propellants. This variant also used the M-10 as the first stage. Aerojet General's Aerobee-Hi sounding rocket, then still in development but scheduled to fly in August 1955, constituted the second stage. The third stage would again be an Atlantic Research solid-propellant rocket using ammonium perchlorate as oxidizer and polyvinyl chloride as fuel. When the Stewart Committee voted for Vanguard as the IGY choice, it favored the M-15 configuration even though it would take longer to develop. However, the committee recommended that NRL consider a different third stage, of the Sergeant type.[25]

In mid-August 1955 when the Army submitted to the Stewart Committee its revised proposal in support of Project Orbiter, the Navy put together a revised proposal of its own with a Thiokol third stage (Thiokol having built the motor for the Sergeant missile, the type of rocket the committee favored). Milton Rosen, representing the sea service, also gave the committee a revised estimate of when it could launch the first satellite. Rosen had originally stated that it would take thirty months, which proved astonishingly accurate. But Martin thought it could do the job in eighteen months if awarded the contract, and under the pressure of competition with the Army, Rosen endorsed this overly optimistic estimate against his better judgment. He bolstered the projection in his August memo with enclosures: a telegram from Thiokol promising delivery of a solid-propellant third stage nine months after a contract was let, a GE guarantee to deliver the first-stage engine in nine months, a promise by Aerojet to deliver Aerobee-Hi engines eleven months after it signed a contract, and an agreement from Martin "to put a satellite in being in approximately 18 months" if government and industry clearly understood "the part each was to play in the program execution." These assurances allowed the decision in favor of Vanguard to stand. In a memo to the secretaries of the Army, Navy, and Air Force, Deputy Secretary of Defense Reuben Robertson ratified the final decision of the Stewart Committee on September 9, 1955.[26]

As Project Vanguard developed, only Thiokol among these four contractors did not play a role. For the second stage of the launch vehicle, the Aerobee-Hi did not meet the requirements later specified, so Aerojet had to go back to the drawing board for an engine that proved unexpectedly trouble-some, although the firm had raised some red flags even for producing Aer-obee-Hi in the eleven months to which it had agreed. Long after Vanguard was relegated to the history books, Rosen had occasion to comment on a pending NASA publication that said of NRL's launch vehicle, "The first stage was derived from the Viking sounding rocket, and the second stage was de-rived from the Aerobee sounding rocket." Rosen, without indicating his own part in the process, wrote:

> Th[is] myth, as I call it, was generated by the NRL people who were trying to sell the Vanguard project to the Stewart committee, in order to mitigate the impression that much of the project was new devel-opment, and indeed, at the time there was some hope that some of the Viking and Aerobee technology could be transferred. The truth of the matter is that Vanguard was almost entirely a new design; the engines were new, the structure was new, the guidance was new. All that was transferred was the valuable experience that the NRL, Martin, and Aerojet engineers derived from Viking and Aerobee designs and operations. The myth that the Vanguard was derived from the Viking and Aerobee sounding rockets continues to haunt those who perpe-trated it.[27]

On the other hand, Rosen added in 2002, "I said then and I say now: without Viking there would have been no Vanguard."[28]

In any event, it took from September 9 until October 6, 1955, for the Naval Research Laboratory to receive official direction from the chief of naval research to begin carrying out Project Vanguard. The objectives of the project were "(a) To put an object into orbit around the earth; (b) to prove that the object is in an orbit; and (c) to conduct at least one scientific experi-ment using the object." And, according to a congressional committee report, the intent was to put this "satellite in orbit during the IGY," which ended in December 1958.[29]

NRL appointed John Hagen as director of Project Vanguard. Born in Nova Scotia on July 1, 1908, Hagen had earned a B.A. at Boston University in 1929, an M.A. at Wesleyan in 1931, and a Ph.D. in astronomy at Georgetown in 1949. Between his master's and doctorate, Hagen had gone to work at NRL in 1935, devoting much of his research to improving radar techniques but also helping to develop an automatic indicator for the ground speed of aircraft. In 1954 he became head of NRL's Atmosphere and Astrophysics Division, and the next year he was named director of Vanguard.[30] Rosen served as technical director under Hagen.

NRL did not wait for official direction before getting to work on Vanguard. Knowing speed was essential, it let a contract through its parent organization, the Office of Naval Research, with the Martin Company on September 23, 1955, for the design, construction, and preflight testing of the Vanguard vehicle. On October 1, Martin in turn subcontracted development of the first-stage engine to GE. Once the weight, thrust, and other parameters for the first stage had been calculated, Martin could set targets for the next two stages. It determined that for the second stage to provide the velocity required, given the estimated weight of the vehicle's structure and instrumentation, it needed a thrust of 7,500 pounds and a specific impulse (measure of performance) of 278 lbf-sec/lbm (pounds of thrust per pound of propellant burned per second) at altitude. Even though Aerobee-Hi could not meet these specifications, Martin contracted with Aerojet on November 14, 1955, to develop the second-stage engine.

The third stage had to impart enough additional velocity to the payload to make it reach orbital speed. It could accelerate more easily than the overall launch vehicle because after the shedding of stages 1 and 2, it weighed much less. Also, when it began firing, it was further from the center of Earth and thus less impeded by Earth's gravitational pull. Somewhat paradoxically, therefore, it needed a higher mass fraction (the mass of the propellant divided by the total mass of the stage) than the other stages so it could take full advantage of its greater ability to accelerate. (To give an indication of how critical the third-stage weight really was, one source indicated that the addition of a pound to the first stage reduced the final velocity of the satellite by 1 ft/sec; for the second stage, the reduction produced by one pound of additional weight was 8 ft/sec; but for the third stage the reduction in speed rose tenfold to 80 ft/sec.) Thus team engineers decided to put the guidance and control equipment in the second stage rather than the third, and to use spin on the third stage to provide stability while it was burning. This in turn dictated a solid propellant with an internal-burning charge, as the spinning would make liquid propellants slosh, complicating their effective use. Martin determined that the third stage needed a weight of 484 pounds, a mass fraction of 0.816 (compared with less than 0.7 for the first two stages), a specific impulse at altitude of 245, and a total impulse of 97,600 pounds per second. These parameters ruled out a Sergeant-type motor.

Several companies submitted proposals to furnish this stage. Since Martin and NRL expected developmental difficulties, Martin awarded a contract to the Grand Central Rocket Company in February 1956 and, in a parallel move, the Navy issued one in April 1956 to the Allegany Ballistics Laboratory (ABL), now operated by the Hercules Powder Company, to develop third-stage motors. Having two contractors would improve the chances of getting a viable motor in time for the Vanguard launches.[31]

Even before Martin began awarding these subcontracts, it dealt a "disappointment, even a shock" to NRL. The Navy laboratory had learned unofficially that it had won the battle with the Army for the right to launch the IGY satellite, and early in September it had issued a letter of intent to Martin for work on the launch vehicle. Before doing so, the Navy had conferred with the Department of Defense, hoping thereby to avoid competition with other projects for Martin's rocket engineers. On September 14, Martin learned that it would win the contract to design, develop, and test the Titan I missile (as it became), with a letter contract signed with the Air Force in October and a final contract in January 1956. Since Vanguard's priority at the Department of Defense was low—the IGY satellite program, it was stipulated, must not interfere with military missile projects—while the much larger Titan program enjoyed a high military priority, Martin split up the experienced Viking team, sending only some of its Viking engineers to work on Vanguard. To be sure, the situation was not as dire as it sometimes has been portrayed. Martin's project manager for Vanguard, its chief engineer, and its chief of flight testing were all veterans of the Viking team. But NRL had counted on the entire Viking team transferring to Vanguard.[32]

Of more consequence, perhaps, for NRL perceptions and morale, Martin situated its Vanguard engineers in a loft of its aircraft plant that was extremely hot in summer, cold in winter, and subject to sparrow droppings. Rosen, despite being described as "intense" and "hard-driving," had gotten along with the Martin engineers on the much smaller Viking project, but NRL's disappointment at Martin's splitting up the Viking team and operating from a balcony may have contributed to a less amicable working relationship with the Baltimore firm this time around. Another factor was a disagreement about approach. The Air Force allowed Martin considerable freedom to develop Titan I, whereas the NRL team wanted to monitor not only plans but execution. And where Martin was inclined to use shortcuts and empirical data in design and development, NRL favored greater emphasis on analysis. This contrast in approach added friction to what was already a difficult design-and-development effort.[33]

The First-Stage Engine

Even though the General Electric X-405 first-stage engine, as it came to be designated, was based on GE's A-3B, the modifications were numerous and difficult, making this a substantially new engine, as Rosen said. The A-3B had burned liquid oxygen and alcohol, but the X-405 replaced the alcohol with kerosene as part of the effort to raise the specific impulse from 225 to 254 lbf-sec/lbm.[34] Available sources on Project Vanguard and GE do not indicate whether GE solved early problems with kerosene as a rocket fuel entirely on its own, but it could have learned of solutions to those problems from an Air Force program with North American Aviation.

In January 1953, Lt. Col. Edward Hall and others from Wright-Patterson Air Force Base insisted to Sam Hoffman of NAA that he convert from alcohol to a hydrocarbon fuel for a 120,000-pound-thrust Navaho engine. Hoffman protested, as the standard kerosene the Air Force used was JP-4, whose specifications allowed a range of densities. It also clogged a rocket engine's slim cooling lines with residues. The compounds in the fuel that caused these problems did not affect jet engines, but they simply would not work easily in rockets' power plants. To resolve these problems, Hoffman initiated the Rocket Engine Advancement Program, which resulted in the development of the RP-1 kerosene rocket fuel, without JP-4's high levels of contaminants and variations in density. This fuel went on to power the Atlas, Thor, and Jupiter engines. The specifications for RP-1 were not available until January 1957, but that was still before the actual delivery date of the X-405 engine for test vehicle 2 of Vanguard.[35]

GE's X-405 may thus have been a direct or indirect beneficiary of NAA's work.[36] In any event, other changes from the A3–B were an increase in the chamber pressure from some 450–500 pounds per square inch up to 616 psi and an extension of the burn from 50 to 146 seconds. The specific impulse rose to roughly the requisite 254 lbf-sec/lbm as well. Implementing these improvements brought the usual problems.

The higher pressure necessitated a slightly thicker chamber wall. It also pushed up the heat transfer rate from 0.4 to 0.6 $BTU/in^2/sec$. This was offset by faster propellant flow through the regenerative cooling passages, but to raise the flow rates, the injectors had to be modified. The hydrogen peroxide in the turbopump system also had to flow more quickly to provide the higher pressure. Despite these issues, GE had the first production engine (P-1) ready for delivery on October 1, 1956. But then damage to the lining of the combustion chambers occurred during static testing of the P-2 and P-3 engines. Failures of two chamber liners in the P-4 engine delayed delivery until the problem could be solved. A redesign proved effective but required careful attention to injector specifications to preclude local hot spots and combustion instability. In fact, GE ended up testing fifteen injectors with six variations in design between January and April 1956 before finding one that would work. This was "rule-of-thumb" engineering, but it produced a comparatively uncomplicated engine with a minimum of relays and valves that never experienced a burn-through in flight.[37]

The Second-Stage Engine

The development of Aerojet General's second-stage engine, designated AJ10–37, saw many more problems than GE's first stage. The stringent velocity requirements of stage 2 imposed severe weight limitations and called for a high specific impulse. The engineers selected unsymmetrical dimethyl hydrazine (UDMH) and inhibited white fuming nitric acid (IWFNA) as the propellants because they were hypergolic (self-igniting on contact, eliminating ignition issues), had a high loading density (reducing tank size), and delivered the requisite specific impulse. Another perceived advantage of hydrazine and acid was a comparative absence of combustion instability in experimental research.[38]

The history of the evolution from the aniline–nitric acid propellants used in the WAC Corporal[39] and the first Aerobee sounding rocket[40] to the UDMH and IWFNA used in Vanguard is complicated but worth relating because of what it illustrates about propellant chemistry and the number of institutions contributing to it. The basic aniline-RFNA combination worked as a self-igniting propellant pairing. But aniline is highly toxic and rapidly absorbed through the skin. A person who comes into contact with a significant amount of it faces a swift death from cyanosis. Moreover, aniline's high freezing point means it can be used only in warm weather. RFNA is highly corrosive to propellant tanks, so it has to be loaded just before firing, and when poured it gives off dense concentrations of nitrogen dioxide, which is also poisonous. The acid itself burns the skin as well.

Two chemists at JPL discovered as early as 1946 that white fuming nitric acid (WFNA) and furfuryl alcohol with aniline, while just as poisonous and corrosive, at least did not produce nitrogen dioxide. But WFNA turned out to be inherently unstable over time. Chemists in the rocket business throughout the country also found this complicated substance difficult to analyze. By 1954, however, those at the Naval Ordnance Test Station in Inyokern, California, and at JPL had thoroughly investigated nitrogen tetroxide and nitric acid and come up with conclusions that were to be used in the Titan II. Meanwhile, chemists in various places—among them JPL, the NACA's Lewis Flight Propulsion Laboratory, the Naval Air Rocket Test Station in New Jersey, the Air Force's Wright Air Development Center in Dayton, and Ohio State University—reached a fundamental understanding of nitric acid by 1951, with their findings published by 1955. The Naval Air Rocket Test

Station was apparently the first installation to discover that small percentages of hydrofluoric acid both reduced the freezing point of RFNA/WFNA and inhibited corrosion with many metals. Thus was born inhibited RFNA and WFNA, for which the services and industry representatives under Air Force sponsorship drew up military specifications in 1954. In this way the services cooperated to solve a common problem despite their continuing competition for roles and missions.

During the same period, chemists were also looking for replacements for aniline or chemicals that could be mixed with it to make it less troublesome. Hydrazine seemed to be a promising candidate, and in 1951 the Navy's Bureau of Aeronautics, through its Rocket Branch, issued contracts to a firm named Metallectro and to Aerojet to work with hydrazine derivatives and see if any were suitable as rocket propellants. The two firms found that UDMH rapidly self-ignited with nitric acid, leading to a military specification for UDMH in 1955.[41]

Despite the severe weight limitations on stage 2, Vanguard project engineers decided on a pressure-fed propellant delivery system rather than stage 1's pump-fed system. The pumps produced angular momentum as their turbines rotated, and for stage 2 this would be hard for the roll-control system to overcome. Concerns about reliability and development problems led to a decision to use heated helium gas as the pressurant. Aerojet convinced the Martin Company, Rosen, and Hagen to employ stainless steel instead of aluminum propellant tanks, because steel had a better strength-to-weight ratio. Aerojet argued that the lighter metal would, paradoxically, have had to be 30 pounds heavier than steel to handle the pressure.

Moreover, a "unique design for the tankage" placed the sphere containing the helium pressure tank between the two propellant tanks, serving to divide them, thus saving the weight of a separate bulkhead. A solid-propellant gas generator augmented the pressure of the helium and added its own chemical energy to the system at a low cost in weight. Initially Aerojet built the combustion chamber entirely of steel, and it logged 600 seconds of burning without corrosion. However, it was too heavy. So Aerojet engineers developed a lightweight chamber made up of spaghetti-type aluminum regenerative-cooling tubes wrapped in stainless steel, cutting the weight by more than 20 pounds. This chamber was apparently the first built of aluminum tubes to be tested with nitric acid and UDMH.[42]

Despite Aerojet's experience in this area, in 1956 problems arose with welding stainless steel tanks. They were resolved when, at Martin's recommendation, Aerojet made improvements in tooling and inspection. The California rocket engine firm also had to try several types of injectors before it found the right combination of features. One with 72 pairs of impinging jets did not deliver sufficient exhaust velocity, so Aerojet engineers added 24 nonimpinging orifices for fuel at the center. This raised the exhaust velocity above specifications.[43]

Despite the use of inhibited white fuming nitric acid, the lightweight aluminum combustion chamber, which could be lifted with one hand, experienced gradual erosion. It took engineers "weeks of experimenting" to find out that a coating of tungsten carbide substantially added to the chamber's life. There also were problems with valves for flow control, requiring significant redesign of the valves and rerouting of the "plumbing."[44]

A final problem lay in testing an engine that had to ignite at an altitude upwards of 30 miles. At the outset of the project, there was no vacuum chamber large enough, but apparently Aerojet acquired or built one, because NRL propulsion engineer Kurt Stehling wrote, "Several tests were made at Aerojet with engine starts in a vacuum chamber." In any event, to preclude problems with near-vacuum pressure at altitude, the engineers included a "nozzle closure" that kept pressure in the combustion chamber until exhaust pressure from ignition blew it out.[45]

The Third-Stage Motors

One firm the Martin Company selected to develop the third stage for Vanguard was a relative newcomer to the rocket business. Charles Bartley had left JPL in 1951 and founded the Grand Central Rocket Company in 1952. Initially located in Pacoima, in the San Fernando Valley, it moved eastward in 1954 to Mentone, just past Redlands in San Bernardino County. Bartley served as the president until September 1958, when he left the firm to organize two companies that produced solid-propellant equipment such as sounding rockets and small propulsion devices to eject pilots from aircraft. At Grand Central he hired Lawrence Thackwell from Thiokol and both Larry Settlemire and John I. Shafer, who, like Thackwell, had worked for him at JPL. Thus he brought considerable talent and knowledge from JPL to the new company. Ultimately Grand Central passed through interim owners to Lockheed, becoming Lockheed Propulsion Company in 1961.[46]

In the meantime, the company under Bartley's direction contracted for and developed the third-stage motors that flew on all Vanguard launches except the final one (excluding test launches without a third stage). The motor used Thiokol's LP-3 polysulfide rubber as the binder and fuel (28 percent of the propellant) and ammonium perchlorate as the oxidizer (71 percent of the propellant). It was configured with a five-pointed, internal-burning star. The case was very thin stainless steel. The propellant burned at a low chamber pressure, making the thin case possible, but Cooper Development Corporation of California, which made the case, had to spin and roll it in extremely thin sheets, then weld and heat-treat it properly for it to work. Coating the nozzle with aluminum oxide provided protection against over-heating. Since ignition of this stage occurred more than 200 miles above the earth, it required a high-energy igniter plus sealing of the throat to keep in some pressure until firing began. The igniter used the same propellant as the motor.

The propellant yielded a specific impulse at altitude of 239 lbf-sec/lbm, slightly below the design figure of 245, and the metal parts initially weighed 55 instead of the specified 53 pounds. To compensate for the underperformance of the propellant, a product improvement program decreased the weight of the parts so as to raise the mass fraction and achieve the necessary velocity. By measures that included reducing the thickness of the liner between the propellant and the case, the engineers lowered the weight by five pounds. There was also evidence of combustion instability, but it was not intense enough to be significant. Testing of the motor revealed problems with cracking of the propellant at the star points, as had occurred at JPL with the Sergeant test vehicle. Placing polyvinyl acetate cement at the star points inhibited the cracking. As of June 1957, Grand Central Rocket had completed development of stage 3 and satisfied the specifications to reach orbital velocity.[47]

The alternate third stage was fated to be flown on only one Vanguard mission, the last of the series, because of problems with its technical development. But it was more innovative than Grand Central's motor, and it (or variants of it) later found use in military missiles as well as both the Delta and the Scout launch vehicles. The firm that produced this motor came from a different lineage than the other solid-propellant firms discussed up to this point.

The Hercules Powder Company, which had operated the government-owned Allegany Ballistics Laboratory since the end of World War II, came into existence in 1912. An antitrust suit forced its parent company, E. I. du Pont de Nemours, to divest some of its holdings, one of which became Hercules. The spinoff firm began as an explosives company that produced more than 50,000 tons of smokeless powder during World War I and then began to diversify into other uses of nitrocellulose. During World War II the firm supplied large quantities of double-base propellants for tactical rockets, using an extrusion process. After the war Hercules began using a technique for casting such propellants. Chemists poured a casting powder consisting of nitrocellulose, nitroglycerine, and a stabilizer into a mold and added a solvent of nitroglycerine plus a diluent and a stabilizer. With heat and the passage of time this yielded a much larger grain than could be produced by extrusion alone.

Wartime research by John F. Kincaid and Henry M. Shuey at the National Defense Research Committee's Explosives Research Laboratory, operated by the Bureau of Mines and the Carnegie Institute of Technology in Bruceton, Pennsylvania, south of Pittsburgh, had produced this process. Kincaid and Shuey along with other propellant chemists had developed it further after transferring to ABL, and under Hercules management ABL continued work on cast double-base propellants. Flight testing of a JATO using this propellant followed in 1947. The casting process allowed Hercules to produce a propellant grain as large as the castable composite propellants that Aerojet, Thiokol, and Grand Central were developing in this period but with a slightly higher specific impulse—albeit a greater danger of exploding rather than releasing the exhaust gases at a controlled rate.[48]

Although the Navy had let the contract to Hercules for this motor, designated JATO X241 A1, it had delegated responsibility for technical coordination to the Glenn L. Martin Company. The first propellant Hercules' Allegany Ballistics Laboratory used for the motor was a cast double-base formulation with insulation material between it and the case. This propellant yielded a specific impulse of about 250 lbf-sec/lbm, higher than Grand Central's propellant and higher than the specification of 245 lbf-sec/lbm for both motors. A key feature of the motor was its case and nozzle, composed of a laminated fiberglass made from epoxy resin. ABL had subcontracted work on the case and nozzle to Young Development Laboratories, which in 1956 developed a method of wrapping threads of fiberglass soaked in epoxy resin around a liner made of phenolic asbestos. (A phenol is a compound used in making resins to provide laminated coatings or form adhesives.) After curing, this process yielded a strong, rigid shell with a strength-to-weight ratio 20 percent higher than the stainless steel Aerojet was using for its propellant tanks.[49]

In 1958, with its third-stage motor still under development, Hercules acquired the fiberglass winding firm. Its founder, Richard E. Young, was a test pilot who had worked for the M. W. Kellogg Company on the Manhattan Project. Kellogg had designed a winding machine in 1947 under Navy contract, leading to a winding laboratory that built a fiberglass nozzle. In 1948 the operation moved to Rocky Hill, New Jersey. There Young set up labs under his own name and tackled the problem of strength-to-weight ratios in rocket motors through developing lighter materials. He and the firm evolved from nozzles to cases, seeking to improve a rocket's mass fraction, which was as important as specific impulse in achieving high velocities. In the mid-1950s, ABL succeeded in testing small rockets and missiles using cases made with Young's Spiralloy material.[50]

This combination of a cast double-base propellant and the fiberglass case and nozzle created a lot of problems for Hercules engineers. By February 1957, ABL had performed static tests on about twenty motors, with fifteen of those firings producing failures of insulation or joints. Combustion instability arose in about a third of the tests. Hercules installed a plastic paddle in the combustion zone in an attempt to interrupt the acoustic patterns (resonance) that caused the instability. This did not work as well as hoped, so the engineers developed a suppressor of thicker plastic. They also improved the bond between the insulator and the case, then cast the propellant in the case instead of just sliding it in as a single piece. Despite these modifications, nine cases failed during hydrostatic tests or static firings. The culprits were high stress at joints and "severe combustion instability."[51]

In February 1958, in addition to X241, ABL began developing a follow-on third-stage motor designated X248 A2. Perhaps the lab did so in part to reduce the combustion instability, for the new motor's propellant was 3 percent aluminum, which upon burning produced particles in the combustion gases that suppressed (damped) high-frequency instabilities. But another motivation was increased thrust. The new motor was the one that actually flew on the final Vanguard mission on September 18, 1959. As of August 1958, ABL had already developed a modification of this motor, X248 A3, for use as the upper stage in a Thor-Able lunar probe. By this time, also, ABL was testing the motors in an altitude chamber at the Air Force's Arnold Engineering Development Center in Tullahoma, Tennessee, and was experiencing problems with ignition and with burn-throughs of the case.[52]

The X248 solid rocket motor consisted of an epoxy-fiberglass case filled with the case-bonded propellant. The nozzle was still made of epoxy fiberglass, but its coating was now a "ceramo-asbestos." By November 11, 1958, wind-tunnel static tests showed that the X248 A2 filament-wound exit cone was adequate. The motor had a specific impulse at altitude of 256 lbf-sec/lbm, and its other problems had been overcome. The X248, wrote Stehling, offered "considerable improvement in reliability and performance over the X241 contracted for originally" and succeeded in launching the Vanguard III satellite weighing 50 pounds, whereas Grand Central Rocket's version of the third stage could orbit only about 30 pounds of payload.[53]

Aerodynamics

As finally designed, the Vanguard launch vehicle was 71 feet 2 inches long. Its first stage was 45 inches in diameter, tapering to 32 inches for the second and third stages. Unlike the Viking, it was finless with an integral-tank construction (the tanks serving as structural support), and it had a weight at liftoff of 22,600 pounds. These dimensions made for a slender rocket. As such, it had a low bending moment, meaning that it could bend relatively easily under the force of winds or other air "loads." NRL and Martin consequently arranged for a wide variety of wind tunnel tests at low speeds, transonic speeds (roughly Mach 0.8 to 1.2), and supersonic speeds, using different tunnels for each range since there was no single tunnel that could provide accurate data over the entire velocity scale from Mach 0.0 to 3.5. (Speeds above three and a half times the speed of sound would occur at such high altitudes that the density of the atmosphere and the resultant air loading were too low to be a concern.)

Table 1.1 shows the wind tunnels used for Vanguard testing and the speed ranges tested. There were two tests at supersonic speeds because of configuration changes between the first and second test. A second series of low-speed tests was occasioned by concern about the "variable Reynolds number effect at high angles of attack."[54]

The tests at MIT were ordered because Vanguard's relative slenderness made it more subject to the von Kármán vortex-shedding effect than other

Table 1.1. Wind Tunnel Testing for Vanguard

Test Date	Wind Tunnel	Speed Range
10/28/55	Aberdeen Proving Ground Supersonic Wind Tunnel #3	Mach 2.0–4.0
3/23/56	University of Maryland Low-Speed Wind Tunnel	150 mph at dynamic pressure of 57.55 lb/ft^2
5/18/56	University of Maryland Low-Speed Wind Tunnel	50–200 mph
9/17/56	Naval Air Missile Test Center, Point Mugu, California, Supersonic Wind Tunnel	Mach 1.6–3.5
10/29/56	Wright Air Development Center, Wright-Patterson AFB, Ohio, Transonic 10-Foot Wind Tunnel	Mach 0.8–1.2
12/10/56	MIT Aero-elastic Wind Tunnel	40–75 mph

Sources: "Project Vanguard Report . . . 1 June 1957," 2-13 to 2-15, 2-102, NHRC; Hagen, "The Viking and the Vanguard," 130; Martin, "Vanguard Satellite Launching Vehicle," 18, NHRC.

rockets while on the launchpad. Named after the famous aerodynamicist who discovered it, the von Kármán effect could result from a steady wind that produced oscillatory forces or vortices; these would form periodically on the sides of the cylindrical fuselage and then disengage, imparting forces that could destroy the rocket on the pad if they were of the proper magnitude and frequency. Analysis and testing had to take into account conditions before and during propellant loading, as the amounts of propellants affected the proclivity of the long, thin rocket to bend and possibly break apart. A preliminary theoretical analysis indicated that wind velocities between 10 and 18 ft/sec (about 7 to 12 mph) could produce serious enough vibrations to destroy the vehicle under certain propellant loads. The results of the wind tunnel tests at MIT were quite complex and their scope had to be limited, but they provided the Vanguard team with rough parameters for determining safe launch conditions. They also led engineers to develop mechanisms—removable rubber spoilers on the second stage—to diminish the effects. Technicians mounted twelve of these spoilers, with drag cones attached. The spoilers suppressed the vortices, while the drag cones ensured that the spoilers fell off the vehicle after liftoff, when the oscillations were no longer a problem.[55]

Structure

The Vanguard structure was mainly monocoque, meaning that the outer skin bore the major portion of the bending forces, with the pressurized internal tanks also serving as structural supports. The materials for the first stage were principally aluminum and magnesium alloys. The second stage contained the propellant and helium tanks, which were made of stainless steel, as was the motor casing for the third stage except in the final Vanguard configuration where the ABL fiberglass case was used.[56]

Guidance and Control System

The final major concern of Vanguard's designers was the guidance and control system in stage 2. Experience with Viking had demonstrated that fins for stabilization could actually cause destabilization if they were not lined up properly. Besides, they added weight. Martin engineers had designed a gimballing system that was sensitive enough not to require fins, but the Navy's budget for Viking did not permit implementing the system. It was used later on Vanguard, which could then dispense with fins.[57] Some Deltas and Titans subsequently followed the Vanguard precedent and also did not use fins.

altitude, by which time the second and third stages had tilted to a completely horizontal attitude. The third stage provided 50 percent of the orbital velocity—about 18,000 mph relative to Earth's surface—the remaining 3 percent coming from geophysical effects including Earth's rotation, since Vanguard was launched to the east.[61]

During coasting flight, another major component of the flight control system came into play: the coasting time computer. Provided by Electronic Communications, this device ascertained the velocity of the vehicle when the second-stage engine burned out and computed the proper time for the third-stage motor to ignite. An integrating accelerometer provided acceleration data throughout the Vanguard's flight up until second-stage cutoff. From the information it supplied, the computer calculated the vehicle's velocity at stage-2 burnout and energized a timing motor. This in turn signaled the spin-up of the third stage and its separation via Atlantic Research's two small solid-propellant retro-rockets. Fifteen seconds after the spinning of the third stage began, a delay fuse fired, initiating ignition of the rocket motor.[62]

Flight Testing

The original Vanguard schedule as of November 1955 called for six test vehicles to be launched between September 1956 and August 1957, with the first satellite launching vehicle to lift off in October 1957—by happenstance, the month of Sputnik. Had the project remained on schedule, it is conceivable that NRL could have launched a satellite at about the same time as the Soviet Union. As it was, problems with the first- and second-stage engines caused delays. On December 8, 1956, already more than two months late, Viking rocket number 13 was launched to test the Vanguard launch complex and the telemetry system, develop familiarity with range safety, and the like. The test was largely successful, but by February 1957 even Viking number 14 had not flown. Rescheduled for the end of April, that vehicle actually launched on May 1 with a prototype of the Grand Central third stage tested as the Viking's second stage for spin-up, separation, ignition, propulsion, and trajectory (the actual second stage not flying on this mission). Although the sequence of spinning up and separating for the Grand Central rocket was "untried and complicated," the test was successful and almost in accord with the revised timetable.[63]

However, the third test vehicle (designated TV-2 because the first test with the Viking rocket was TV-0), involving a prototype Vanguard first stage with inert second and third stages, could not be launched until October 23, 1957.

This was almost three weeks after Sputnik 1 and four months behind the June liftoff date stipulated back in February. For guidance, TV-2 contained only some of the guidance-and-control and other electronic components of a complete Vanguard vehicle. The old Viking system still provided much of the control. Before the test vehicle left the Martin plant, the contractor's crew found that the roll jet and pressurization systems were not performing according to specifications and had to be partially redesigned. Despite numerous hardware difficulties along the way, TV-2 finally launched successfully and met all of its objectives. The first stage operated as designed, and conditions appeared favorable for successful separation of the first and second stages.[64]

By the time of this launch, pressures were mounting on John Hagen. The project director had been up on Capitol Hill to brief the staff of the formidable Senator Lyndon B. Johnson and his Senate Preparedness Subcommittee, which was scrutinizing the nation's missile and satellite programs in the wake of Sputnik. Hagen had also briefed President Eisenhower on both TV-2 and TV-3, the second of which was now scheduled for launch on December 4. Already in July, before the furor attending the Soviet satellite, the Navy had decided to launch TV-3 and TV-4 with minimal satellites weighing 3.4 pounds instead of an instrumented nose cone for the original 21.5–pound satellite. Hagen had so indicated to Eisenhower, pointing out that there was no guarantee they could be put into orbit. Despite this warning, the president's press secretary informed reporters that during December the Vanguard project would launch a test vehicle with a satellite on board. The press, of course, seized upon the mere test as America's answer to Sputnik.[65]

TV-3 would be the first Vanguard launch with all stages operational. Its mission was not just to launch a small satellite but to test and evaluate all three stages. More particularly, it would be the first test flight of the problem-ridden second stage as well as the complete guidance and control system. It was rare in those days for any rocket to have three successful tests in a row, as Vanguard had. The odds were certainly against a fourth. Still, TV-3 had passed its preflight "functional, instrumentation, and restrained-firing tests" and, following delays for various reasons, sat poised on the launchpad on December 6, 1957, ready to send the grapefruit-sized satellite into orbit to join its larger Soviet cousin. Then the odds caught up with Vanguard. The first stage ignited, but the test vehicle rose only slowly, "agonizingly hesitated a moment . . . and . . . began to topple [as] an immense cloud of red flame from burning propellants engulfed the whole area." The press did not

Figure 1. Vanguard test rocket TV-3 exploding on the launchpad. A malfunction in the first stage caused the vehicle to lose thrust after two seconds, topple over, and burn. U.S. Navy photo courtesy of NASA.

react charitably but called the vehicle "Kaputnik, Stayputnik, or Flopnik," while Americans, in one historian's words, "swilled the Sputnik Cocktail: two parts vodka, one part sour grapes."[66]

In the somber wake of the colossal failure in public relations, technicians and engineers from GE and the Martin Company pored over records from ground instrumentation, films of the failed launch, and the two seconds of telemetered data from the toppling inferno that was to have orbited America's first satellite. Not surprisingly, the Martin people blamed the problem on an "improper engine start" caused by low fuel tank pressure, whereas those from GE said there was no improper start and blamed a loose fuel-line connection. The Martin engineers turned out to be correct, although there were more problems than low fuel tank pressure. To solve that one, GE negotiated a specification change with Martin to increase the minimum pressure in the fuel tank by 30 percent. But telemetry data also showed a high-pressure spike on engine start that the GE engineers had not caught in testing because of low-response instrumentation. The pressure spike had ruptured a high-pressure fuel line, causing destruction of the rocket. The remedy was to increase from 3 seconds to 6 seconds the period when oxygen was injected into the combustion chamber ahead of the fuel. The modified engine worked without problems in fourteen static and flight tests following this disaster.[67]

Meanwhile, waiting in the wings was a backup (BU) version of TV-3, virtually identical to the vehicle that had failed to launch. Following repairs to the launchpad, on February 5, 1958, stage 1 of TV-3BU fired properly and the test vehicle lifted off successfully to the cheers of the Vanguard team. However, 57 seconds into the launch, at an altitude of about 1,500 feet, the control system malfunctioned. Investigation later showed that a broken wire or connection had caused the vehicle to go out of control and break apart at a point between the first and second stages because of structural loads on the fuselage.[68]

TV-4 contained modifications made to the stage-1 engine following the failure of TV-3, but it did not yet incorporate the tungsten-carbide coating in the aluminum combustion chamber of the stage-2 engine. And it was still a test vehicle. On March 17 the slender Vanguard launch vehicle lifted off on the first actual flight test of the troubled second stage. It performed well enough this time to place the small 3.4–pound Vanguard I satellite in a highly stable orbit. The strapdown guidance system proved itself by producing an error of less than one degree in the satellite's angle of injection. It did this despite problems with the roll-control jets and a rough start by the second-stage engine. The mission was a huge, if belated, success. The Vanguard team had developed a complex, high-performance launch vehicle in a mere two years, six months, and eight days—only eight days longer than Rosen's initial estimate, though considerably longer than the revised one submitted to the Stewart Committee the second time around.[69]

The remaining Vanguard launches went as shown in table 1.2. Midway through the launching cycle, on November 30, 1958, Project Vanguard transferred to NASA. Project personnel remained at NRL for a time, but NASA delayed further launches until a committee of rocket experts not associated with the project looked at it. The committee proposed some slight changes in test procedures and circuitry, and the project personnel continued with the remaining four launches.[70]

Table 1.2. Vanguard Launches

Vehicle/Date	Description and Results
TV-5 4/28/58	Final test vehicle, like a satellite launching vehicle but with additional instrumentation. Attempted to launch a 21.5–pound satellite. First two stages satisfactory, with stage 2's below-normal thrust offset by stage 1's better-than-normal performance. Stage-2 cutoff sequence interrupted electrically, preventing coasting-flight control system from igniting stage 3. Upper stages coasted to 358 miles of altitude but, having insufficient speed to go into orbit, fell into the ocean.
SLV-1 5/27/58	First nontest satellite launching vehicle. Attempted to launch a 21.5–pound satellite. Normal flight through stage-2 cutoff, when a pressure switch apparently malfunctioned, disrupting the pitch gyro and sending stage 3 to an altitude variously estimated between 1,850 and 2,200 miles but not into orbit.
SLV-2 6/26/58	Stage-2 engine cut off after eight seconds of burning, probably because corrosion of the oxidizer tank caused clogging of filters in the inhibited white fuming nitric acid lines.
SLV-3 9/26/58	Attempted, after flushing the oxidizer tanks of stage 2, to launch a 23.3–pound satellite. Stage-2 performance was below normal, and the satellite narrowly missed orbital speed. Again the problem seemed to be clogging, but this time of a fuel filter, not an oxidizer filter.
SLV-4 2/17/59	Placed the 23.3–pound Vanguard II satellite into orbit.
SLV-5 4/13/59	Suffered violent yaw on stage-2 ignition, apparently from flame oscillation, causing stages 2 and 3 with the satellite to tumble in the pitch plane and fall into the ocean.
SLV-6 6/22/59	In stage 2, with a hydraulic system changed by Aerojet and separation reprogrammed to occur earlier, a previously reliable regulating valve failed after ignition. Helium pressure could not vent, and an explosion sent the vehicle into the Atlantic about 300 miles downrange.
TV-4BU 9/18/59	Test vehicle backup, equipped with ABL's X248 A2 third stage. Launched 52.25–pound X-ray and environmental satellite Vanguard III into orbit.

Sources: Stehling, *Project Vanguard*, 223–42, 269–81; Green and Lomask, *Vanguard*, 283–87; U.S. Congress, "Project Vanguard," 66; memo, Chief of Naval Research to Director, Advanced Research Projects Agency, September 25, 1958, in RG 255, box 1, ARPA folder, NA; Newell, "Launching of Vanguard III," NHRC; and seven Space Activities Summary documents: for SLV-1, SLV-2, SLV-3, and SLV-6 in Vanguard folders 006637–006639 and 006642, NHRC; for SLV-5 in "Vanguard Launch Vehicle 5," folder OV-106121–01, NASM; for Vanguard II in "Vanguard II (Feb. 17, 1959)," NHRC; for Vanguard III in "Vanguard III Satellite Launch Vehicle 7," NASM.

Analysis and Conclusions

Given this inconsistent performance, what, finally, can be said about Project Vanguard, as well as its predecessor, the Viking rocket? Were they successes or failures? What did they contribute to the evolution of launch vehicle technology? Viking was a small program that contributed more than often is realized. Its early use of aluminum was a step toward that metal's extensive use in later missiles and launch vehicles. Viking's gimballing was an advance beyond the swiveling technology used on the MX-774 test vehicle,[71] and designs that Martin drew up during the Viking project led to a further evolution of the technology in Vanguard that prepared the way for extensive use of it on many other missiles and launch vehicles. Viking also afforded experience that helped with Project Vanguard.

Given the limited expectations initially invested in Project Vanguard, it succeeded much better than often has been thought. It did put a satellite into orbit before the end of the IGY, although the grapefruit-sized Vanguard I was smaller than the satellite NRL had intended to orbit. After the end of the IGY, the project put two other satellites into orbit that were of the requisite size, one of them more than double the originally intended size. The Vanguard team did this despite the low priority the Department of Defense allotted to the project and the injunction not to interfere with high-priority missile projects. It also launched the three satellites in the midst of the enhanced expectations that came with the launch of Sputnik and the public clamor it produced. Another obstacle was the splitting up of the Viking team by Martin after it won the Titan I contract. Developing a substantially new launch vehicle of three stages at a time when the technology was still not mature and doing this in a comparatively short time were also notable achievements.[72] They were, however, diminished by the salesmanship that had won Project Vanguard the approval of the Stewart Committee and the Department of Defense. Giving the impression that the project would use stages that were nearly developed and saying it could do so in eighteen months created goals that the team could not meet, as Rosen sensed.

Measured against the contributions of Vanguard to launch vehicle technology, rather than NRL's own initial marketing, the project appears much more successful. The Air Force's Thor-Able launch vehicle used the Thor intermediate range ballistic missile as a first stage and modified Vanguard second and third stages, the last being the original third stage developed by Grand Central Rocket Company. The air arm had better success with the second stage than did the Vanguard Project, for two reasons. One was special cleaning and handling techniques for the propellant tanks that came into being after Vanguard had taken delivery of many of its tanks. Also, Thor-Able did not need to extract maximum performance from the second stage as Vanguard did, so it did not have to burn the very last dregs of propellant. This residue contained a disproportionate amount of scale from the tanks, but the Air Force could close the valves before the scale entered the fuel lines in the regenerative cooling jacket.[73] Thus the important Thor-Able system not only benefited from two stages of Vanguard and took over their technology; it also learned from problems that Vanguard experienced and avoided them.

In January 1959, Rosen proposed to Abe Silverstein, NASA's director of space flight programs, that the Thor-Able be evolved into what became the Delta launch vehicle. Drawing on his experience with Vanguard, Rosen suggested substituting the ABL third stage for the one from Grand Central Rocket, designing more reliable control electronics than Vanguard's, replacing the aluminum combustion chamber in the Vanguard second stage with a stainless steel one, and adopting Bell Telephone Laboratories' radio guidance system then being installed in the Titan ballistic missile. Silverstein commissioned Rosen to develop the Delta launch vehicle along those lines, and it became highly successful.[74]

A variant of the ABL third stage for Vanguard, known as the Altair I (X248 A5), later became the third stage for Delta and the fourth stage for the all-solid-propellant Scout launch vehicle. A follow-on, also built by Hercules Powder Company (at ABL), became the third stage for Minuteman I. And the fiberglass casing for the ABL third stage was a feature of these later stages and found many other uses in missiles and rockets.[75] The strapdown guidance and control system, finally, although it had imperfect electronics, was another contribution to missile technology and to later launch vehicles.

These were significant legacies to American rocketry. Clearly, if they are taken into consideration, Vanguard was an important success. Impressively, the industry-Navy team made this contribution with some 15 people at NRL and a total team of only 180. Its cost was $110 million. "It wasn't first in orbit, but it did its job and lived up to its name of being the vanguard of many space projects to follow."[76]

2

The Thor-Delta Family
of Space-Launch Vehicles, 1958–1990

The Thor missiles did not remain in operational use very long,[1] but even before the Air Force retired them in 1963, it had begun to use Thor's airframe and propulsive elements, including its vernier engines, as the first stages of various launch vehicles. With a series of upratings and modifications, Thor remained in use with such upper stages as Able, Able-Star, Agena, Burner I, Burner II, and Burner IIA until 1980. In addition, NASA quickly chose Thor as the first stage of what became its Thor-Delta (later called just Delta) launch vehicle family, which has had an even longer history than the Air Force's Thor series of launch vehicles did. The Delta launch vehicles initially drew upon Vanguard upper stages, as did the Thor-Able used by the Air Force.

After the initial launches of Thor-Able in 1958 to test reentry vehicles, followed by flights to test the radio-inertial guidance system for the Titan ICBM in 1959, Thor with its various upper stages went on to launch a great variety of spacecraft, including reconnaissance satellites and other classified payloads for the Air Force. Also included were many satellites to gather data about Earth, and an array of communications, navigation, and meteorological satellites, plus some payloads consisting of scientific experiments for DoD. Beginning in 1960, Delta with an even greater variety of upper stages launched planetary exploration probes as well as weather, communications, and scientific satellites. Starting with the passive Echo communications satellite, ranging through Telstar, Pioneer, Nimbus, and Landsat to the Cosmic Background Explorer, and including the early Intelsat communications satellites, Delta provided a reliable and comparatively inexpensive launch platform for a large number of missions and a variety of customers.[2]

Throughout its history, Delta evolved by uprating existing components or adopting new ones that had proved themselves. It used this low-risk strategy to improve its payload capacity through the Delta II at the end of the period covered by this history. But it did not stop there, evolving through a Delta III, first launched (unsuccessfully) in 1998, and a Delta IV that finally had its (successful) initial launch on November 20, 2002. The unsuccessful first (and second) launch of the Delta III and numerous delays in the launch of the Delta IV because of both software and hardware problems suggested, however, that the design of new launch vehicles was still not something engineers had "down to a science" even in the twenty-first century.[3]

Thor-Able

The initial Thor engine used for space ballistic-missile tests had 150,000 pounds of thrust, with the two vernier engines adding 1,000 pounds each. The main engine for this version was the Rocketdyne MB-3 Basic system, followed by a Block I. The Air Force used it with Thor-Able and some of the Thor–Able-Star launches. For some Able-Star missions, the Thor used an uprated MB-3 Block II or Block III engine with 170,000 pounds of thrust. The newer system had slightly higher chamber pressure and specific impulse with a hypergolic ignition system replacing the earlier pyrotechnic one. The vernier engine used with the Block II and III main Thor engines also used hypergolic ignition. The pumps for the main engine fed the verniers until engine cutoff, following which tank pressure fed them for nine seconds after the main engines ceased to fire.[4]

The use of the Able upper stage grew out of the Air Force's desire to test more advanced reentry concepts than the blunt heat-sink type without interfering with ICBM development. In November 1957 the nonprofit Rand Corporation, an Air Force think tank, suggested in a briefing to Lt. Gen. Donald L. Putt, the Air Force's deputy chief of staff for development, that a Thor first stage and a Vanguard second stage would provide a vehicle with numerous applications. Louis Dunn of Space Technology Laboratories (STL) urged the same combination after an analysis of available components for a two-stage test vehicle. The Air Force was able to award direction of the project to STL despite restrictions on the lab's activities, as this was not a production project, with only three flights then scheduled. STL developed and produced the autopilot, guidance, and telemetry systems, with Aerojet providing the second-stage engines, evidently under subcontract to Douglas, the Thor contractor.[5]

Accurate details about the Able stage are maddeningly sparse. We know that the Ables used for the reentry tests rode on a Thor taken from the research-and-development production line and listed as using an MB-1 Basic engine, although it provided the MB-3's thrust of 150,000 pounds. Designated AJ10–40 (as compared with the Vanguard second stage's AJ10–37), the Able engine was a modified Vanguard propulsion unit that still used a regeneratively cooled combustion chamber with aluminum-tubular construction. Able kept the propellants (IWFNA and UDMH), tanks, helium pressurization system, and propellant valves from the Vanguard. New were a control compartment, skirts and structural elements for mating the upper stage with the Thor, a tank-venting and pressurization-safety system, some electrical components, and a roll-control system. Perhaps learning already from Vanguard's experience, STL anticipated that transport of the tanks would dislodge scale, so it flushed them before use. The engine produced a thrust of about 7,500 pounds for roughly 120 seconds.[6]

On its first reentry test, the Thor-Able failed because of a faulty turbopump, the second Thor failure attributed to that problem. The next Thor-Able launch, on July 9, 1958, was successful, although the nose cone could not be recovered before it sank off the African coast. A repeat of this scenario occurred on July 23, but in both cases, telemetry data showed that GE's melamine ablative heat shield for the Atlas could tolerate four times the heating load of the heat-sink warhead used on the Thor missile. Three mice, carried as bioastronautical test specimens on the flights, were also lost when the reentry vehicles sank beneath the waves of the Atlantic.[7]

The next use of Thor-Able was to launch three separate lunar probes. On March 27, 1958, the secretary of defense announced that the Advanced Research Projects Agency (ARPA) would manage several programs to launch spacecraft, with the three Air Force lunar probes assigned to the Ballistic Missile Division (BMD). Again, STL won the contract to oversee modification of the second and (this time) third stages from Vanguard plus the probes in what came to be called Operation or Project Mona (also Pioneers 0, 1, and 2). The Thor engines used for these probes were MB-3s, but the missile's inertial guidance system was too heavy for this mission. Instead, engineers used the Bell Telephone Laboratories radio guidance system developed as a backup for the primary system in the Thor missile. They placed it in the second stage, with both the Thor and Able having autopilots, as had been the case in the Project Able reentry test vehicles (there, without the Bell Labs guidance system). Spin would control the third stage, as in Vanguard. Aerojet designated the Able engine AJ10–41 for the mission, but it was evidently basically the same engine, modified from Vanguard, that had been used in the reentry tests.[8]

Although the Air Force considered using other solid-propellant motors for Project Mona, including the Vanguard third stage produced by Grand Central, it decided on ABL's X248 in its A3 configuration, which had an asbestos-phenolic steel composite exit cone in place of the asbestos-phenolic filament-wound cone developed for Vanguard. Both this and the Vanguard version (X248 A2) underwent extensive testing at the Air Force's Arnold Engineering Development Center, and between the launches of Pioneer 0 and Pioneer 1, ABL increased the insulation outside the propellant grain in an area where there was an internal-burning slot. According to one source, the X248 delivered an average thrust of 2,500 pounds, only slightly lower than the later Altair I (X248 A5) version used as a third stage for the Delta and as a fourth stage for the Scout launch vehicles.[9]

Interestingly, the STL project manager for these launches was George E. Mueller, who later managed the Apollo program for NASA. The project did not schedule any test flights for Thor-Able launch vehicles used for the lunar probes. On August 17, 1958, the first vehicle, Pioneer 0, lifted off from Cape Canaveral and exploded 77 seconds into the launch, another victim of a turbopump gearbox failure in the Thor stage. On the first day of its existence, October 1, 1958, NASA took over direction of the project but delegated authority for it back to the Air Force, which launched Pioneer 1 on October 11. This time the Thor stage performed satisfactorily. The Able stage ignited and, at the proper time, so did the third stage—its first chance to perform in space, since Pioneer 1 and 2 both launched before Vanguard 3. Although the X248 operated roughly as designed, its total impulse was slightly (less than 0.5 percent) below intended performance. This left the spacecraft only a little bit below the speed needed to reach the Moon. The guidance system had also set the trajectory some 3 degrees too high, and the integrating accelerometer had shut down the second-stage engine just below the planned velocity. Pioneer 1 did reach an altitude of 71,700 miles and returned some useful data about the Van Allen Belt and other phenomena.[10]

For the third attempt to launch a lunar probe, engineers modified the launch vehicle to correct for the trajectory deviation on the second launch. They eliminated the arrangement previously used, so that the integrating accelerometer no longer determined second-stage engine cutoff. For flight 3, the radio-inertial control system would send this command using new Doppler equipment. It appeared that there had also been problems with separation between the second and third stages, with "a nozzle blocking condition" at third-stage ignition caused by a "structural asymmetry of the compartment below the third stage nozzle." Since spin rockets rotated both the second and third stages before separation and the spin rate of the second stage had decreased during separation, interference may have arisen between the two stages. To correct this problem, engineers added four additional spin rockets plus two forward-firing retro-rockets to the second stage. This would ensure the proper spin and would separate the two stages by about a foot before third-stage ignition, which would be delayed by one second. All of these modifications went for naught on the third lunar mission, launched on November 8, 1958. Stages 1 and 2 functioned satisfactorily, and the spin rockets provided the proper velocity of 2.1 revolutions per second. But then the third stage failed to ignite, perhaps because of a broken ignition wire, and Pioneer 2 never reached the Moon.[11]

For the next series of launches, the Thor-Able configuration reverted to just the two stages in what was called the Precisely Guided Re-entry Test Vehicle (PGRTV). These launches used a Thor with an MB-3 Basic engine and an Able with a propulsion system designated AJ10–42 to test the Bell Labs radio-inertial guidance system, built for the Titan missile. Testing of both AVCO and GE ablative materials on nose cones also continued. There were six PGRTV launches from Cape Canaveral between January 23 and June 11, 1959. On the first, an electrical malfunction at main-engine cutoff precluded ignition of the second stage, called Able 2 in this series. On the next five launches, both stages performed satisfactorily. Some of the nose cones were lost, but those that were recovered and the telemetered data from the others showed that the radio-inertial guidance system from Titan worked well, and the missions provided information for selection of ablative materials for ICBM heat shields. The series also demonstrated that the Thor-Able launch vehicle could function satisfactorily for multiple missions.[12]

The final Thor-Able missions, launched between August 7, 1959, and April 1, 1960, used the three-stage configuration plus a spacecraft with its own injection rocketry. Leading off was the launch of the Explorer 6 satellite, NASA's first in that series, by Thor-Able 3. The stage-1 Thor had an MB-1 engine rated, according to one source, at 153,248 pounds of thrust at sea level. Stage 2 used an AJ10–101A engine, a modified version of the AJ10–40 with thrust at altitude listed at 7,664 pounds and a specific impulse of more than 270 lbf-sec/lbm, while stage 3 used the X248 A3 motor with specific impulse listed at more than 250 lbf-sec/lbm and average thrust at 3,150 pounds. The onboard equipment for the radio-inertial guidance system used in the Able stage on earlier missions was too heavy, at 120 pounds, for this mission, which required the placement of the 140-pound Explorer 6—the most sophisticated American payload to date—in a highly elliptical orbit extending from a perigee of 115 miles to an apogee of 24,618 miles. STL, which continued to be the contractor for these missions, designed a new three-axis attitude control system weighing only 33.5 pounds. It included circuit boards and three integrating gyros that were heat regulated. Although it would fly "open loop" on this mission and not actually control the vehicle, this system was capable of sending signals to an electronic assembly that would activate servos to gimbal the engine and correct the attitude in pitch and yaw. Separate electronic channels selectively operated four pneumatic jets to control roll. For this mission, ground transmitters and receivers would control the second stage and the satellite, but for following missions the onboard equipment would provide some actual control. The launch was successful, and Explorer 6 yielded much of value, including the first comprehensive "mapping" of the Van Allen belt, the first televised photo of cloud cover, and the discovery of a system of large electrical currents in the outer atmosphere.[13]

The next two Thor-Able missions, lumped together as Thor-Able 4, were the attempted launch of the Navy's Transit 1A navigation satellite and the successful launch of the Pioneer 5 space probe. The Transit 1A mission used the Bell Labs system for guidance, an X248 A7 for the third stage, and in stage 2 a new retro system. Project engineers had discovered in laboratory tests that the retro-rockets, installed during Project Mona to ensure separation before stage-3 ignition, posed a significant danger of contaminating the payload, as their exhaust products could condense on the satellite. So they substituted a cold-gas system in which high-pressure nitrogen was supposed to furnish the thrust for stage separation. On September 17 the Thor-Able launched on its mission to place the 265–pound satellite in orbit, but the third stage failed to ignite. Examination of flight data showed that the new retro system did not operate. This probably allowed the second stage to bump the third repeatedly and sever the ignition wires. But before the satellite returned to the atmosphere and burned up, it transmitted information that helped the Navy conclude it would be possible to "establish correction factors for refraction of signals through the ionosphere" so that such a satellite could provide "a highly accurate global all-weather means of fixing precisely the position of ships and possibly aircraft." Such a system would allow Polaris submarines to update their inertial systems every few days from single passes of a Transit satellite.[14]

Initially planned as a probe to fly past the planet Venus, Pioneer 5 experi-

Figure 2. Thor-Able 3 launch on August 7, 1959, from the Atlantic Missile Range. The payload was Explorer 6, for meteorological study and other purposes. Courtesy of NASA.

enced technical difficulties in development by STL and became instead a so-
lar-orbiting investigator of magnetic fields, radiation, solar phenomena, and
cosmic dust. The Air Force's development plan called the Thor-Able for the
mission "essentially the same" as for Able 3, but the launch vehicle for Pioneer 5 reportedly had an MB-3 Thor and an X248 A4 third stage, although it
is not clear that the new configurations changed the thrust parameters. The
STL space guidance system on the Able second stage was scheduled to operate closed-loop (actually controlling the vehicle) for this launch, with a fuel-
injector pressure switch designed to cut off the first-stage engine. It would
then initiate separation of the second stage, while a relay activated by the
same signal would trigger second-stage ignition and the "uncaging" of the
gyros so they could begin to function. At some point in the sequence, radio
commands from the ground would take over control of the vehicle, including cutoff of the stage-2 engine. Timers would ignite spin rockets for stage 3
and ignite the X248 motor, which would burn to depletion of its propellants.
Radio command would also initiate separation of Pioneer 5 from the third
stage. Except for the closed-loop operation, the second-stage control system
was the same as for Able 3, including a Burroughs J-1 digital computer on
the ground at the Cape.[15]

The Thor-Able 4 vehicle successfully launched the 94.8–pound Pioneer 5
into its solar orbit on March 11, 1960. When the third stage burned out, the
velocity actually exceeded what was necessary. The probe's transmissions
from interplanetary space confirmed the existence of magnetic fields that
scientists had previously only suspected and yielded other significant data.
Pioneer 5 ceased to transmit on June 26, 1960, at what was then a record
distance of 22.4 million miles from Earth, later exceeded by Mariner 2.[16]

The final Thor-Able launch, on April 1, 1960, again used the three stages,
this time with a reversion to the Bell Labs guidance system and a redesigned
retro system. Its payload, the 269.5–pound meteorological satellite Tiros 1,
would return 22,952 cloud-cover images during its 78–day useful life, preparing the way for NASA to provide daily cloud-cover images to weather
services around the world following the launch of Tiros 4 (by a Thor-Delta)
on February 8, 1962.[17]

Of the sixteen Thor-Able launches, all stages worked satisfactorily on ten,
for a 62.5 percent success rate—sufficiently good for this early period that
the Air Force could refer to Thor-Able as "an extremely capable and reliable vehicle combination." Long before the final Thor-Able launch, however,
ARPA had issued an order on July 1, 1959, calling for development of the
Able-Star upper stage, derived from the Able but offering two and a half

times the total impulse plus the ability to shut down, coast, and restart in space. This feature permitted a more precise selection of orbit for a satellite than was possible before. Soon after Able-Star became operational in January 1960, it replaced the Able as an upper stage.[18]

Thor–Able-Star

The Air Force not only continued to use STL for systems engineering and technical direction of development for the Thor–Able-Star launch vehicles, it also chose STL to design and produce the new second stage. This latter responsibility amounted to an oversight role, however, because in the fall of 1959 the Air Force Ballistic Missile Division awarded Aerojet a contract to do the actual design and production of Able-Star's propulsion system and electronics. Aerojet in turn contracted with Space Electronics Corporation, a subsidiary of the Glendale, California–based Pacific Automation Products, to design and produce the upper stage's electronic systems, which included the onboard guidance and control equipment. Aerojet apparently liked the work its subcontractor did, because it soon acquired Space Electronics and merged the firm with its own Spacecraft Division to form a subsidiary known as Space General Corporation. A radio guidance system that could override the onboard equipment was built by Bell Labs, perhaps the same system used with many Thor-Able flights.[19]

Aerojet could develop the Able-Star engine (AJ10–104) in a matter of months both because it was derived from the Able engines and because it was simple. The Air Force and/or STL told the Aerojet engineers to make the system rugged, with only those subsystems and components needed to meet the requirements for restart, attitude control during coasting periods, and longer burning time than the Able provided. As an Aerojet publication states, "The objective during its development was to achieve maximum flight capability through limited redesign, overall simplification and optimum utilization of flight-proven components."[20]

Aerojet designed and built the combustion chamber to be "practically identical" to the ones used on the Able upper stages, so it remained an aluminum, regeneratively cooled, pressure-fed device. For reasons that are not stated but may have involved the Air Force's desire to have the same propellants for Agena and Able-Star, the latter switched from the IWFNA used in Able to IRFNA (inhibited red fuming nitric acid) as the oxidizer, keeping UDMH as the fuel. Helium under pressure continued to feed the propellants to the combustion chamber, where the injector had concentric rings of orifices that mixed the hypergolic IRFNA and UDMH in an impinging-stream pattern. There were three helium containers made of titanium to supply the pressurizing gas. Experience had suggested that a nozzle closure diaphragm, previously used to ensure high-altitude starts, was not needed. Another change from Able was an optional nozzle extension that altered the expansion ratio from 20:1 to 40:1. Rated thrust rose from 7,575 to 7,890 pounds with the nozzle extended, and rated specific impulse likewise climbed from over 260 to almost 275 lbf-sec/lbm.[21]

Space Electronics' onboard guidance system used a strapdown inertial reference system with three gyros sensing movement in the roll, pitch, and yaw axes. These plus a sequencer and a programmer fed into an "electronics package," whose "logic elements" (computer) signaled engine gimbals and a gas-jet roll control system to adjust the attitude of the vehicle. A nitrogen pressure vessel provided pitch and yaw control as well as roll control following the first burn of the Able-Star while it was coasting in preparation for a second burn to inject the spacecraft into the proper trajectory. The radio guidance system could override the programmer, but the programmer was necessary during liftoff and on a long coast, when the Able-Star would be beyond radio control in its orbit around Earth. The radio link did cut off the engine's first burn. The programmer signaled the second burn, and an integrating accelerometer signaled cutoff when the vehicle reached the programmed velocity. Before the second burn, the pressurized nitrogen provided additional impulse via "positive acceleration jets" to settle the propellants to the rear of their tanks in zero gravity so the helium could force them into the combustion chamber for hypergolic ignition.[22]

Although Aerojet's design and development of Able-Star were quick, they were not problem-free. The virtually identical combustion chambers for the Able engines required only 115–second test firings. Able-Star, with its longer tanks and increased burn time, had to undergo five 300–second static test firings. In November 1959 a new design for the injector manifold, apparently occasioned by the conversion to IRFNA as oxidizer and coolant, caused a burn-through of the injector plate and adjacent cooling tubes. In a piece of cut-and-try engineering, Aerojet made unspecified adjustments to the designs, and later that month two combustion chambers operated successfully for the full 300 seconds.[23]

The Air Force used the Able-Star upper stage mated with two different configurations of the Thor for the Transit 1B through 4A navigation satellites, the Courier 1A and 1B communications satellites, and the ANNA 1B (ANNA being Army, Navy, NASA, Air Force) active geodetic satellite, together with a number of secondary (piggyback) payloads on these missions. The Transit 1B launch on April 13, 1960, marked the first programmed restart of a rocket engine in space, but although the coast attitude-control system worked, a malfunction in the Able-Star ground guidance system resulted in an elliptical rather than a circular orbit. STL attributed this rather vaguely to "noise in the ground guidance system data [that] resulted in angular ambiguities which were not correctly resolved" but said the orbit was still "useful." For Transit 2A, launched on June 22, the orbit was again elliptical. This time the fault lay with sloshing of the propellants in the stage-2 tanks, which produced roll forces. To overcome the roll, the attitude-control system used up the supply of nitrogen gas. This left no source of control forces during the coast period, no acceleration to settle the propellants before engine restart, and no control over roll during the second burn. The Able-Star nevertheless did restart and placed Transit 2A in a usable orbit. The two Transit satellites in orbit at the end of 1960 were "completely successful." Experiments to determine position using the two spacecraft proved that "fixes" from them were accurate to "within one-quarter of a mile."[24]

To correct the sloshing problem, engineers and technicians placed anti-slosh baffles in both Able-Star propellant tanks. On November 30, 1960, a totally different problem prevented placing Transit 3A in a circular orbit. A premature cutoff signal shut the Thor engine down early. The range safety officer commanded the destruction of stage 2 to protect South America, but the remaining Thor stage landed in Cuba. Investigation suggested the inadvertent signal was due to a leak or plug in a pressure transmission line, a faulty pressure switch (or switches), or a malfunctioning pressure sensor. STL urged modifications to make the circuit more reliable through new switches and better quality control. For Transit 3B, the Thor stage worked fine on February 21, 1961, but the navigation satellite and a secondary payload, the Lofti (low-frequency trans-ionosphere) satellite, wound up in a lower and more elliptical orbit than intended. The two satellites failed to separate from one another or the Able-Star stage, and they reentered the atmosphere after only 37 days in orbit. The culprit was a malfunctioning "counter" in the second-stage programmer, which failed to signal three things: (1) the restart of stage 2, (2) the spinning and (3) the separation of the payloads. On June 28 a vehicle substantially the same as Transit 3A's successfully launched Transit 4A into a nearly circular orbit. It also placed in orbit two secondary payloads, which failed to separate from one another but were "fully functional."[25]

Sometime before mid-October 1961, STL transferred responsibility for this program to the Aerospace Corporation. Then on November 15 a Thor–

Figure 3. A Thor–Able-Star unsuccessfully launching the Transit 3A satellite on November 30, 1960. Official U.S. Air Force photo, courtesy of the 45 Space Wing History Office, Patrick AFB, Fla.

Able-Star launched Transit 4B into a near-circular orbit along with a secondary payload designated Transit Research and Satellite Control (TRASC) to research a way of stabilizing satellites in space through natural gravitational forces. This was the last Able-Star launch slated for the Transit program, with future launches planned on Scout vehicles to be launched from the West Coast over the Pacific Missile Range. A Blue (military) Scout vehicle in fact launched the first prototype of an operational Transit satellite, designated 5A, on December 18, 1962, and in January 1964 the overall Transit system became operational.[26]

Meanwhile, still under STL technical direction, a Thor–Able-Star attempted to launch a Courier delayed-repeater communications satellite on August 18, 1960. The launch vehicle was essentially identical to the one that launched Transit 2A, with baffles added to limit propellant sloshing. Unfortunately, a failure in the stage-1 hydraulic system led to loss of control, structural breakup, and explosion. After modifications to the Thor hydraulic system, the Courier 1B satellite was correctly launched on October 4 into a near-circular orbit at an altitude of slightly more than 500 miles. There was no evidence of propellant sloshing on the flight. The purpose of the satellite was to test a method of relieving crowded communications lines by delayed relay of information. The experiment was a success, with large amounts of data transmitted between Puerto Rico and New Jersey. However, the delay of up to two hours before the message was repeated while the satellite completed its orbit was unsatisfactory for telephone transmission and military purposes. The future lay with much higher, geosynchronous orbits, where a satellite was positioned so that it hovered above a given location on the rotating Earth, making nearly instantaneous relay of messages a reality.[27]

Between September 28, 1963, and August 13, 1965, a series of seven Thors with MB-3 Block 2 and 3 engines offering 170,000 pounds of average thrust launched classified Navy and Air Force payloads, with one failure on April 21, 1964. According to Mark Wade's Thor–Able-Star Internet site, most of the launches included a Transit satellite, often in conjunction with other payloads, although why these satellites should have been launched on Thor–Able-Stars rather than Scouts is not clear. All of the launches in this series were from Vandenberg AFB. They concluded the Thor–Able-Star launches, of which there were twenty in all, with a success rate of 75 percent.[28]

Thor-Agena

Even before the first launch of Thor–Able-Star, the Air Force had begun using an Agena upper stage with the Thor, and for more than a decade this combination became a preferred choice for a great many missions, often classified, including those to place a family of reconnaissance satellites in orbit under what began as the WS-117L program. The initial Thors in this combination had MB-1 Basic engines. In three versions from A to D (without a C), the Agena also operated with uprated Thors, Atlases (for which it was originally designed), and Titan IIIs to orbit a great many military and not a few NASA spacecraft.[29]

The Air Force began the development process for the Agena in July 1956. On October 29 the service selected Lockheed Missile Systems Division as prime contractor for development of both the WS-117L reconnaissance satellite system and an associated upper stage that became the Agena. The Missile Systems Division had just moved from Burbank near Los Angeles to a new facility in Sunnyvale, California, near both the NACA's Ames Aeronautical Laboratory and Stanford University. The engine for the Lockheed upper stage was a modified version of the Hustler propulsion unit (model 117) that Bell Aerospace had developed for the B-58 bomber's air-to-surface missile, designated the Powered Disposable Bomb Pod. The Air Force cancelled the missile project, but its engine, with 15,000 pounds of thrust, had by then undergone preliminary flight rating tests, in May 1957. It burned IRFNA and JP-4 and was available that autumn for application to the Agena when Lockheed contracted with Bell for the engine.[30]

The Lockheed Aircraft Company had been in the business of building airplanes since 1926. During the Korean War it had expanded into the missile field, with its Missile Systems Division formed in 1954. Soon after it got the Agena contract, the division also won one with the Navy for Polaris. Lawrence D. Bell had founded Bell Aircraft in 1935. Initially known primarily for its helicopters, the firm became recognized during and after World War II for the XP-59 Airacomet jet airplane and the X-1, X-2, and X-5 experimental research aircraft. When they developed the Hustler rocket engine, Bell engineers also produced a 12,000–pound rocket engine for the Air Force's Rascal air-to-surface missile. The Rascal engine's extreme complexity led them to seek a simpler system for the Agena. In the course of Agena's development through the D-model, Bell Aircraft was purchased in 1960 by Textron and became that firm's Bell Aerospace Division.[31]

Despite their pursuit of simplicity, for the Agena engine Bell engineers still designed and tested 26 components consisting of 1,860 individual parts. Since it is unusual to find such a listing, it is worth noting that the engineers, designers, and technicians involved in the project included specialists in pumps, valves, gas turbines, tanks for propellants, flow systems, and combustion devices. A separate group of specialists did the testing, while systems engineers oversaw the entire effort and ensured that the various components would work together. By the time Lockheed selected it for the new Agena (A-model) upper stage, the Bell model 8001 (XLR81BA-3) was practically an "off-the-shelf" engine because of the work on the Hustler version. One addition was gimballing to provide control in pitch and yaw. The gimbals necessitated relocation of a turbine exhaust duct so that it would not interfere with the motion of the engine. A third noteworthy change was a nozzle closure to ensure that the engine started in space after first-stage cutoff. Finally, the burn time increased from 60 to 100 seconds. Like the Hustler engine, the model 8001 burned IRFNA and JP-4. The two propellants yielded a specific impulse (presumably at altitude) of 265.5 lbf-sec/lbm. A small solid-propellant starter cartridge began operation of the gas turbine that drove the propellant pumps.[32]

The Agena stage with the Bell 8001 engine flew only once, on February 28, 1959, for the launch of Discoverer 1 by a Thor–Agena A. "Discoverer" was the name publicly released for the secret Corona reconnaissance satellites, which had been separated from the WS-117L program by this time. WS-117L retained the Sentry program, later called Samos, which also developed reconnaissance satellites. But to conceal its real purpose, Discoverer was billed to the public as a scientific and engineering program. On the outcome of this first Agena launch into a polar orbit from Vandenberg AFB, accounts differ, some claiming the launch itself was successful and others not, but in any event, the satellite's telemetry system failed.[33]

Before this launch, in 1958 Lockheed awarded Bell a second contract, to improve the Agena engine's performance. Bell changed the fuel to UDMH, which ignited spontaneously upon contact with IRFNA, still the oxidizer. This of course eliminated the need for an ignition system but did still require a solid-propellant starter cartridge to begin the flow of propellants into the combustion chamber by powering the propellant pumps. With the new fuel, the specific impulse rose from 265.5 to 277 lbf-sec/lbm. The new engine had a longer thrust chamber, which provided a more efficient expansion of the gases produced in combustion. And the duration of combustion was longer, 120 seconds instead of 100. Now called Bell model 8048, the engine used a cavitating venturi tube to maintain the propellant flow from the turbopumps to the combustion chamber at a constant level. (Venturis are essentially pipes featuring converging and diverging orifices. They enable measurement and regulation of the flow rate of fluids.) A final change was an increase in the nozzle expansion ratio from 15:1 to 20:1. Taken together, these modifications increased by 500 pounds the weight of payload the Agena could put into orbit.[34]

Bell delivered the 8048 engine to Lockheed during the fall of 1958, and it first flew on April 13, 1959, for the launch of Discoverer 2. Including this mission, Thor–Agena A launches totaled fourteen by September 13, 1960, with the Agena failing three times and the Thor twice. This yielded only a 64 percent success rate overall, but 79 percent for Agena alone, although not all of the satellite operations were themselves successes even when both stages of the launch vehicle worked.[35]

Meanwhile, late in the spring of 1959 Lockheed awarded Bell a third contract, this time to double the burn time of the engine and provide a dual-start capability so the upper stage could enter a coasting or "parking" orbit and then restart to inject a satellite into another orbital path. To enable restart, Bell added a second solid-propellant starter cartridge on the model 8081. But this left the problem of nozzle closure, as the closure used on earlier engines would already have been breached by the first start on separation from the Thor stage. Bell found the solution at the Arnold Engineering Development Center. Model 8048 had been the first engine tested in a high-altitude test chamber at Arnold, which simulated altitudes of 80,000 to 100,000 feet. In 1959, tests of model 8081 in this chamber showed that use of an "oxidizer lead," in which the IRFNA preceded the UDMH into the combustion chamber, ensured a start in space without the nozzle closure. Doubling the burn time of course required larger propellant tanks; while the diameter of the upper stage remained constant at 5 feet, the length increased from 14 feet 3 inches to 20 feet 8 inches. This became the engine for Agena B, which first flew on October 26, 1960, in a failed attempt to launch Discoverer 16.[36]

Information is limited, but it appears that between October 26, 1960, and May 15, 1966, there were as many as 48 launches of Thor–Agena B vehicles and about 39 were successful. If these two statistics are correct, they yield an 81 percent success rate. Not all of them carried Discoverer (Corona) satellites, but most did, ranging from the failed launch of Corona 16 to the successful orbiting of Corona 56. Besides other classified launches for the Department of Defense, the Thor–Agena B placed some NASA satellites in orbit. Among the latter was a Canadian satellite for studying the ionosphere, Alouette 1, which NASA launched on September 28, 1962. Designed by Britain's de Havilland Aircraft Company, Alouette 1 reportedly was the first spacecraft developed by a country other than the Soviet Union or the United States. Also included in the Thor–Agena B launches were the meteorological satellites Nimbus 1 and 2, placed in orbit on August 28, 1964, and May 15, 1966. Nimbus 1 transmitted some 27,000 exceptionally clear images during a period of 27 days, after which the solar array panels no longer could track the sun and the mission had to be terminated. Nimbus 2 transmitted data for longer than 32 months between 1966 and 1969, including satellite images that appeared on many televised news and weather broadcasts.[37]

By the time of Nimbus 2, the thrust-augmented Thor was available as a booster for the Agena B, allowing NASA to design a heavier payload. Whereas Nimbus 1 weighed a little less than 830 pounds, Nimbus 2 tipped the scales at more than 910 pounds. The thrust-augmented Thor came into the launch vehicle inventory in 1963. It consisted of a Thor with about 170,000 pounds of thrust and three Thiokol TX-33–52 (Castor I) solid-propellant, strap-on rocket boosters, which increased liftoff thrust to 331,550 pounds. These Castor I's were attached to the outside of the Thor stage, hence the term "strap-on," which did not mean that the booster literally was strapped on, only that it was connected to the rest of the Thor or other launch vehicle by some sort of hardware such as explosive bolts. These bolts would explode upon actuation and separate the strap-ons from the next stage. The TX-33–52 is sometimes designated as the Sergeant booster, and indeed it was a descendant of the line beginning with the RV-A-10 motor that continued with the Sergeant. But the TX-33–52 belonged to the Castor series that was developed further as the second-stage motor for the Scout launch vehicle and used polybutadiene-acrylic acid as its binder (first used in the Minuteman, not the Sergeant) with ammonium perchlorate as the oxidizer and aluminum as the principal fuel. Besides Nimbus 2, the thrust-augmented Thor and Agena B launched a classified Air Force mission and Explorer 31. The Explorer launch on November 29, 1965, also carried Alouette 2. The two satellites constituted a joint project of NASA and Canada's Defence Research Board to continue studies of the ionosphere.[38]

Much earlier, in the fall of 1959 Bell began designing the engine for the Agena D, which became the standard Agena propulsion unit. This Bell model 8096 had the same duration as the 8081 but included a nonregeneratively cooled nozzle extension made of titanium with an expansion ratio that was more than doubled to 45:1. Bell engineers chose an extension cooled only by the radiation outward of the heat. A design analysis determined that use of the fuel to cool the nozzle posed too many complexities. The titanium did require intersecting molybdenum hoops and stringers to prevent distortion of the nozzle shape by the pressure of the hot exhaust gases. Simulated altitude testing at the Arnold Engineering Development Center showed that the nozzle extension withstood temperatures of more than 2,000°F. It, the higher expansion ratio, and a redesigned injector raised the specific impulse of the 8096 engine above 290 lbf-sec/lbm, bettering the 8081 model by 10 lbf-sec/lbm. The newer engine, using the same propellants, yielded a thrust of 16,000 pounds.[39]

Most of the Agenas apparently used essentially the same guidance and control system, basically a strapdown gyroscopic reference system built by Minneapolis-Honeywell with two hermetic integrating gyros that sensed changes in the pitch and yaw axes plus a miniature integrating gyro for sensing roll deviations. Two horizon sensors aided in determining the correct attitude of the vehicle. An electronics system including a sequence timer operated a nitrogen cold-gas-jet system for attitude control during coasting periods, and for roll control during engine operation when gimballing provided the pitch and yaw control. Supplementing the onboard equipment was a Bell Labs radio guidance system for making adjustments from the ground and for starting or resetting the onboard timer. Bell itself built an integrating accelerometer called a "single-axis velocity meter" that detected acceleration and fed the information to a "binary counter" (computer) programmed to shut down the Agena engine at the proper velocity for both the first and second burns. A backup stabilization system used Earth's magnetic field to determine the proper attitude of the Agena in orbit if the primary system failed.[40]

The first launch of a Thor–Agena D seems to have occurred on June 28, 1962, with a favorable outcome for Corona 45. The first launch of a thrust-augmented Thor also featured an Agena D, but the Thor failed in the attempt on February 28, 1963. The last Corona launch using a Thor–Agena D combination was apparently the successful orbiting of Corona 145 on May 25, 1972, by which time both Thor and Agena D had evolved further. Meanwhile, NASA had made use of Thor–Agena D combinations to launch Orbiting Geophysical Observatory 2 on October 14, 1965; the Pageos 1 satellite on June 23, 1966, in support of the National Geodetic Satellite Program; Nimbus 3 on April 14, 1969; Orbital Geophysical Observatory 6, the last of the series, on June 5, 1969; the Space Electric Rocket Test (SERT) on February 3, 1970, an overall unsuccessful mission because of shorts in the ion thrusters being tested but a successful launch; and Nimbus 4 on April 8, 1970. The two Nimbus missions and the SERT mission employed the long-tank, thrust-augmented Thor first stage, known as Thorad for short.[41]

Already in January 1966 the Air Force's Space Systems Division (SSD) had announced contracts for a stretched or long-tank Thor to replace the thrust-augmented Thor (TAT). Douglas would provide the longer Thor. Thrust augmentation would continue to come from Thiokol, but Thorad's three strap-ons would be Castor IIs rather than the Castor I motors used on the earlier first stage. The Thorad was more than 70 feet long, as compared with 56 feet for the TAT. The added length came mainly in the form of extended

Figure 4. Agena target docking vehicle as seen from the *Gemini 8* spacecraft. Agena was not only a target in Project Gemini but also an extremely important upper stage for a great many satellite launches. Courtesy of NASA.

tanks that increased the burn time of the first stage. For the Castor II (TX-354–5), basically developed (as TX-354–3) in 1964 for the Scout second stage, among other applications, Thiokol kept the steel case used on Castor I but substituted carboxy-terminated polybutadiene for the polybutadiene-acrylic acid used as the binder for the earlier version, keeping aluminum and ammonium perchlorate as fuel and oxidizer. This increased the specific impulse from less than 225 for Castor I to more than 235 lbf-sec/lbm for Castor II and the total impulse from 1.63 to 1.95 million pounds—which in turn enlarged the payload capacity for the Thorad by 20 percent over the thrust-augmented Thor.[42]

In April 1967 the Air Force contracted with Lockheed for an uprated Agena, with Bell developing the improved engine. Bell engineers changed the propellants to high density acid (HDA) and a combination of UDMH and silicon oil (called USO). HDA contained 44 percent nitrogen tetroxide instead of the 14 percent in inhibited red fuming nitric acid. The added nitrogen tetroxide increased engine performance but also raised the combustion temperature. The silicon oil provided a film coating for the injector face and the walls of the combustion chamber to protect them from burn-through. Most other aspects of the new engine, designated Bell model 8096–39, remained unchanged, but the new propellants increased the specific impulse from about 290 to 300 lbf-sec/lbm. Tests on the uprated engine began in 1971, and it was flight qualified by 1974.[43]

Available trustworthy data appear to be inadequate to tally the number of successful TAT and Thorad launches with Agena D as the second stage. The last use of TAT–Agena D may have been on January 17, 1968, with the launch of a classified Air Force mission. This was also the 150th Thor–Agena launch. After the May 25, 1972, launch of Corona 145, the Agena D continued to serve as an upper stage on Atlas and Titan boosters until 1987.[44]

Thor–Burner I and II

Another series of upper stages used with the Thor first stage included Burner I, Burner II, and Burner IIA. Burner I actually bore little relation to Burners II and IIA. Information about it is sparse, but sources refer to it as the Altair, a derivative of the Vanguard third stage developed by Hercules Powder Company at the Allegany Ballistics Laboratory. The version used to launch Vanguard III used the X248 A2 engine. (An X248 A3 was the third stage engine used in many of the Thor-Able launches.) But the sources for Burner I also state that it used the Altair powerplant employed as the fourth stage of the Scout launch vehicle. This was the X248 A5, which provided propulsion for the Delta third stage as well. There appear not to have been any highly significant changes in the motor from the A2 to the A5 version, with the near-vacuum specific impulse remaining at about 255 lbf-sec/lbm at 40°F and the average thrust being about 2,600 pounds.[45]

The first launch of the Thor–Burner I occurred on January 18, 1965, and the last on March 30, 1966. There apparently were only four such launches, all from Vandenberg AFB into sun-synchronous orbits, one of them a mission failure. The spacecraft were classified at the time but apparently were Block 4A Defense Satellite Applications Program meteorological satellites to inform the U.S. military of weather conditions for launching reconnaissance satellites and other defense purposes such as mission planning during the conflict in Vietnam. In 1973 the program became the Defense Meteorological Satellite Program and soon was no longer classified.[46]

Burner I got so little use because of the development of Burner II. Conceiving a need for a guided upper stage that was low-cost and could be used with more than one first-stage vehicle, on September 2, 1961, the Air Force's Space Systems Division awarded study contracts to the Boeing Company and Ling-Temco-Vought leading to development of what became Burner II. As a result of its initial work, Boeing won a fixed-price contract worth $6.5 million on April 1, 1965, to provide one ground-test version and three flight versions of the new upper stage. By September 15, Maj. Gen. Ben I. Funk, commander of SSD, could announce the development of the new stage, which became the smallest maneuverable upper-stage vehicle in the Air Force inventory.[47]

The primary propulsion for Burner II came from the Thiokol Star 37B motor, a spherical design promoted by a NASA engineer of Cajun heritage from Louisiana named Guy Thibodaux and others. Originally designed as the main retro-rocket for the Surveyor spacecraft, the motor needed only increases in propellant and structural strength to adapt to the needs of an upper stage. The spherical shape, having the greatest volume per surface area of any solid configuration and also the smallest mean curvature among all convex forms, allowed the motor to contain more propellant than a cylinder of comparable volume while also lowering the stress on the case, yielding a high mass fraction of 0.91. The propellant used a carboxy-terminated polybutadiene binder with aluminum as the fuel and ammonium perchlorate as the oxidizer, configured with an eight-point star and a steel case. This yielded a specific impulse of about 265 lbf-sec/lbm and an average thrust of 9,680 pounds during a burn time of slightly over 42 seconds.[48]

The overall Burner II stage was 65 inches in diameter and 68 inches high, excluding a nose shroud. Besides the 37–inch-diameter motor, it included a hydrogen-peroxide reaction control system for correcting the trajectory in pitch and yaw, a system of eight cold-nitrogen-gas thrusters for control in roll (plus pitch and yaw when the stage was coasting), and a strapped-down inertial guidance-and-control system. Using data from a gyro reference unit, a programmer and associated electronics equipment directed both the nitrogen and hydrogen-peroxide reaction control jets on the Burner II. The guidance and control system on Burner II also provided steering commands to the autopilot on the Thor booster before separation and directed ignition of the upper-stage motor.[49]

Between September 15, 1966, and February 17, 1971, Thor–Burner II vehicles launched four Block 4A, three Block 4B, and three Block 5A Defense Satellite Applications Program weather satellites from the Western Test

Range. During this same period, the Thor–Burner II also launched scientific satellites as part of the Department of Defense's Space Experiments Support Program managed by SSD. The first such launch occurred on June 29, 1967, at Vandenberg AFB. It placed an Army Sequential Correlation of Range satellite and a Navy Charged Particle and Auroral Measurements satellite in orbit. On February 16, 1971, the second such launch by a Thor–Burner II placed a radar calibration and drag sphere for the Naval Research Laboratory in orbit, followed on June 8, 1971, by the successful launch of a celestial infrared measurements satellite and some spacecraft attitude sensing devices.[50]

The Block 5B versions of the Defense Satellite Applications Program weather satellites were about twice as heavy as the 5A, necessitating increased thrust for Burner II. Consequently, the Air Force's Space and Missile Systems Organization (SAMSO) contracted with Boeing for an uprated Burner II that became Burner IIA. (The Air Force Systems Command created SAMSO on July 1, 1967, to bring the Ballistics Systems Division and Space Systems Division into a single organization headquartered in Los Angeles at the former SSD location.) Boeing did the job with a minimum of modifications by adding a Thiokol Star 26B motor to form a second upper stage, with the control system moved to the new stage. The new motor was again spherical, with an identical propellant mass fraction to that of Star 37B but a lighter case made of titanium. The propellant was almost the same, and the smaller (26–inch-diameter) motor added 7,745 pounds of thrust on average over a burn time of almost 18 seconds.[51]

With the Burner IIA, a Thor first stage launched five Block 5B and two Block 5C meteorological satellites from October 14, 1971, to May 24, 1975, in what became the Defense Meteorological Satellite Program (DMSP). A final Thor–Burner IIA launch on February 18, 1976, failed because the Thor ceased firing prematurely. This unsuccessful launch attempt marked the last use of the Burner IIA. It was not the end of the DMSP program, however, because a Thor coupled with a Thiokol Star 37S motor in a titanium case (rather than the steel used on the Star 37B) launched four improved Block 5D weather satellites between September 11, 1976, and June 6, 1979. The new motor yielded an average thrust of 9,790 pounds of thrust over a burn time exceeding 42 seconds. It was not a part of an upper stage per se but an integral element of the satellite, making for weight savings. Then Atlas Es and Titan IIs launched ten more DMSP satellites between 1980 and 1999. The last of these satellites came under the control of the National Oceanic and Atmospheric Administration rather than the Department of Defense after a merger designed to reduce government expenses by allowing the civilian agency to oversee all U.S. meteorological satellites.[52]

Thor-Delta

A final major use of Thor as the first stage of a launch vehicle was in the Thor-Delta, subsequently referred to as the Delta launch vehicle. Whereas the other Thor-based launch vehicles were primarily Air Force assets sometimes used by NASA, Delta was a NASA-developed space-launch vehicle used on occasion by the Air Force until near the end of the period covered by this book, when the Air Force began to make extensive use of Deltas II and III, with a Delta IV on the horizon at the beginning of the twenty-first century. Originally conceived by NASA in 1959 as an interim vehicle to lift medium payloads by using existing technology, with modification only as needed for specific missions, Delta has enjoyed a remarkably long career.[53]

As partly explained already, the idea for Thor-Delta seems to have come from Milton Rosen. Rosen was then working at NASA Headquarters in the Office of Space Flight Development headed by Abe Silverstein. His immediate supervisor was Abraham Hyatt, who had become assistant director for propulsion following a decade of work at the Navy's Bureau of Aeronautics. Hyatt assigned Rosen, his chief of propulsion development, the task of putting together a national launch vehicle program including all vehicles in use or planned by NASA and/or the Department of Defense. Once Rosen completed the report, NASA sent it to President Eisenhower in January 1959. As Rosen later wrote, "It was from working on this report that I became aware of the need for a modest-sized, highly versatile launch vehicle that would be useful for a large number of scientific missions." He realized that it needed to be "reliable, which meant it should call for a minimum of new development, and finally, it would have to be available as soon as practicable."[54]

Rosen decided that the best vehicle for meeting this set of requirements was the Thor-Able, incorporating the Thor missile as the first stage and the second and third stages of Vanguard. He recalled proposing what became the Delta to Silverstein near the end of January 1959 with "several important changes" from the Thor-Able that resulted from his "experience with Vanguard." One was use of the Hercules X248 in place of Grand Central's third stage, a step the Air Force had already taken for its series of unsuccessful Pioneer launches in late 1958. Rosen also suggested use of the Bell Labs radio guidance system, which again the Air Force and STL had adopted for the Pioneer launches. Further suggestions in Rosen's proposal were "repackaging of the control electronics" used in the Vanguard second stage and "replacement of the aluminum thrust chamber in the Aerojet second stage by one made of stainless steel" to avoid Vanguard's erosion problems.[55]

Rosen said "Silverstein approved the project" and told him to develop the launch vehicle. "I started by calling my friend, Elmer Wheaton, Vice President of Engineering at the Douglas Aircraft Company," Rosen recalled. At Rosen's suggestion, Wheaton sent the new agency a letter of proposal on February 3, 1959. "Within 10 weeks," according to Rosen, "NASA had a signed contract with the Douglas Aircraft Company for the development, production and launching of 12 Delta rockets." NASA announced the contract for $24 million on April 29. By using components already proven in flight, NASA and Douglas eliminated the need for developmental flights, and their contract set a goal for initial reliability of 50 percent with a final rate of 90 percent, a very ambitious proposition in 1959.[56]

Obviously, Delta owed a great deal to the Navy's (and now NASA's) Vanguard upper stages and the Air Force's Thor. Perhaps even more important than the hardware transfers were the personnel from the Vanguard program, including Rosen. They brought their experience to decision-making positions at the new Goddard Space Flight Center within NASA as well as at NASA Headquarters. Finally, the experience of the Air Force and STL with Thor-Able clearly contributed significantly. At Goddard, Vanguard veteran William R. Schindler headed a small technical group that provided direction and technical monitoring for the new launch vehicle. On November 24, 1962, NASA converted this technical direction to formal project management for Delta. Even though Douglas, the prime contractor for the entire vehicle, was also the airframe contractor on the Thor, NASA bought Thors from the Air Force for the first Deltas and furnished them back to Douglas. These first stages used either the MB-3 Basic or the follow-on Block I engines with about 150,000 pounds of thrust. NASA and its contractors removed their AC Spark Plug inertial guidance systems, and NASA procured Bell Labs radio guidance systems through the Air Force. The systems used on the initial Deltas were in the series 300, which used vacuum tubes and a large number of relays.[57]

The second-stage engine in the initial Delta configuration was Aerojet's AJ10–118, derived from the Vanguard second stage but, as Rosen suggested, with a stainless steel combustion chamber. Apparently the first twelve of these engines had a 20:1 expansion ratio for the nozzle and unsymmetrical dimethyl hydrazine as the fuel with inhibited white fuming nitric acid (IWFNA) as the oxidizer, although NASA and Aerojet soon converted to inhibited red fuming nitric acid (IRFNA). Delta had become the only program in the country using IWFNA, and NASA changed in part to conform to the usual practice but also because the denser IRFNA gave increased performance for a given volume. With IRFNA, the second stage delivered an average thrust at altitude of 7,575 pounds and a specific impulse of more than 265 lbf-sec/lbm.[58]

The third stage for the first dozen Deltas included the Hercules X248–A5 Altair I, built at Allegany Ballistics Laboratories, procured from the Navy, and furnished to Douglas. This apparently was a slightly uprated version of the X248–A3 used on the Able upper stage in the Pioneer launches, itself derived from the upper stage used to launch Vanguard III. Its cast double-base propellant in a fiberglass/resin case yielded an average thrust of 2,590 pounds at 40°F.[59]

On May 13, 1960, an attempt to launch the spherical, passive reflector Echo satellite with the first Thor-Delta failed when the third stage propellants did not ignite because a small chunk of solder in a transistor broke loose in flight, shorting out a semiconductor that had passed all of its qualification tests. A similar but less costly problem with another transistor on the third Delta launch led NASA to change the specifications and testing of such components. Meanwhile, on August 12, 1960, the second Delta mission successfully placed Echo 1 in orbit, and the remainder of the original twelve Deltas all had successful launches of a variety of payloads from the Tiros 2 through 6 weather satellites to the Telstar 1 communications satellite—the first commercial spacecraft launched by NASA (on July 10, 1962, seventy days before Tiros 6 on September 18, the last of the original dozen Delta launches). Thus, NASA and Douglas easily surpassed their goal of an initial 50 percent success rate.[60]

It would take far too many pages to recount each modification to the basic Delta launch vehicle. The initial Delta could launch 100 pounds of payload into geostationary (also called geosynchronous) transfer orbit (GTO). (These transfer orbits were not themselves geostationary but were orbits from which upper stages or satellite kick motors could place the spacecraft in geosynchronous orbits.) Starting in 1962, Delta evolved through a series of models with designations such as A, B, C, D, E, J, L, M, M-6, N, 900, 904, 2914, 3914, 3910/PAM (for Payload Assist Module), 3920/PAM, 6925 (Delta II), and 7925 (also a Delta II), the last of these introduced in 1990. Payload capabilities climbed, at first gradually and then more rapidly, so that the 3914 introduced in 1975 could lift 2,100 pounds to geostationary transfer orbit and the 7925, 4,010 pounds.[61]

Over the course of the three decades from 1960 to 1990, the Delta pro-

Figure 5. An early Delta vehicle carrying Telstar 1 aloft from Cape Canaveral on July 10, 1962. Telstar 1 was a communications satellite, the first commercial spacecraft launched by NASA. Courtesy of NASA.

gram increased the capabilities of the booster and upper stages, lengthened and enlarged the tanks of the two liquid-propellant stages, enlarged and upgraded third-stage motors, improved guidance systems, and introduced increasingly large and numerous strap-on solids to provide so-called zero-stage boost. During this period, the program generally followed Rosen's initial approach of introducing only low-risk modifications or ones involving proven systems. This enabled on average a launch every sixty days, with a reliability over the thirty years of 94 percent (189 successes out of 201 attempts, the last one during the period of this book occurring on November 26, 1990), which comfortably exceeded the ultimate goal of 90 percent reliability set by the early program.[62]

The first improvement in the guidance system occurred in 1962, when the program introduced the Bell Labs series 600 radio-inertial guidance system on Delta B. Fully transistorized unlike the series 300, it also had fewer relays. The onboard portion of the system in stage 2 was lighter by about 150 pounds (pounds that became available for payload) and more accessible to ground maintenance because the stage was less packed with components. In the propulsion systems, after earlier incremental changes, the Delta D in 1964 marked a plateau. The Thor stage was using the MB-3 Block III engine with about 172,000 pounds of average thrust (presumably including 1,000 pounds for each of the verniers) without longer tanks. Already on Delta C in 1963, the stage-2 propulsion had advanced to the AJ10–118D (used also on a Delta B the same year), and the third-stage motor became ABL's X258. The principal changes to stage 2 were the lengthening of the propellant tanks by 3 feet (introduced on Delta B) and the change from IWFNA to IRFNA. This did not increase the average thrust, which remained at 7,575 pounds above 55,000 feet in altitude.[63]

The X258 or Altair II was initially a Scout fourth-stage motor developed by Hercules in 1963 under the joint sponsorship of the Bureau of Naval Weapons and NASA. Hercules shifted from the composite double-base propellant used for the X248 to a composite-modified double-base propellant somewhat similar to the one used on the second stage of Polaris A2. Like the A2 motor, Altair II used ammonium perchlorate and HMX along with nitrocellulose, nitroglycerin, and aluminum. The X258 had a slotted-cylinder internal-burning grain configuration and a fiberglass/resin case that yielded a mass fraction of 0.877, actually slightly lower than the X248–A5's and considerably lower than the Polaris A2 stage-2 motor's. Nevertheless, for a considerably shorter burning time (down from 43 to 24 seconds) the newer motor provided considerably more average thrust (5,888 pounds, up from 2,590 in the X248) and an increase in specific impulse from about 255 to upwards of 280 lbf-sec/lbm. This added performance increased the payload that Delta could launch into GTO, from 150 pounds for Delta B to 180 pounds for Delta C.[64]

What made Delta D especially significant was its addition of three Thiokol Castor I solid motors to augment the liftoff thrust from the launchpad, creating the first thrust-augmented Delta. These were the same strap-on motors used with the first thrust-augmented Thor. Sources differ on how much they augmented the first-stage thrust, but one source puts the figure at 53,000 pounds each for a total thrust at liftoff of 331,000 pounds including that provided by the Thor. This again raised the payload the Delta could put into geosynchronous transfer orbit, to 230 pounds.[65]

Figure 6. The 1965 launch of Intelsat I, first of a long series of Intelsat communications satellites, from Cape Kennedy by a Delta D with three Thiokol Castor I solid-propellant rocket motors strapped to the main stage. This was the first thrust-augmented Delta, raising the payload the vehicle could lift to geosynchronous transfer orbit from 180 pounds for the C model to 230 pounds for the D version. Courtesy of NASA.

Although thrust augmentation continued on subsequent models of Delta, the D-model flew only twice, to launch Syncom 3 on August 19, 1964, and Intelsat I on April 6, 1965. Syncom 3 was the third and last of a series of active repeater communications satellites launched into GTOs by Deltas, the first two by Delta Bs in 1963. Although successfully launched, Syncom 1 stopped communicating after its apogee rocket fired. Syncoms 2 and 3 provided data, facsimile, telephone, and television transmissions, with Syncom 3 providing coverage of the 1964 Olympic Games from Tokyo. Also called Early Bird, Intelsat I was the first of a long series of Intelsat communications satellites. Linking Europe and North America, it transmitted telephone, black-and-white television, teletype, and facsimile data.[66]

Delta E in 1965 brought a further uprating of the vehicle. There were initially no significant changes to the strap-on solid motors or the Thor, but an increase in the diameter of the second-stage tanks almost doubled the amount of propellant, extending the burn time from 174 to 398 seconds. In addition, the Delta E offered a choice of third-stage motors, the existing X258 or the FW-4 built by the comparatively new United Technology Corporation (UTC) of United Aircraft, which developed the solid rocket motors for the Titan III. The FW-4, also used on the fourth stage on the Scout Standard Launch Vehicle, employed a propellant consisting of poly-butadiene-acrylic acid-acrylonitrile (PBAN, also used on the SRMs for Titan III), ammonium perchlorate, and aluminum. Designed with a cylindrical-port/transverse-slot internal burning cavity, a fiberglass case, and a nozzle expansion ratio of 50:1, it had a propellant mass fraction of 0.911 and an average thrust of between 5,510 and 5,950 pounds, depending on the ambient temperature. Its specific impulse was more than 285 lbf-sec/lbm. The FW-4 was thus roughly equivalent to the X258 in performance, though there were some slight differences in capability to match with given payloads, and it added the benefit of a second source of supply. With either third stage, Delta E could lift 330 pounds to GTO, 100 more than Delta D.[67]

The fiberglass for the FW-4, developed by the Owens-Corning Company of Santa Clara, California, included a low-density silica material that reportedly weighed 35 percent less than comparable materials used by competitors. The FW-4 also employed an igniter made of aluminum that itself burned during combustion of the propellant and contributed to the propulsion. Because the third stage used spin for control, however, development of the motor was not problem-free. In early tests, centrifugal forces from the spinning caused the deposit of considerable aluminum oxide from the burning process on the motor chamber, degrading performance. UTC engineers

Figure 7. A Delta E rocket launching from Kennedy Space Center on December 5, 1968, carrying Highly Eccentric Orbit Satellite 1 for the European Space Research Organization. Courtesy of NASA.

shifted to a finer aluminum powder in the propellant, which reduced the deposits.[68]

Delta J in 1968 brought a new third stage motor, the Thiokol Star 37D, very similar to the Star 37B on the Burner II. Its average thrust of 9,680 pounds was significantly greater than the previous Delta third-stage motors.' The year 1968 also saw the introduction of the long-tank Thor with the same basic configuration as the Thorad and three Castor II strap-on motors on Deltas L and M. Since the Delta could be launched with one of several third-stage motors or with no third stage at all, the flexibility of the system had become very great by this time, although specifics about each configuration and each launch are scattered and not necessarily consistent. With Delta M, however, the vehicle was now able to launch 785 pounds into geosynchronous transfer orbit. By the end of 1968, Delta in its various configurations had reached its sixty-third launch of mostly weather, scientific, and communications payloads with only four failures, a roughly 94 percent success rate. It had begun launching not just from Cape Kennedy but from the Western Test Range in 1966, permitting safer and more straightforward launches into polar orbits than had been possible from Florida, where complicated and inefficient dogleg trajectories were necessary to avoid flight over Caribbean islands.[69]

Another upgrade, started in 1969, provided for three, six, or nine strap-on Castor II solid motors, each with an average thrust of some 52,130 pounds. To support as many as nine strap-ons, engineers strengthened the aft structure of the first stage in a "universal boat tail" configuration. Second-stage propulsion now used what is often called the Titan III Transtage engine. Although the new engine for Delta, designated AJ10–118F in the same numerical series as the earlier Aerojet engines derived from Vanguard, was similar to the earlier Titan Transtage engine produced by Aerojet and designated AJ10–138, they were not identical. In both, the combustion chamber was of fiberglass impregnated with resin and had an ablative lining for cooling. And both, like the Titan II engines, used a half-and-half mixture of hydrazine and UDMH as the fuel, igniting hypergolically with a nitrogen-tetroxide oxidizer. This replaced the IRFNA used on earlier versions of the AJ10–118 and increased the specific impulse by about 11 percent to almost 295 lbf-sec/lbm. The F version of the engine had a thrust of between 9,235 and 9,606 pounds, well above the roughly 7,575 of the earlier versions, and it was capable of as many as ten starts in orbit. The new engine completed its preliminary design in 1970 but did not fly until July 23, 1972.[70]

A major upgrade on what one source describes as the Delta 0900 was the replacement of the radio-inertial guidance system used until this time with a strapped-down inertial guidance system that also replaced the autopilots in the first two stages. The original system, designed by Bell Telephone Laboratories and often referred to as a Western Electric Company design, was now old enough that critical components were hard to locate when they had to be replaced. Engineers developed the inertial measurement unit for the new system by converting the abort gyro package from the Apollo lunar mod-

Figure 8. A Delta 900 launch vehicle with strap-on solid-propellant boosters. This particular vehicle is preparing in December 1972 to launch a Nimbus E spacecraft. Courtesy of NASA.

ule. Information on this abort sensing assembly is scattered and incomplete, but United Aircraft's Hamilton Standard Division apparently developed the strapped-down inertial system for the lunar module under contract to TRW, which in turn was a subcontractor for Grumman, the firm that produced the module itself. The strapped-down inertial sensor unit consisted of "three floated, pulse-rebalanced, single-degree-of-freedom, rate-integrating gyros;

three pendulous, fluid-damped accelerometers and associated pulse torquing electronics," as well as related equipment. The Delta Inertial Guidance System, as it came to be called, did not use the digital computer from the lunar module but instead adopted the one Teledyne had developed for the Centaur. This computer with a 4,096–word memory provided navigation, guidance, and steering for both first and second stages of the Delta once it was introduced about 1970.[71]

In 1970 NASA signed a new memorandum of agreement with the Air Force that, among other things, gave the civilian agency control over the Thor production line at the McDonnell Douglas Corporation (formed by the 1967 merger of Douglas Aircraft and McDonnell Aircraft). NASA no longer had to procure Thors from the Air Force, and in this agreement, the civilian agency also achieved its own control over procurement of Castor strap-on motors from Thiokol. Even before the agreement went into effect, on January 23, 1970, a Delta N-6 successfully launched the Tiros M (also known as the Improved Tiros Operational System or ITOS 1) meteorological satellite. This marked the first use of six strap-on motors. The N-6 launched the 675–pound satellite into a sun-synchronous orbit. Three of the motors fired at liftoff, the other three at about 46,000 feet to help propel the satellite to about 900 miles above Earth. Then on July 23, 1971, a Delta 0900 with the first use of nine strap-ons successfully launched the Earth Resources Technology Satellite (ERTS 1 or Landsat 1) weighing 2,070 pounds into a polar orbit at an altitude of about 560 miles. This launch was also the first use of the digital inertial guidance system and the AJ10–118F second stage.[72]

In 1971–72, McDonnell Douglas implemented a number of other changes to the Delta. One introduced the so-called straight-eight configuration in which, for the first time, Delta had a constant diameter of 8 feet from nose to base; before, there had been various taperings from the Thor to a narrower upper stage or stages and, often, payload. A second major modification was the introduction into the Thor of the RS-27 engine, a repackaging of the Rocketdyne H-1 engine developed for the Saturn I. The H-1 was a derivative of work from 1957 to 1959 on the X-1 engine by the Experimental Engines Group at Rocketdyne to make the Thor MB-3 engine simpler, more reliable, and lighter. On September 1, 1958, the Army Ballistic Missile Agency awarded Rocketdyne a contract to uprate the Thor-Jupiter engine for a large booster that became Saturn I. Only later known as the H-1, this engine required only half a year for delivery of the first production version to ABMA. It had only 165,000 pounds of thrust, however, less than the Thor MB-3 Block II, but the H-1 was to go through versions of 188,000, 200,000, and

Figure 9. A January 3, 1971, photo of a Delta N-6 launch vehicle carrying an ITOS meteorological satellite, with strap-on solid-rocket boosters at the base of the vehicle. This and other Delta photos together illustrate the great variety of configurations used in Delta launches. Courtesy of NASA.

205,000 pounds as the Saturn project evolved, with Saturn I using the first two, and Saturn IB the 200,000– and 205,000–pound versions.[73]

Development continued well beyond the initial delivery, with the preliminary flight-rating test for the 188,000–pound version not occurring until September 28, 1962. Beginning in 1963, the testing process included the inducing of combustion instability (by setting off small bombs in the combustion chamber), which showed that the injectors inherited from the Thor and Atlas could not recover and restore stable combustion. Rocketdyne engineers rearranged the injector orifices and added baffles to the injector face to solve the problem. Cracks in liquid-oxygen domes and splits in regenerative-cooling tubes also required reengineering, the latter during uprating from 165,000 to 188,000 pounds of thrust. Embrittlement by sulfur from the RP-1 in the hotter environment of the higher-thrust engine required a change of materials from nickel alloy to stainless steel for the tubular walls of the combustion chamber. There were other problems, but the Saturn personnel resolved them in time for the launches of Saturn I and IB from late 1961 to early 1968.[74]

Late in 1971, McDonnell Douglas awarded a contract to Rocketdyne to develop, test, and begin production of the RS-27, using hardware from the H-1 modified to be compatible with the Thor airframe. There followed many subsequent contracts to provide the RS-27 for the Delta. Although production of the engine began with turbopumps and combustion chambers identical to those in the H-1, the RS-27 generated about 207,000 pounds of thrust at sea level as compared with roughly 205,000 for the most powerful H-1s. Rocketdyne completed the preliminary design before the end of 1971, with the first flight occurring on January 18, 1974. Both the H-1 and the RS-27 were single-start engines with no provision for adjusting the thrust level. Both burned liquid oxygen and RP-1 kerosene in regeneratively cooled combustion chambers. Where the H-1 had a rated thrust duration of 155 seconds, the RS-27 was rated for 227 seconds.[75]

No doubt the longer duration was attributable to a further extension of the propellant tanks on the Thor-Delta, with the first stage now 73.6 feet long as compared with the 70.5 feet of the Thorad. One other change was the elongation of the Star 37D to the Star 37E configuration to increase the propellant load from 1,440 to 2,290 pounds without changing the type of propellant or the grain configuration. A change of the case material from steel to titanium yielded a slight improvement in the propellant mass fraction from 0.911 to 0.926. The average thrust increased from 9,680 pounds to 15,472. With the Star 37E in the third stage, the overall length of the vehicle rose to 116 feet. This Delta model 2914 could lift 1,593 pounds of payload to geosynchronous transfer orbit. The first launch with this overall configuration was on April 13, 1974, when the Delta successfully placed the 1,258–pound Westar 1 domestic communications satellite into transfer orbit, from which the apogee kick motor transferred it to a geosynchronous orbit on April 16.[76]

But already in 1972–73 Deltas had flown with the extended long-tank

Figure 10. A Delta 1910 launch vehicle lifting off from Cape Canaveral on June 21, 1975, carrying the Goddard Space Flight Center's eighth Orbiting Solar Observatory. Courtesy of NASA.

Thor, as the longer version was called. This version featured a new structure. Earlier Thors had used stringers welded into the aluminum skin of the propellant tanks with intersecting ribs. Now McDonnell Douglas used an "isogrid" construction with the tank walls made of an aluminum-alloy plate machined to a thin skin with integral ribs that formed a triangular pattern. The result was a thinner skin with fewer ribs and no stringers. This meant that the new tanks were simpler to make, stronger, and also lighter than before.[77]

Then, for the Westar 1 launch, the Delta 2914 also featured TRW's TR-201 second-stage engine undergoing its maiden flight. Design of this engine began in October 1972. Its combustion chamber was made of quartz phenolic, with cooling by ablation as in the AJ10–118F, though the TR-201 weighed only 298 pounds as compared with the Aerojet engine's dry weight of 1,204 pounds.[78] Both engines burned nitrogen tetroxide with a fifty-fifty mixture of hydrazine and UDMH, igniting hypergolically. But the lighter TRW propulsion unit yielded 9,900 pounds of thrust compared with a maximum of 9,606 for the Aerojet unit. The newer system did have only five restarts available instead of ten.[79]

In another upgrade of the Delta, Castor IV strap-on motors with higher thrust replaced the Castor IIs. The Castor IV was actually less sophisticated, with its binder reverting to polybutadiene-acrylic acid from the carboxy-terminated polybutadiene used in the Castor II. The reason for the shift may have been cost: the Castor IV, 29.8 feet long and 40 inches wide, contained much more propellant than the 19.8–foot-by-31–inch Castor II, and CTPB was more expensive than PBAA. Both motors had steel cases with aluminum as fuel and ammonium perchlorate as oxidizer. Although the specific impulse and mass fraction of the Castor IV were slightly lower than the Castor II's, the burning time increased from 37 to 54 seconds and the average thrust rose from 52,130 to 85,105 pounds. The Delta model 3914 was the first to use the Castor IV. Flown on December 13, 1975, this vehicle succeeded in launching the 1,018.6–pound RCA Satcom 1, the first in a series of three RCA commercial communications satellites, into geosynchronous transfer orbit.[80]

In 1980 Delta adopted a new third stage, the Payload Assist Module (PAM), which used a Star 48 motor. Thiokol began developing the motor in 1976 for use with the space shuttle. It was an offshoot of the propulsion unit for Minuteman stage 3, which the firm began producing in 1970 essentially using the original Aerojet design. The purpose of the PAM on the shuttle was to propel satellites from a low parking orbit, about 160 miles above Earth, to a higher final orbit. It used the same hydroxy-terminated polybutadiene (HTPB)-aluminum-ammonium perchlorate propellant as Thiokol's Antares IIIA rocket motor, a third-generation third-stage propulsion unit for the Scout launch vehicle, as well as the firm's Star 30 apogee boost motor. Thiokol made the Star 48 motor case of titanium and used the recently developed advanced-composite called carbon-carbon, for the nozzle's exit cone.[81]

Figure 11. A Delta 3920 launch vehicle with a PAM upper stage lifting off from Kennedy Space Center on September 21, 1984, carrying Galaxy-C, a communications satellite. Courtesy of NASA.

Another major addition to Delta's arsenal of engines and motors before the introduction of the Delta II was Aerojet's AJ10–118K propulsion unit, developed for the Air Force's improved Transtage injector program and used in a third stage for the Air Force's Titan III, as well as in the second stage of the Japanese N-II and the same stage of Delta. First flown on a Delta in 1982, the 118K version of the engine burned the same nitrogen tetroxide and Aerozine-50 storable, hypergolic propellants as the AJ10–118F and TR-201 engines. Pressurized helium fed the propellants to the combustion chamber of an even lighter engine than the TR-201. At 220 pounds, the 118K was about 75 pounds lighter than the TRW engine. It had an even higher specific impulse, at about 320 lbf-sec/lbm, than the 300–plus of the TR-201, presumably achieved through a nozzle expansion ratio of 65:1, compared with 46:1 for the older engine. Although the vacuum thrust of 9,800 pounds for the Aerojet engine was slightly lower than the TR-201's 9,900, its lower weight probably more than offset the difference. Like all of Delta's engines (as distinguished from solid-propellant motors), this one used gimbals for steering.[82]

Had it not been for the space shuttle *Challenger* disaster on January 28, 1986, the addition of this engine might have constituted the last chapter in Delta's history. With shuttles scheduled to carry most satellites into orbit, it appeared that NASA would discontinue orders for the reliable Delta. In the wake of *Challenger*, however, an enduring need for the Delta as well as other expendable launch vehicles became clear. On January 21, 1987, the Air Force awarded a contract to McDonnell Douglas for the first of the Delta IIs, and more contracts from both NASA and the Air Force followed. The primary payload for the Delta II was to be the Navstar Global Positioning System (GPS) navigational satellite, previously slated for launch by the space shuttle. By October 1987 the space agency itself had revised its plans to include a mixture of expendable and reusable launch vehicles, with the orbiters used only for missions requiring either human participation or the landable vehicles' other special capabilities.[83]

The Delta II improved in two steps. The first was the Delta 6925, which used an RS-27 engine in a Thor with the propellant tanks stretched again, from 73.6 to 85.6 feet, in what was called the extra-extended long-tank Thor. The diameter remained at 8 feet, but the burn time jumped from 227 to 265 seconds without a change in the average thrust. A complementary uprating was the shift from the Castor IV to the Castor IVA strap-on motors, an effort begun at Goddard Space Flight Center in the early 1980s. Tested and qualified in 1983, the new motors were not introduced then because of the impending phase-out of the Delta in favor of the space shuttle. With the post-*Challenger* resurrection of expendable launch vehicles, McDonnell Douglas proposed the Castor IVAs as a low-risk improvement to Delta II. The new strap-ons kept the old dimensions, case, and nozzle throat material. But they used an HTPB-aluminum binder with a higher loading of solids. This increased the average thrust for the same size motor from 85,105 to 98,187 pounds. The second and third stages for the Delta II 6925 did not change.[84]

Figure 12. A Delta II 6925 launch at Cape Canaveral, February 14, 1989. Official U.S. Air Force photo, courtesy of the 45 Space Wing History Office, Patrick AFB, Fla.

Although little information seems to be available about it, the inertial guidance system changed for Delta II. It consisted of a Delta strapdown inertial measurement system with three gyroscopes and four (instead of three) accelerometers. The guidance computer was a Delco rather than a Teledyne product. This system guided and controlled the first two stages, with the third stage remaining spin-controlled.[85]

With the changes in the first and strap-on stages, the Delta II 6925 now had a payload capability to transfer orbit of 3,190 pounds, up from 2,800 pounds for the Delta 3920/PAM, an increase of almost 14 percent. The first use of the 6925–model Delta II occurred on February 14, 1989, with the successful launch of the first of the Block II operational series of Navstar GPS satellites. One anomaly was higher vibration in the second-stage inertial guidance system, although this posed no serious problem on this flight. Investigation revealed the problem lay with the slope in the fairing on that part of the fuselage, which engineers effectually modified in a minor example of cut-and-try engineering. Thereafter the Delta II 6925 successfully launched the other eight Block II GPS satellites by October 1, 1990.[86]

The second step in Delta II development was model 7925. The leading "7" indicated two changes from the "6" designator. One was an uprating of the Thor engine to the RS-27A model, with a nozzle expansion ratio increased from 8:1 to 12:1. At a cost in sea-level thrust (down from 207,000 pounds to 200,000 pounds) and specific impulse at sea level (from almost 265 to about 255 lbf-sec/lbm), the stage gained in thrust at altitude (from 231,000 to 237,000 pounds) and in altitude specific impulse (from about 295 to more than 300 lbf-sec/lbm). To offset the losses at sea level, the Castor IVAs gave way to strap-on graphite epoxy motors (GEMs) developed by Hercules Aerospace and produced at facilities in Utah, which more than compensated, although their greater heat required an upgraded thermal protection system to shield the Thor stage from their exhaust.[87]

The graphite epoxy motors were almost 6 feet longer than the Castor IVAs, and their filament-wound cases made of graphite fiber and an epoxy resin were lighter than the Castor IVA cases constructed of steel. Like the Castor IVAs, the newer motors used an HTPB propellant. With a slightly higher nozzle expansion ratio, the GEMs also had a higher specific impulse, giving them an average thrust of 98,900 pounds at sea level, compared with 97,700 for the Castor IVAs. Multiplied by nine strap-on motors, the increase was 10,800 pounds, more than making up for the decrease of 6,000 pounds in thrust at sea level resulting from the higher nozzle expansion ratio of the RS-27A. Where the Castor IVAs' burn time was 56.2 seconds, the GEMs burned for 63 seconds, raising overall thrust. With no change in the second and third stages, the Delta II model 7925 could carry 4,010 pounds of payload to GTO, an increase of more than 25 percent over model 6925.[88]

The first model 7925 Delta II successfully launched the initial Block IIA Navstar GPS satellite on November 26, 1990, virtually at the end of the period covered by this book. The Air Force Space Command declared the satellite ready for use on January 15, 1991, one day before the air campaign began during Operation Desert Shield/Desert Storm against Iraq. It provided three-dimensional velocity, timing, and position data for ground forces, ships, and aircraft during the conflict and permitted navigation in the featureless desert, guiding tanks, artillery, aircraft, and missiles.[89]

The Delta II model 7925 went on to launch many more Block IIA GPS satellites in the course of the 1990s, and the protean launch vehicle continued to evolve through the Delta III and Delta IV configurations. (In a sense, Delta IV can be considered a new launch vehicle because it had an entirely new first stage. However, many Deltas had featured new stages, and the design of the Delta IV emphasized reliability and low cost, as had previous rockets in the Delta family.) But already by the end of 1990 the Delta had shown a remarkable ability to evolve as new technologies emerged and to increase its capabilities to launch ever larger payloads. As it changed, it retained a high level of reliability, making it one of the great success stories in the history of launch vehicles.[90]

Summary and Conclusions

Despite the hurried and difficult gestation of the Thor missile, the basic rocket became, from 1958 on, the first stage for a large variety of launch vehicles. The early ones built upon the even rockier start of the Vanguard launch vehicle. But from the initial Thor-Able, an amalgam of Thor and Vanguard stages, the Air Force and NASA developed an important family of medium launchers that carried everything from reconnaissance satellites to scientific, weather, and communications satellites into orbit and sent even a few space probes on their trajectories into the solar system. Using many liquid- and solid-propellant upper stages, the Thor and Delta launch vehicles were not so much innovators as borrowers of new technology. Consequently, their various models experienced fewer birth pangs than other missiles and rockets.

Nevertheless, the Able-Star did have to use empirical solutions to unforeseen injector-plate burn-throughs, and the second stage of the Delta II 6925 required adjustments of some fairings to avoid vibrations that designers had not expected. Also, when the Thor adopted the RS-27 engine, it took over technology from the Saturn H-1 that had overcome combustion instabilities and other engineering problems only through cut-and-try engineering. Despite such problems, by using mostly components already tested and proven, the Delta achieved a high reliability that made it an enduring member of the launch vehicle family. From an interim launch vehicle in 1959, it became one of the few that lasted into the twenty-first century. It shared this honor with the Atlas family of launch vehicles, the subject of the next chapter.

3

The Atlas Space-Launch Vehicle and Its Upper Stages, 1958–1990

Even before it began its service as a missile, the Atlas had started to function as a space-launch vehicle. In December 1958 as part of Project Score, an entire Atlas (less its two jettisoned booster engines) went into temporary orbit carrying a repeater satellite that could receive messages from Earth and send them back. Then, simultaneously with their role in Project Mercury, modified Atlas missiles began to serve as space-launch vehicles for both the Air Force and NASA in a variety of missions. For this purpose, the basic Atlas was standardized, uprated, lengthened, and otherwise modified in a variety of configurations that were often individually tailored for specific missions. Engineers mated the vehicle with a number of different upper stages, of which the Agena and Centaur were the best known and most important. In these various configurations, Atlas space boosters launched satellites and spacecraft for such programs as Samos, Midas, Ranger, Mariner, Pioneer, Intelsat, the Fleet Satellite Communications System, the Defense Meteorological Satellite Program, and the Navstar Global Positioning System. Following the end of the period covered in this book (roughly 1990), some Atlases even used strap-on solid motors to supplement their thrust at liftoff.[1]

Most of the upper stages used with Atlas were derivatives of other programs. This was true of both the liquid-propellant Agena and a variety of solid-propellant upper stages. The Centaur, however, was in a sense a derivative of Atlas, in that it used the steel-balloon tank structure envisioned by Charlie Bossart and developed for the Atlas missile. Nevertheless, adapting that structure to liquid hydrogen fuel proved to be a major challenge that required a great deal of cut-and-try engineering as well as a major reorganization of the way Centaur was managed. Still, after initial delays, the Centaur worked and went on to contribute technology to the Saturn upper stages and to the space shuttle.

Project Score

The Advanced Research Projects Agency was the sponsor of Project Score (Signal Communications by Orbiting Relay Equipment), in which a B-model Atlas zoomed into orbit on December 18, 1958, controlled by a General Electric radio-inertial guidance system. Fewer than a hundred people were let in on the secret that the missile was being orbited with 150 pounds of communications relay equipment, although many employees at Cape Canaveral had their suspicions because of deviations from the normal pattern for missile test flights. Two hours after the tank section of the missile was in low Earth orbit, President Eisenhower announced that the United States had orbited an object weighing some 8,800 pounds. (The main section of the Atlas weighed almost 8,700 pounds; pods installed by the Army Signal Corps along the missile's side carried the 150 pounds of communications equipment.) Among the transmissions from the satellite was a Christmas message recorded by the president. Reportedly, this was the first time a human voice had been broadcast from space. The satellite also relayed messages transmitted from ground stations in Georgia, Arizona, and Texas. Expected to stay in space only twenty days, the vehicle and its equipment remained in orbit for more than a month before reentering the atmosphere on January 21, 1959, and burning up over the Pacific Ocean near Midway Island. While the Soviets' Sputnik 3, launched on May 15, 1958, had weighed an impressive 2,925 pounds, the Project Score satellite almost tripled that weight.[2]

Atlas-Able

There were three attempts to launch Atlases mated with Able upper stages.[3] These efforts were outgrowths of the Thor-Able lunar probes and employed similar hardware and instrumentation in the upper stages. Stimulated by a Space Technology Laboratories (STL) proposal to the Advanced Research Projects Agency in June 1958, NASA conceived the project in November of that year, initially to place a probe in a Venus orbit. The new space agency worked with the Ballistic Missile Division (BMD) and STL to obtain the launch vehicles, with BMD responsible for conducting the program under NASA's overall management. By June 1959 if not before, NASA had decided on a somewhat less ambitious lunar-orbiting payload. Atlas's thrust, greater than Thor's, allowed a heavier and more sophisticated satellite to be launched into an orbit around the Moon. It weighed 362.5 pounds, about half of which consisted of a hydrazine liquid-propellant rocket designed by STL for final velocity corrections and injection into lunar orbit. This was a much larger rocket than the 5–pound solid-propellant version used in the successful Thor-Able 3. The sole function of the still comparatively small liquid-propellant thruster on this mission was injection to ensure a satisfactory orbit.[4]

For reasons that are not clear from available sources but probably related to availability, the first Atlas used in this program was a C-model modified to mate with the Able second stage. The third stage used an Allegany Ballistics Laboratory X248, known as the Altair. This configuration, designated Atlas-Able IVa, underwent a static firing test without the payload at Cape Canaveral on September 24, 1959. After the conclusion of the test, a ruptured propellant line led to an explosion that damaged Launch Complex 12 severely.[5]

Fortunately, a backup Atlas D was available from Project Mercury. Equipped with Able upper stages and the satellite, as Atlas-Able IVb it launched successfully from Pad 14 at the Cape on November 26, 1959. Unfortunately, the payload shroud, designed for jettison after 175 seconds of flight, fell off about 45 seconds after liftoff. Exposed to heavy dynamic pressures, the third stage and satellite disconnected from stage 2 and exploded. Examination of flight data indicated that, as the launch vehicle rose into the thinner atmosphere, the pressure differential inside and outside the shroud caused the premature jettisoning. On the next two missions, engineers corrected this design defect by the simple expedient of drilling small holes in the shroud to equalize the pressure as the vehicle ascended.[6]

Two more Atlas Ds with Able upper stages, known as Atlas-Able Va and Vb, launched with Pioneer satellites aboard. The Atlas Ds used a modified General Electric radio guidance system, with stages 2 and 4 employing the guidance system STL had designed for Thor-Able 4. (Here, stage 4 appears to refer to the satellite itself, with its liquid propellant rocket.) On September 25, 1960, Atlas-Able Va launched with the Atlas stage performing as designed until vernier engine shutoff, which did not occur 5 seconds after sustainer engine cutoff as planned. The oxidizer valve position switch (a backup device for the thrust chamber pressure switch, which failed to function) initiated stage-2 separation, but soon after the Aerojet engine on stage 2 started, an oxidizer leak developed and the Able stages went off course before stabilizing for about 45 seconds and then tumbling. As a result, the STL ground guidance system commanded stage-2 cutoff some 8 seconds early. Stage 3 ignited and operated for roughly 40 seconds, then the satellite separated and its hydrazine-powered engine fired. But the deviation in trajectory during the second-stage burn could not be corrected sufficiently to place the satellite in orbit around the Moon.[7]

Investigators attributed the oxidizer leak to a "random structural failure" without pinpointing its cause, while the guidance and control problem was attributed to failure of an actuator. The first problem did prompt removal and examination of the Atlas-Able Vb second-stage combustion chamber, but this did not reveal any inherent structural weakness that could lead to another oxidizer leak. Since a faulty explosive bolt at stage-2 separation could have damaged the stage-2 combustion chamber, engineers modified the cages around the bolts. In further corrective action, redesign of the method of initiating stage-2 separation made a thrust valve switch the primary device, with the thrust chamber pressure switch relegated to backup. Engineers also subjected the pressure switch to further acceptance testing, which included actuation in a vacuum while experiencing severe vibration. Although no evidence existed that vacuum conditions in space had contributed to the problems on the previous launch, the Atlas-Able team added a closure to the upper stages to ensure that combustion occurred under pressure. And among other changes, engineers and technicians reinforced portions of the actuators involved in pitch and yaw control.

With these modifications, Atlas-Able Vb launched on December 15, 1960, from Cape Canaveral. The first 66.7 seconds of flight were normal, but then the first- and second-stage axial accelerometers registered anomalies, suggesting some violent occurrence affecting the upper stages. A subsequent decrease in liquid-oxygen pressure was followed by structural failure of the entire vehicle. Examination of recovered parts of Atlas-Able Vb by a review group (including representatives from STL, Aerospace Corporation, Rocketdyne, Aerojet, Convair, the Space Systems Division, and NASA) did not reveal any propellant leakage or combustion in the second stage. The group speculated that the mishap was due to a failure of the liquid oxygen tank in the Atlas, possibly caused by dislodging of some component in the upper stages that could have struck the tank. The Air Force described the three failures of the Atlas-Able vehicles as random incidents with no discernable pattern. Air Force spokespersons denied that there were any structural problems with the Able upper stages. The Atlas in combination with other upper stages was to achieve a distinguished record of successes, but at the time of the Atlas-Able launches, Atlas had only two successful launches in seven missions. Atlas never again launched with an Able upper stage, so every attempt with this combination was a failure.[8]

Atlas-Agena

The Atlas-Agena combination was not without its problems, but it was certainly far more successful than Atlas-Able.[9] For launching the Agena, as for Project Mercury, the Atlas D had to be strengthened in its upper section by increasing the thickness of the skin to handle the greater loads the capsule or upper stage created. Guidance while the Atlas engines were burning continued to be provided by the GE radio-inertial system. An Atlas D–Agena A first got off the launching pad at Cape Canaveral on February 26, 1960, but the Agena stage failed to separate. The entire launch vehicle reentered the atmosphere about 2,500 miles downrange, burning up in the process. This was an attempt by the Air Force to launch a Midas (Missile Defense Alarm System) satellite designed to provide early warning of enemy missile launchings through use of infrared scanning to detect the rockets' exhaust plumes.

A more successful launch occurred with another Atlas D–Agena A on May 24, 1960, placing the Midas 2 satellite in orbit. It was the first early warning satellite to achieve orbit, but although it was designed to operate for forty months, its telemetry system ceased to function two days after launch. This was the last launch of the Atlas D–Agena A from Cape Canaveral. A Samos (Satellite and Missile Observation System) reconnaissance satellite was the other payload for the Agena A atop an Atlas D. On October 11, 1960, the first attempt from the Navy's Point Arguello launch facility south of Vandenberg AFB failed to achieve orbit, even though the launch vehicle functioned properly up to the point of inserting the Agena, which also served as part of the spacecraft, into a nearly circular orbit. At the launch, an umbilical cable did not separate from the Agena at the proper time before liftoff. In pulling out, it damaged the control system, preventing the gas jets that were supposed to provide control from functioning. A second attempt on January 31, 1961, succeeded in placing Samos 2 in orbit from Point Arguello. It marked the Air Force's last use of Agena A in conjunction with an Atlas. Because of the upper stage's small size and lack of restart capability, on January 16, 1959, the Air Force had issued a change to its Agena contract with Lockheed, directing the contractor to study and test a modified second stage with increased tank capacity and the ability to stop and restart its engine in space. This led to the Agena B, used with the Atlas D launch vehicle.[10]

The first Atlas D–Agena B succeeded in launching Midas 3 into a polar orbit from Vandenberg AFB on July 12, 1961. This was the heaviest U.S. satellite yet orbited. Unfortunately, the satellite's solar array failed to provide auxiliary power, so the infrared telescope returned data for only five orbits. Meanwhile, in early 1960 NASA had selected the Atlas D–Agena B combination for its Project Ranger to launch a spacecraft to the surface of the Moon. During the final two minutes of descent, the spacecraft was to televise the lunar surface at a resolution expected to be vastly better than photographs taken from Earth. The launch vehicle and spacecraft on Ranger 1 were to perform tests in preparation for the actual lunar missions in the program. The objective was to place Ranger 1 in a highly elliptical Earth orbit where it could collect data on the space environment—solar plasma, magnetic fields, cosmic rays. On August 23, 1961, the Atlas and Agena boosted the Ranger into a low Earth orbit. But the Agena, which had executed a second burn on the Midas 3 mission, failed to restart on Ranger 1 and could not inject the spacecraft into its higher orbit. The vehicle stayed in orbit for three days. Researchers checked out some of the systems on board and collected some data before it reentered the atmosphere. This prompted NASA's assistant director for lunar and planetary programs to say that "although the design orbit was not achieved, the flight constituted a fairly good test of the spacecraft."[11]

Analysis of telemetry data revealed the cause of the Agena stage's failure to restart: the malfunction of an oxidizer manifold pressure switch, which was supposed to send electric current to a solenoid, thereby opening the fuel valve. Overheating, presumably from the Sun's rays, produced the malfunction. Design engineers had obviously failed to foresee this problem in what was still the comparatively new environment of space. They solved it in another of the many instances of cut-and-try engineering in this period by adding heat shields, redundant switches, and circuits to subsequent Agena stages. In addition, they found switches that were better able to withstand high temperatures. There were no more failures of this particular type on the Agena B.[12]

Although Agena B achieved a success rate of 83 percent through 1966 on seventy-one launches by both Thor and Atlas first stages, it registered another failure (of a different sort) on its second Ranger mission. As with Ranger 1, the mission was for Agena on a second burn to carry the spacecraft from a parking orbit into a highly elliptical Earth orbit, where the Ranger 2 was supposed to study the space environment over a five-month period. Launched from Cape Canaveral on November 18, 1961, the Atlas D booster performed its portion of the mission, and the Agena B's first burn positioned the spacecraft in the intended parking orbit. Once again, there was no second burn to insert the vehicle into its higher orbit. This time, analysis of the tracking data suggested that a roll gyro in the inertial reference system either failed in flight or was nonfunctional at launch time. This caused the Agena to tumble in orbit, which in turn made the propellants flow away from the intake lines to the combustion chamber, preventing the second burn. Ranger 2 reentered the atmosphere on November 19, having transmitted little useful data.[13]

Two failures of Agena B on as many Ranger missions induced General Ritland at SSD to assemble an investigating board to ensure that the upper stage would succeed on Ranger 3. The board determined that a faulty electric-power relay rendered the gyro inoperative. Procedural and equipment changes ensued, but on January 26, 1962, a malfunction of the Atlas guidance system caused by faulty transistors sent Ranger 3 past the Moon into a heliocentric orbit. This failure did not lead to any hardware changes but did produce new quality-assurance procedures. On April 23, 1962, both the Atlas D and the Agena B functioned properly and sent Ranger 4 to lunar impact, only to have that spacecraft fail to carry out its intended operations.[14] Meanwhile, beginning with the launch on March 7, 1962, of Samos 6 by the Air Force, the Atlas Ds used for space launches began to be equipped with the baffled injectors and the hypergolic ignition system used with the booster engines on the Atlas Es and Fs.[15]

Soon after this, NASA inaugurated a program to use the same basic Atlas D–Agena B launch vehicle to send a spacecraft to the vicinity of Venus. Its mission was to gather data on the Venusian surface and the environment surrounding the planet. The agency did this in the knowledge that, including Air Force missions, as of June 23, 1962, the last thirteen Agena launches had been successful. On July 22 Mariner R-1 launched from Cape Canaveral, but a guidance equation error caused an incorrect trajectory, forcing the range safety officer to destroy the launch vehicle and its payload 290 seconds into the mission. However, Mariner 2 did have a successful mission. Launched on August 27, it passed close enough to Venus to gather significant data about its atmosphere and surface in the first completely successful interplanetary mission by any country.[16]

Despite the success of the mission, the Atlas D had not performed altogether satisfactorily. Shortly before the cutoff of the two booster engines, one of the two vernier engines on the Atlas failed for reasons that remained unclear. This did not pose a problem until the jettisoning of the boosters eliminated their roll control. Without that source of stability, the launch vehicle began to roll. The one functioning vernier opposed the roll and put the vehicle into a roll in the opposite direction. The Atlas-Agena-Mariner

Figure 13. An Atlas-Agena launching Mariner R-1 from Cape Canaveral Launch Complex 12, July 22, 1962. Courtesy of NASA.

was spinning at a rate of about once a second when the malfunctioning vernier started to work again, arresting the rotation. Separation and ignition of the Agena occurred satisfactorily, but because of the Atlas's roll problem, at ignition the Agena's pitch was two degrees below the planned trajectory. The horizon sensors in the second-stage guidance system did not correct the error for 15 seconds, but the Agena halted the first burn when its velocity meter sensed the preset speed. The Agena-Mariner coasted in orbit, and the restart of the Agena engine took place satisfactorily. At engine cutoff, the Mariner separated and headed for Venus.[17]

Coming on top of the other problems with Atlas-Agena vehicles, this comparatively minor malfunction led to a briefing at NASA Headquarters on October 1, 1962, in which an unidentified presenter pointed out that, on the six Atlas-Agena launches for NASA up to that time, there had been two Agena and three Atlas failures. The launch vehicles had performed properly on Ranger 4, even though the spacecraft had failed. But Mariner 2, the sole success, had succeeded despite a malfunction on the Atlas D. True, there had only been minor problems with the Agena B since Ranger 2, so Air Force–NASA cooperation had apparently achieved reliability there. And the two organizations were now proceeding with a standardized upper stage known as Agena D (for which Lockheed had received an Air Force letter contract on August 25, 1961). Atlas, however, remained a problem. Testing and checkout of NASA's six Atlas vehicles had shown that none of them was flightworthy. The Air Force and General Dynamics claimed that each of the three Atlas failures was random and that such problems were to be expected. Engineers analyzed the failures when they occurred and took corrective action when warranted. This situation was not acceptable to NASA.[18] The space agency pointed out that its uses of Atlas-Agena were quite different from the Air Force's. Whereas the Air Force did not require a high degree of guidance-and-control accuracy to place a satellite in Earth orbit, NASA required great precision for missions to the Moon and the planets.[19]

As it happened, on May 17, 1962, General Dynamics had already proposed to the Air Force that a standard Atlas space-launch vehicle be developed in lieu of individually tailoring Atlas Ds for each specific mission. On June 5, personnel from Marshall Space Flight Center—responsible since mid-1960 for managing NASA's Agena program as well as the Centaur upper stage—met in Huntsville with representatives of NASA Headquarters and the NASA organization at the Cape to review the proposal and make suggestions to the Air Force, which contracted with General Dynamics for the standardized Atlas, known as Space-Launch vehicle 3 (SLV-3). Despite NASA's urgings, the Air Force did not elongate the propellant tanks of SLV-3, which used the General Electric Mod 3G radio-inertial guidance system developed for its use on the Eastern Test Range. But standardization did apparently result in greater reliability. Whereas the Atlas D as a launch vehicle (LV-3A) had a total of 43 successful launches out of 53 attempts for a success rate of 81 percent, SLV-3 was successful on 49 of 51 space launches, or 96 percent, most of them with Agena upper stages.[20]

Even after the Agena D came into service in combination with Atlas first stages in 1963—by which time technical direction of Agena had passed from Marshall to Lewis Research Center—both NASA and the Air Force continued to use the Agena Bs still in inventory. For example, the Atlas D–Agena B combination remained the launch vehicle for the finally successful Ranger project lasting until March 21, 1965. Once problems with the spacecraft were resolved, Ranger 7 demonstrated for Project Apollo that the lunar surface would be suitable for human landings, and Rangers 7 through 9 transmitted thousands of photographs with a resolution as much as a thousand times better than those taken from Earth. Similarly, on September 5, 1964, an Atlas D and Agena B successfully launched NASA's OGO (Orbiting Geophysical Observatory) 1 into a highly eccentric orbit to measure Earth's atmosphere and magnetosphere as well as solar emissions and other phenomena. Then on June 6, 1966, an Atlas SLV-3 and an Agena B launched OGO 3 into another eccentric orbit, where it collected data within the magnetosphere and in interplanetary space. Meanwhile the Air Force had launched its final Atlas D–Agena B mission from Vandenberg AFB on July 19, 1963, placing Midas 9 in orbit. The combination had by then launched over a dozen other Air Force payloads.[21]

The Air Force launched the first Agena D coupled with an Atlas (also a D-model) on July 12, 1963, from Vandenberg AFB. Its mission, as with many of the subsequent Atlas–Agena D launches, was classified, but on October 16, 1963, Space Systems Division launched a pair of Vela satellites, the first of a series, from Cape Canaveral on an Atlas D–Agena D. The Vela program, which SSD managed for DoD, placed radiation-detection satellites in near-circular orbits about 70,000 miles above Earth to furnish information on nuclear detonations within Earth's atmosphere or in space as far as 100 million miles away. The final pair of Vela satellites went into orbit aboard a Titan IIIC in April 1970. The series helped to monitor the Limited Nuclear Test Ban Treaty reached with the Soviet Union in 1963. In addition, it provided information on solar flares and other radiation that might affect humans in space.

NASA too used the Agena D extensively. The first such mission, using an Atlas D for the first stage, launched on November 5, 1964, carrying the Mariner 3 space probe past Mars to gather data on the red planet's surface and atmosphere. The two stages of the launch vehicle functioned as intended, but the spacecraft did not. A shroud protecting it during ascent failed to jettison from Mariner 3 at the appropriate time. The shroud was a new model made of fiberglass and magnesium, and the inner fiberglass core had separated from the outer skin. Engineers quickly designed an all-metal shroud to replace it, using a magnesium section with an inner thermal liner. This allowed Mariner 4 to launch aboard another Atlas D–Agena D on November 28. This time the launch vehicle again functioned properly, and the new shroud ejected effectively. This Mars mission was highly successful, sending back the first photos of another planet from a near encounter plus much new data.[22]

Among other comparatively early missions launched by the Agena D, all with Atlas SLV-3s as the first stage, were Orbiting Astronomical Observatory 1, Lunar Orbiters 1–5, and Applications Technology Satellites (ATS) 1–3. Of these, only ATS 2 experienced a problem with Agena D, a failure of the fuel supply system that prevented the upper stage from reigniting to insert the satellite into a circular orbit. It went into a highly elliptical orbit in which its stabilization system could not overcome the inertial forces, producing tumbling that was deleterious to the mission.[23]

In May 1965, Robert C. Seamans Jr., NASA's associate administrator, and Harold Brown, DoD's director of defense research and engineering, agreed to improve the performance of SLV-3. General Dynamics' Convair Division got the contract, resulting in an SLV-3A for the Agena and an SLV-3C for the Centaur. The change, which NASA had requested before SLV-3 came into existence, allowed an increase in weight of the later versions of Agena plus its payloads. General Dynamics lengthened the Atlas by 9.75 feet from the 68.95 feet of SLV-3 to 78.7 feet for SLV-3A. Its longer propellant tanks accommodated 48,000 more pounds of propellant. To compensate for the extension's added weight, Rocketdyne improved the combined thrust of the two booster engines from 330,000 pounds on SLV-3 to 336,000 pounds on SLV-3A, while the sustainer engine's thrust rose from 57,000 to 58,000 pounds. With SLV-3A, the Agena D could place roughly 7,500 pounds in Earth orbit, compared to 6,000 pounds with SLV-3. By 1972 the thrust had increased further, to 370,000 pounds for the boosters and 60,000 for the sustainer. The propellants remained unchanged, with the increased thrust coming from higher propellant flow rates. Guidance for the Atlas stages used with Agena remained radio-inertial.[24]

From 1968 to 1978, the Atlas SLV-3A flew only twelve times, mostly on classified missions. But its first launch on March 4, 1968, was to place NASA's OGO 5 in an eccentric orbit to collect geophysical data for a better understanding of how Earth functioned as a planet and interacted with the Sun. It was the most successful of the five OGO missions, with the onboard systems operating for forty-one months. The last of the SLV-3A flights occurred in early April 1978. The Atlas-Agena carried an electronic eavesdropping satellite named Aquacade. In all, there were approximately 110 Atlas-Agena launches, with the launch of an Agena and a modified Atlas F for Seasat 1 on June 27, 1978, being the last. Agena continued to fly missions on the Titan family of launch vehicles until at least 1987, but by May of 1979, on Thor, Atlas, and Titan boosters, it had proved itself to be a workhorse of space, achieving a reported success rate of about 93 percent.[25]

Atlas-Centaur

Another very important upper stage used with the Atlas was the Centaur. If Agena was the workhorse of space, Centaur was the Clydesdale. Its powerful engines enabled it to carry heavier payloads into orbit than Agena could handle. The Centaur could do this because it burned liquid hydrogen as well as liquid oxygen. Hydrogen offered more thrust per pound of fuel burned per second than any other chemical propellant then available—some 35 to 40 percent more than RP-1 when burned with liquid oxygen.[26] The added performance allowed various versions of Atlas-Centaur to support such NASA missions as landing on the lunar surface in the Surveyor project and orbiting High-Energy Astronomy Observatories as well as placing 35 communications satellites in orbit through 1989, including 23 Intelsat (International Telecommunications Satellite Consortium), 4 Comstar (Communications Satellite Corporation), and 8 FLTSATCOM (Fleet Satellite Communications) satellites. As with other upper stages flying on Atlas vehicles, not all of the Centaur missions were successful, but most were.[27]

Until the vehicle became operational, however, Centaur went through an unusually difficult gestation.[28] The background to its development and the circumstances surrounding that process were complicated. Both Goddard and Oberth had written about the advantages of hydrogen as a rocket fuel, and for the von Braun team in Germany, Walter Thiel recognized the potential of liquid hydrogen. However, he had experienced leaks and problems in handling the fuel when he actually tried to use it at Kummersdorf, writing in 1937, "the extremely low temperature of the liquid hydrogen . . . ; the high boil-off rate . . . ; the danger of explosion; the large tank volume required as a result of the low specific weight . . . ; and the need to use insulated tanks and ducts, create . . . difficulties which will pose strong obstacles to . . . experimental and development activities."[29]

Despite such difficulties, Professor Herrick L. Johnston created a cryogenics laboratory at Ohio State University with research funds available during World War II. There he produced some liquid hydrogen by 1943. The university began experimenting with a rocket burning hydrogen and oxygen from 1947 to 1950, and Aerojet performed similar tests from 1945 to 1949, with some cooperation between the two projects. Both programs developed pumps for liquid hydrogen and operated thrust chambers. JPL also tested a liquid-hydrogen engine during 1948. There was, however, no immediate use of the valuable data these programs yielded.[30]

In 1954 a small group of researchers at the NACA's Lewis Flight Propulsion Laboratory began experimenting with liquid-hydrogen rocket engines. Lewis engineers tested tanks, pumps, heat exchangers, and turbojet engines. The efforts culminated in a joint flight-research project with the Air Force in late 1956 and early 1957 in which one engine of a B-57 bomber operated part of the time with gaseous hydrogen as its fuel. The laboratory developed a predilection for liquid hydrogen, and Lewis associate director Abe Silverstein supported its development and use. He was later a key manager at NASA Headquarters and then director of Lewis Research Center (as the lab was called after the founding of NASA), so these developments provided an important background to the later development of Centaur.[31]

Meanwhile, a more immediately important development in the use of hydrogen as a fuel came in 1956–58 with the Air Force's highly secret Suntan project. This was an attempt to produce an airplane fueled with hydrogen that could outperform the U-2 reconnaissance aircraft, which itself was secret. The project involved contracts with Lockheed for two prototype aircraft, designated CL-400, and with the Pratt & Whitney division of United Aircraft for a study and, if feasible, the development of a hydrogen-powered engine. At Wright Air Development Center in Ohio, the Air Force put Lt. Col. John D. Seaberg, an aeronautical engineer who had worked at Chance-Vought until recalled to active duty during the Korean War, in charge of overseeing work on the airframe, airplane systems, and liquid-hydrogen fuel tanks. Pratt & Whitney designed a new engine, designated model 304, and a centrifugal pump to feed the liquid hydrogen to the engine. The Air Force also funded two liquid-hydrogen plants near Pratt & Whitney's isolated test center in West Palm Beach, Florida, where it could do research on hydrogen in an area distant from population centers. The Air Force effectively cancelled Suntan in June 1958 before the project reached fruition, but by then the various efforts to work with liquid hydrogen had laid the technological groundwork for Centaur.[32]

Before the defense establishment made use of this technology, however, it had to be nudged by a proposal from Convair's Krafft Ehricke. Called to service in a German Panzer division on the western and then the eastern front during World War II, the young man was still able to earn a degree in aeronautical engineering at the Technical Institute of Berlin. He was fortunate enough to be assigned to Peenemünde in June 1942, where he worked closely with Thiel. Though he came to the United States as part of von Braun's group and moved with it to Huntsville, Ehricke was a much less conservative engineer than von Braun. He transferred to Bell Aircraft in 1952 when it was working on the Agena upper stage and other projects. Then in 1954, when he believed interest at Bell had shifted away from space-related efforts, he heeded a call from Bossart to work at Convair.[33]

At the San Diego firm, Ehricke initially served as a design specialist on Atlas and was involved with Project Score. By 1956 he was beginning to study possible vehicles for orbiting satellites, but he could find no support for such initiatives until after the Soviet Union successfully orbited Sputnik 1. Then General Dynamics managers asked him to design an upper stage for Atlas. He and some other engineers, including Bossart, decided on liquid hydrogen and liquid oxygen as the propellants. And Ehricke worked with Rocketdyne to develop a proposal entitled "A Satellite and Space Development Plan." This featured a four-engine stage with the hydrogen and oxygen fed to the engines by pressure rather than pumps, neither Rocketdyne nor Ehricke being aware of the pumps Pratt & Whitney had developed. In December 1957, James Dempsey, vice president of the Convair Division, sent Ehricke and another engineer named William H. Patterson off to Washington, D.C., to pitch the design to the Air Force.[34]

The air service did not act on the proposal, but in 1958 Ehricke proposed it to the new Advanced Research Projects Agency, established by the Department of Defense on February 7, 1958. For a time, ARPA exercised control over all military and civilian space projects before relinquishing the civilian responsibility to NASA in October 1958. Thereafter, for a year, ARPA remained responsible for all military space projects, including their budgets. The new agency made Ehricke aware of Pratt & Whitney's hydrogen pumps and encouraged Convair to submit "a proposal using two pump-fed engines" of a type the engine contractor had suggested, each with 15,000 pounds of thrust. Convair submitted that proposal in August 1958. Near the end of that month, ARPA then issued order number 19–59 for a high-energy, liquid-propellant upper stage to be developed by Convair-Astronautics Division of General Dynamics, with liquid-oxygen/liquid-hydrogen engines to be developed by Pratt & Whitney.[35]

In October and November 1958, at ARPA's direction, the Air Force followed up with contracts to Pratt & Whitney and Convair for the development of Centaur, but NASA's first administrator, Keith Glennan, requested transfer of the project to the civilian space agency. Deputy Secretary of Defense Donald Quarles agreed in principle, but ARPA and the Air Force resisted the transfer until June 10, 1959, when NASA associate administrator Richard E. Horner proposed that the Air Force establish a Centaur project director, locate him at the Ballistic Missile Division in California, but have him report to a Centaur project manager at NASA Headquarters. The Air Force would provide administrative services and NASA would furnish technical assistance. DoD agreed to this suggestion, and the project transferred to NASA on July 1, 1959. Lieutenant Colonel Seaberg from the Suntan project became the Air Research and Development Command project manager for Centaur in November 1958, located initially at command headquarters on the East Coast. Seaberg remained in that position after the transfer to NASA but moved his location across country to BMD. Milton Rosen became NASA project manager at the civilian agency's headquarters. Also in November 1958, Ehricke became Convair's project director for Centaur.[36]

As if all of this were not complex enough, in the fall of 1958 NASA engineers had conceived of using the first-stage engine of Vanguard as an upper stage for Atlas to be known as Vega. NASA intended that it serve as an interim vehicle until Atlas-Centaur was developed. Under protest from Dempsey that Convair already had its hands full with Atlas and Centaur, on March 18, 1959, NASA contracted with General Dynamics to develop Atlas-Vega. With the first flight of the interim vehicle set for August 1960, Vega at first became a higher priority for NASA than Centaur. As such, it constituted an impediment to Centaur development at Convair until NASA cancelled it on December 11, 1959, in favor of the DoD-sponsored Agena B, which had a development schedule and payload capability similar to Vega's.[37]

Besides Vega's competition for resources up to this point, another hindrance to development of Centaur was liquid hydrogen's physical characteristics. Its very low density, extremely low boiling point (−423°F), low surface tension, and wide range of flammability made it extremely difficult to handle. Ehricke had some knowledge of this from his work with Thiel, but the circumstances of the contract with the Air Force limited the amount of testing he could do to ensure that designs would accommodate the peculiarities of the fuel he had selected.[38]

One factor was funding. When ARPA accepted the initial proposal and assigned the Air Force to handle its direction, the stipulations were that charges by Convair-Astronautics not exceed $36 million, that a first launch attempt occur by January 1961, and that the project not interfere with Atlas development. At the same time, Convair was to use off-the-shelf equipment and Atlas tooling and technology as much as possible. Funding for the Pratt & Whitney contract was $23 million, bringing the total initial outlay for the project to $59 million, not including the costs of a guidance and control system, Atlas boosters, and a launch complex for the initial six launches called for by the contract. Ehricke argued that the limited funding restricted ground testing until it was too late. Another constraint was the lack of DoD's highest priority, known as DX, which meant that subcontractors who were also working on projects with a DX priority could not give the same level of service to Centaur as they provided to the higher-priority projects.[39]

With such limitations, Convair and Pratt & Whitney proceeded with designs for the Centaur structure and engines. The structure of the Centaur stage followed the "steel-balloon" pattern of Atlas, with the same 10–foot diameter. The lightness of the resulting airframe seemed necessary for Centaur because of liquid hydrogen's low density, which required that the hydrogen tank be much larger than the oxygen tank. Conventional designs with longerons and ring frames would have yielded a less satisfactory mass fraction than did the pressurized tanks with thin skins—initially only 0.01 inches thick. The elliptical liquid oxygen tank was on the bottom of the stage, with a cylinder housed inside the lower part of the tank to carry the engine thrust evenly to the rest of the vehicle. To create the shortest possible length and the lowest weight, the engineers on Ehricke's project team made the bottom of the liquid hydrogen tank concave so that it fit over the convex top of the oxygen tank. This arrangement saved about 4 feet of length and roughly 1,000 pounds of weight but created two other difficulties in the process. One was the necessity for higher pressure in the oxygen tank than in the hydrogen tank so that the bulkhead between the two would continue to be curved in the hydrogen direction. The other was more vexing, caused by the small size of the hydrogen molecules and their extreme coldness. The skin of the oxygen tank had a temperature of about −299°F, which was so much "warmer" than the liquid hydrogen at −423°F that the hydrogen would gasify from the relative heat and boil off. To prevent that, the engineers devised a bulkhead between the two tanks that contained a fiberglass-covered Styrofoam material about 0.2 inches thick in a cavity between two walls. Technicians evacuated the air from the pores in the Styrofoam and refilled the spaces with gaseous nitrogen. They then sealed the opening by welding. When they filled the tank with liquid hydrogen, the upper surface of the bulkhead became so cold that it froze gaseous nitrogen in the cavity, creating a vacuum by a process called cryopumping (nitrogen being denser in the solid than in the gaseous and liquid states).[40] The vacuum and Styrofoam were supposed to provide insulation for the liquid hydrogen.

Because testing was limited under the tight initial budget, it was not until the summer and early fall of 1961 that Centaur engineers and managers learned of major heat transfer across the bulkhead. This was so great—more than fifty times the expected amount—as to make the vehicle incapable of carrying out its assigned missions, which required the Centaur engine to stop and, after a coasting period in a parking orbit, restart. The bulkhead, it turned out, had very small cracks through which the highly diffusive hydrogen was leaking and destroying the vacuum, permitting heat transfer and causing fuel to boil off. The boil-off built excessive pressure, which had to be vented, lest the hydrogen tank explode. But the venting depleted the fuel below the level needed for the second engine burn. General Dynamics had used Atlas manufacturing techniques for the materials on Centaur, including those for the bulkhead. Atlas quality control could detect leaks in bulkheads down to about one ten-thousandth of an inch. Inspections revealed no such leaks, but the engineers learned in the 1961 testing that hydrogen could escape through even finer openings. Very small cracks that would not be a problem in a liquid-oxygen tank caused major leakage in a liquid-hydrogen tank.[41]

A year before Convair-Astronautics knew of this, on July 1, 1960, NASA Headquarters had assigned responsibility for Centaur to Marshall Space Flight Center, with Seaberg's Centaur Project Office remaining at BMD in California. Hans Hueter became director of Marshall's Light and Medium Vehicles Office, charged with managing the Centaur and Agena upper stages. During the winter of 1959–60, following cancellation of the Vega program, NASA also established a Centaur technical team composed of experts at various NASA centers and headquarters to recommend ways the upper stage could be improved. In January 1960, Navy Cdr. William Schubert became the Centaur program chief at NASA Headquarters.[42]

From December 11 to 14, 1961, John L. Sloop visited General Dynamics/Astronautics (GD/A) to look into Centaur problems, particularly the heat transfer across the bulkhead. Sloop had been head of Lewis Laboratory's rocket research program from 1949 until 1960, when Abe Silverstein brought him to NASA Headquarters. There in 1961 he became deputy director of the group managing NASA's small and medium-sized launch vehicles. After his visit he wrote, "GD/A has studied the problem and concluded that it is not practical to build bulkheads where such a vacuum [as Ehricke's team designed] could be maintained." The firm also believed "that the only safe way to meet all Centaur missions is to drop the integral tank design and go to separate fuel and oxidizer tanks." Sloop disagreed, saying that "if a decision must be made now, I recommend we stick to the integral tank design, make insulation improvements, and lengthen the tanks to increase propellant capacity."[43]

Subsequent events justified Sloop's optimism. In "a program of designing and testing a number of alternate designs to the original intermediate bulkhead," the Centaur team found that adding nickel to the welding of the double bulkhead (and elsewhere) significantly increased the single-spot shear strength of the metal at $-423°F$.[44]

Many other problems plagued Centaur development. Several of them involved the engines. After enduring "inadequate facilities, slick unpaved roads, mosquitoes, alligators, and 66 inches of rain in a single season" while developing the 304 engine for Suntan at West Palm Beach, Florida, Pratt & Whitney engineers "discovered the slippery nature of hydrogen," as did their Convair-Astronautics counterparts on the opposite coast. The extreme cold of liquid hydrogen precluded use of rubber gaskets to seal pipe joints, so the designers had to resort to aluminum that was coated with Teflon and then forced into flanges that mated with the joints. The engineers needed new techniques for seals on rotating surfaces, where carbon impregnated with silver found wide use. Another concern with the cryogenic hydrogen was to keep the liquid from turning to gas before reaching the turbopumps. The engineers attacked that problem by flowing propellants to the pumps before engine start, precooling the system.[45]

The turbopump for the 304 engine lubricated its bearings with oil. To keep it from freezing in proximity to the cold pump, the oil had to be heated, which created a temperature gradient. For Centaur's RL10 engine, the Pratt & Whitney engineers coated the cages holding the bearings with fluorocarbons similar to Teflon and arranged to keep the bearings cold with minute amounts of liquid hydrogen. This had the same effect as lubrication, since the oil's main function was to prevent overheating. The gears in the 304 were made of Waspalloy, a substance of high tensile strength developed for Pratt & Whitney's J48 jet engine, but it bonded in the hydrogen environment. Engineers replaced it with a carbonized steel coated with molybdenum disulfide to provide dry lubrication. This solved the bonding problem but subjected some unlucky engineers to observing tests of the new arrangement through binoculars from an observation post having only a screen door. On the third shift, alligator croakings and other strange noises produced uneasiness in young observers unused to late-night swamp sounds.[46]

The first component tests of the RL10 combustion chamber, which consisted of stainless steel regenerative-cooling tubes brazed with silver, took place in May 1959. As with many other initial tests of combustion chambers, there were signs of burn-through, so the engineers changed the angle at which the hydrogen entered the tubes and aligned the tubes more carefully so that they did not protrude into the exhaust stream. Engine firings two months later showed that the adjustments had eliminated the burn-through but that a cone-shaped chamber produced inefficient combustion. Engineers changed to a bell shape and conducted a successful engine run in September 1959, less than a year from the date of the initial contract.[47]

In a major innovation, the design of the RL10 took advantage of the cold temperature and high specific heat of liquid hydrogen to dispense with a gas generator to drive the turbopump. As the cryogenic fuel passed from its tank into the tubes of the combustion chamber to cool them, it absorbed heat, causing it to vaporize and expand. This provided enough kinetic energy to drive the turbine that operated both the liquid hydrogen and liquid oxygen pumps. It also drove the hydraulic actuators that gimballed the engine. The process was called the "bootstrap cycle," but it still used hydrogen-peroxide boost pumps, fitted into the propellant lines as they exited from the tanks, to start the process. Hydrogen peroxide also powered attitude-control rockets and so-called ullage-control jets, which accelerated the Centaur in a parking orbit and thereby forced the liquid hydrogen to the rear of the tank so it could be properly pumped into the engines for ignition by a spark generator.[48]

Before the RL10, in its two-engine configuration for the Centaur, underwent its first test in an upright position on a test stand, it went through 230 successful horizontal firings. It produced 15,000 pounds of thrust and achieved a specific impulse of 420 lbf-sec/lbm at an expansion ratio of 40:1 through its exhaust nozzle. As required by its assigned missions, it reliably started, stopped, and restarted so that it could coast in a parking orbit until it reached the optimum point for injection into an intended orbit (or, for interplanetary voyages, trajectory). Then on November 6, 1960, two RL10s, upright on a test stand at the Pratt & Whitney facility in Florida for the first time, fired simultaneously and successfully—but only briefly because of a problem with the timer on the test stand. When engineers tried to repeat the test the next day, just one engine fired. The other filled with hydrogen and oxygen until the flame from the first engine caused an explosion that damaged the entire propulsion system beyond repair.

A tape recording of the countdown suggested faulty operation of a test-stand sequencer, so engineers did not suspect a problem with the engine itself. They repaired the test stand and readied another pair of engines on January 12, 1961. Before proceeding, they installed a shutoff valve on the hydrogen tank, separated the exhaust systems of the two engines by a greater distance, and put a blast wall between the two engines for good measure. This time, there was no problem with the sequencing, yet the explosion recurred. In the vertical position, the engineers learned, gravity affected the mixing of the oxygen and hydrogen in a different way than in the horizontal position. So in a further instance of trial-and-error engineering, they adjusted the method of hydrogen feed. They also devised a method of measuring the density of the mixture to ensure the presence of enough oxygen for ignition. With these changes, the two engines fired simultaneously in the vertical test stand on April 24, 1961.[49]

Following this correction, the engines completed 27 successful dual firings at Pratt & Whitney and 5 more at the rocket site on Edwards Air Force Base in California. They then passed the flight rating test from October 30 to November 4, in which they completed 20 firings equivalent in duration to six Centaur missions.[50]

To protect the liquid hydrogen in its tank from boiling off while the vehicle was on the launchpad and during ascent through the atmosphere, engineers had designed four jettisonable insulation panels made of foam-filled fiberglass. These were about a centimeter thick, attached to the tank by straps around their circumference. To keep air from freezing between the tank and the insulating foam, thereby bonding the panels to the tank, designers came up with a helium purging system. To reduce the weight penalty imposed by the 1,350–pound panels, they had to be jettisoned as soon after launch as the atmosphere thinned and the ambient temperature dropped.[51]

Various delays for different reasons—the engine ignition problem, difficulties with elaborate test instrumentation such as a television camera and sensors inside the liquid-hydrogen tank, and others—pushed back the first launch of an Atlas LV-3 with a Centaur upper stage until May 8, 1962, some fifteen months later than planned. The goals of the test flight were to proceed through the boost phase with jettison of the insulation and a nose fairing, followed by Centaur's separation from the Atlas. With only a partial load of fuel, the Centaur was to coast for eight minutes and burn for 25 seconds. The Centaur guidance system would be tested in an open loop (no actual

Figure 14. The Centaur RL10 rocket engine developed by Pratt & Whitney. Courtesy of NASA.

control of the launch vehicle stages), with an autopilot providing the real control.[52]

On the May 8 launch, the two stages rose normally until they approached maximum dynamic pressure at 54.7 seconds into the flight, with aerodynamic buffeting as the vehicle got close to the speed of sound. At that point, an explosion occurred as the liquid-hydrogen tank split open. Initially, engineers decided that the aerodynamic forces had destroyed the insulation and ruptured the tank. About five years later, tests suggested that the real culprit was differential thermal expansion between a fiberglass nose fairing and the steel tank, causing a forward ring to peel off the tank.[53]

Even before this launch, the difficulties with Pratt & Whitney's engine development, resultant schedule delays, and other problems, such as the one with the bulkhead between the hydrogen and oxygen tanks, had led to close scrutiny of the Centaur program, with cancellation a real possibility. Following John Sloop's visit to General Dynamics in December 1961 to look into such problems, he had expressed concerns about the firm's organization. Krafft Ehricke, the program director, had only five men reporting directly to him, and Deane Davis, the project engineer, had direct charge of only two people. Many other people worked on Centaur, twenty-seven of them full-time, but most of them were assigned to six operating divisions not directly under project control. Sloop wrote, "As far as I could tell in three days of discussion, the only people who have direct and up-to-date knowledge of all Centaur systems are Mr. Ehricke and Mr. Davis." Marshall Space Flight Center had "a very competent team of four men stationed at GD/A," and they were well aware of the "management deficiencies" Sloop commented on.[54]

Hans Hueter, von Braun's director for light and medium vehicles, wrote on January 4, 1962, to James Dempsey, president of General Dynamics/Astronautics, expressing his concern about the way the Centaur Program Office was organized in "relation to the line divisions." Hueter mentioned that he and Dempsey had discussed this issue "several times" and reiterated his and others' "impression that the systems engineering is carried on single-handedly by your excellent associates, Krafft Ehricke and Dean Davis." He added, "The individual fields such as propulsion, thermal and liquid behavior, guidance and control, and structures are covered in depth in the various engineering departments but coordination is sorely lacking."[55]

What disturbed both Sloop and Hueter was Astronautics' use of a so-called matrix organization in which Ehricke's program office relied for engineering assistance upon the services of people reporting to other functional department heads. In response to NASA's concerns, Dempsey shifted to a "projectized" organization in which roughly 1,100 employees at Astronautics were placed under the direct authority of the Centaur program director. On February 1, 1962, Dempsey reassigned Ehricke, making him director of advanced systems. Grant L. Hansen became Centaur program director and Astronautics vice president. Trained as an electrical engineer at Illinois Institute of Technology, Hansen had worked for Douglas Aircraft from 1948 to

1960 on missile and space systems, including the Thor, and had experience in analysis, research and development, design, and testing. He came to General Dynamics/Astronautics in 1960 to direct the work of more than 2,000 people on Atlas and Centaur. After February 1962, Ehricke continued to offer Hansen advice and consultation. While Ehricke was imaginative, creative, and "a hell of a good engineer," the company had decided, as Hansen remembered, that he "wasn't enough of a[n] S.O.B. to manage a program like this." Hansen proved to be an effective manager, although it is important to note that he was given authority and an organization Ehricke had lacked.[56]

Several other programmatic changes occurred around this time. On January 1, 1962, NASA (in agreement with DoD) transferred the Centaur Project Office from Los Angeles to Huntsville, Alabama, and converted existing Air Force contracts to NASA covenants. Lieutenant Colonel Seaberg ceased being project manager, and Francis Evans at Marshall Space Flight Center assumed those duties under Hueter's overall direction. By now funding had grown from the original $59 million to $269 million, and the number of Centaur vehicles to be delivered rose from six to ten.[57]

Part of the funding increase had permitted General Dynamics/Astronautics to subcontract with Minneapolis-Honeywell in May 1959 for development of a guidance and control system for Centaur. In its early phases, Centaur's mission was simply to determine whether hydrogen and oxygen could be used in an upper stage. The mission evolved gradually but remained in flux. However, it became clear that the vehicle would need to go into a parking orbit and then restart for injection into another orbit—such as a geosynchronous equatorial one—or a trajectory into space. The guidance and control system on Centaur was designed to function for both the Atlas and Centaur stages, except for attitude and rate sensing from a rate gyro system on Atlas. Centaur's system thus had to provide guidance and control during the boost and sustainer phases of Atlas flight, the initial burn of the two Centaur engines, and up to two restarts, plus adjustments during coasting flight by attitude-control jets using hydrogen peroxide for propulsion.[58]

The necessity for one or more restarts from a parking orbit made an inertial guidance and control system highly desirable. Otherwise, restarts would have been restricted to areas within line of sight of tracking systems. The mission's complexity and its accuracy requirements also dictated use of a digital computer. Honeywell selected a rotating drum computer made by Librascope, a division of General Precision. This L-31 digital computer weighed only 65 pounds, including its input-output system. The basic sensors for the guidance and control system were three pendulous gyro accelerometers mounted on a four-gimbal platform that was stabilized by three sin-gle-degree-of-freedom, floated gyroscopes. Pulses from the accelerometers provided data to the computer, which calculated velocity and acceleration and then integrated the data to ascertain the vehicle's position. With these data points, which it compared with precalculated mission requirements, the computer provided steering commands to both the Atlas and Centaur stages. It also cut off their engines and restarted Centaur's at the appropri-ate times. Except for a clock, the guidance computer itself shut down during long coast periods. Personnel at the Air Force's Central Inertial Guidance Test Facility, Holloman AFB, New Mexico, tested this system on fourteen high-speed sled runs between February 6 and June 18, 1964, long after Min-neapolis-Honeywell had delivered the first guidance system to General Dy-namics/Astronautics on August 11, 1960. At accelerations 7.5 times the force of gravity, the sled tests showed that the system functioned properly in an environment that simulated a launch's G-forces and random vibrations.[59]

Meanwhile, following the launch and explosion of May 8, 1962, the House Subcommittee on Space Sciences, chaired by Rep. Joseph E. Karth (D-Minn.), held hearings on the mishap on May 15 and 18. In a report issued on July 2, the parent Committee on Science and Astronautics was critical, stating that "management of the Centaur development program has been weak and ineffective both at NASA headquarters and in the field."[60]

NASA did not immediately react by making further changes, but there clearly were problems with Marshall's management of Centaur. These came out in the hearings and prompted unfavorable comment in the committee report. Von Braun had spoken of Astronautics' "somewhat bold approach. In order to save a few pounds, they have elected to use some rather, shall we say, marginal solutions where you are bound to buy a few headaches before you get it over with." Hansen had admitted that his firm was inclined to "take a little bit more of a design gamble to achieve a significant improvement, whereas I think they [von Braun's engineers] build somewhat more con-servatively." As the congressional report noted, "Such a difference in design philosophy can have serious consequences."[61]

Ehricke characterized the more conservative design approach of the von Braun team as "Brooklyn Bridge construction." The contrast between that approach and the one at General Dynamics/Astronautics is best revealed by an account that Deane Davis wrote soon after a Marshall visit to GD/A on an unspecified date shortly after Marshall took over responsibility for Centaur in July 1960. A group led by von Braun and including his structures chief, William Mrazek, and Hueter had come to General Dynamics for a tour and briefings on Atlas and Centaur. Mrazek and Bossart had gotten into a discussion on the steel-balloon tanks, with Mrazek (according to Davis) unwilling to admit they could have any structural strength without ribs. Bossart ushered Mrazek to a tank, handed him a 7–pound lead-weighted fiberglass mallet with a rubber cover and a 2–foot handle, and invited him to hit the tank. After a tap and then a harder whack, Mrazek could not find a dent. Bossart urged him to "stop fiddling around. Hit the damned thing!" When Mrazek gave it a "smart crack," the mallet bounced back so hard he could not hold on to it. It flew about fifteen feet, knocking off Mrazek's glasses on the way, and left on the tank only a black smear from the rubber cover, no dent. Davis wrote that Hueter was as amazed as Mrazek by the strength of the tank.[62]

Davis's account is difficult to accept in its entirety because Mrazek had designed the Redstone with an integral-tank structure that was, if hardly as light as Bossart's steel balloon, also not quite Brooklyn Bridge–like. Nevertheless, even in 1962 von Braun was clearly uncomfortable with Bossart's "pressure-stabilized tanks," which he called "a great weightsaver, but . . . also a continuous pain in the neck" that "other contractors, for example the Martin Co., for this very reason have elected not to use." No doubt because of such concerns, von Braun sought quietly to have the Centaur cancelled in favor of a Saturn-Agena combination.[63]

Faced with this situation, on October 8, 1962, NASA Headquarters transferred management of the Centaur program to Lewis Research Center, to which Silverstein had returned as director in 1961 from his position at NASA Headquarters heading the Office of Space Flight Programs. A "sharp, aggressive, imaginative, and decisive leader," Silverstein could be "charming or abrasive," in the words of John Sloop. Deane Davis, who worked with him on Centaur, called him a "giant among giants" and a man he "admired, adored, hated, wondered about—and mostly always agreed with even when I fought him. Which was often." Under Silverstein's direction, the Lewis Center insisted on much more testing than even the Marshall group had done. Everything that could "possibly be proven by ground test" was so tested. Yet Grant Hansen expressed admiration for the group from Lewis and its relations with his engineers at General Dynamics.[64]

Because the RL10 had been planned for use in Saturn as well as Centaur, management of that engine remained at Marshall. The reason given for the transfer of Centaur was that it would allow the Huntsville engineers to concentrate on the Saturn program. A news release quoted NASA administrator James Webb as saying, "This, I feel, is necessary to achieve our objectives in the time frame that we have planned. It will permit the Lewis Center to

Figure 15. Abe Silverstein *(right)* showing a prototype ramjet aircraft model to Edward R. Sharp, director of the NACA Lewis Flight Propulsion Laboratory, November 10, 1951. Silverstein served as NASA Lewis Research Center director from 1961 to 1969 and played a key role in development of the Centaur upper stage. Courtesy of NASA.

use its experience in liquid hydrogen to further the work already done on one of the most promising high energy rocket fuels and its application to Centaur."[65]

Long before technical direction over Centaur passed to Lewis Research Center, engineers from the Cleveland facility had been actively helping to solve the vehicle's engine and structural problems, including use of their

Figure 16. Engine and nozzle from the Centaur upper stage, still under development, at the Propulsion Systems Laboratory, NASA Lewis Research Center (now John H. Glenn Research Center) on October 20, 1960. Courtesy of NASA.

altitude chamber. Many other organizations besides the sled track facility at Holloman had been involved in Centaur development. In 1959 General Dynamics/Astronautics had done some zero-gravity testing in an Air Force C-131D aircraft at Wright-Patterson AFB (and also, at some point, in a KC-135). That same year, the firm had acquired a vacuum test chamber for testing gas expansion and components. With additional funding (to a total of about $63 million) in 1960, GD/A extended its testing program to include vacuum tests at the Air Force's Arnold Engineering Development Center, zero- gravity test flights using Aerobee rockets, and additional static ground testing, including at the Edwards AFB rocket site on test stand 1–1. In 1961, when GD/A's funding rose to $100 million, there were wind tunnel tests of Centaur's insulation panels at NASA's Langley Research Center, additional zero-G testing, construction of a coast-phase test stand to evaluate Centaur's attitude control system, and flight tests of the Centaur guidance system on the Atlas.[66]

At Lewis, Silverstein decided to direct the Centaur project himself, with the assistance of two managers reporting to him personally and with 41 people initially assigned to technical direction. Some forty Marshall engineers helped briefly with the transition. By January 1963 the changeover was mostly complete and Centaur had acquired a DX priority. As costs for the project reached an estimated $350 million, containing them became an issue. Still, Silverstein decided that the first eight Centaurs after the transfer would be test vehicles. The Surveyor series of spacecraft had been assigned as Centaur payloads, and Silverstein determined that none of them would be launched until flight tests had demonstrated Centaur's reliability.[67]

By February 1963 Silverstein had named David Gabriel as Centaur manager, but he placed the project office in the basement of his administrative building so that he could continue to keep tabs on the ongoing development of the troublesome but promising upper stage. Some continuity with the period of Marshall management came in the retention of Ronald Rovenger as chief of the NASA field office at GD/A. From just four NASA engineers, his office rose to a complement of forty. It took until April 1964, but Lewis renegotiated the existing contracts with GD/A into a single cost-plus-fixed-fee document for 14 Centaur upper stages plus 21 test articles. The estimated cost of the agreement was roughly $321 million plus a fixed fee of $31 million, very close to the $350 million estimated at the beginning of 1963. However, Silverstein felt the need for a second contract to cover further modifications resulting from Lewis's technical direction, with the Lewis Center director determining incentive fees himself. Soon the Lewis staff working on Centaur grew to 150 people. Silverstein continued to give the project his personal attention and made a major decision to abandon temporarily the use of a parking orbit and restart for Surveyor. This required a direct ascent to the Moon, considerably narrowing the "window of opportunity" for each launch.[68]

These and other changes under Lewis direction did not immediately solve all of Centaur's problems. There is not enough space here to provide a blow-by-blow account of all the Centaur test flights, which are summarized in table 3.1, beginning with Atlas-Centaur 2 (AC-2).

Data from instrumentation on the insulation panels over the liquid-hydrogen tank on AC-2 showed conclusively that the panels used on AC-1 were inadequate. Engineers designed thicker panels with heavier reinforcement, but the weight rose by almost 800 pounds, making it all the more important to jettison them after about 180 seconds. A drive shaft failure on AC-3 occasioned only a minor redesign. However, avoiding AC-4's problem with liquid hydrogen sloshing away from the bottom of the tank (where it

Table 3.1. Atlas-Centaur Test Flights

Flight	Mission	Objective	Outcome
AC-2 11/27/63	R&D, single burn	Separation of Centaur; Earth orbit; gather data on nose-cone insulation panels	Successful: achieved orbit almost as planned; data gathered
AC-3 6/30/64	R&D, single burn, restart boost pumps	Test jettison of redesigned insulation panels and nose cone; gather data from restart	Jettison successful; failure of drive shaft in hydraulic pump prevented gimballing
AC-4 12/11/64	R&D, two-burn	Restart engines; carry Surveyor model	Partial success: good first burn, but ullage motors unable to keep liquid hydrogen at bottom of tank; weak restart
AC-5 3/02/65	R&D, single burn, separable Surveyor model	Simulate launch of Surveyor	Failed: Atlas fuel valve closed, causing an explosion
AC-6 8/11/65	R&D, single burn	Demonstrate ability to launch Surveyor model similar to actual spacecraft	Successful in separating model and sending it on planned course
AC-8 4/07/66	R&D, two-burn	Perform 25-minute coast in parking orbit, reignite Centaur engine, and send Surveyor model to a target location simulating the Moon	Partial failure: after a hydrogen peroxide leak in parking orbit, not enough remained to power the tank boost pumps
AC-9 10/26/66	R&D, two-burn	Demonstrate restart capability; send Surveyor model on simulated trajectory to Moon	Successful

Sources: Dawson and Bowles, *Centaur*, 90; Green and Jones, "Bugs," 26–35; Richards and Powell, "Centaur Vehicle," 104–6.

had to exit) required investigation and multiple modifications. A slosh baffle helped limit movement of the fuel away from the tank bottom. Screens in the ducts bringing bleed-off hydrogen gas back to the tank reduced energy that could disturb the liquid. On the coasting portion of AC-4's orbit, *liquid* hydrogen had gotten into a vent intended to exhaust *gaseous* hydrogen and release pressure from boil-off. When it exited into the vacuum of space, it created a sideward thrust that tumbled the Centaur and mock-up Surveyor. Fixing this problem entailed a complete redesign of the venting system. Further improvements increased the thrust of both the yaw- and pitch-control engines and those that settled the easily displaced liquid hydrogen to the bottom of the tank during coast. Fortunately, these changes were not necessary before the launch of AC-5 but were implemented for AC-8. The AC-8 mission also incorporated the uprated RL10A-3–3 engine with slightly greater specific impulse derived from a larger expansion ratio for the exhaust nozzle and an increase in chamber pressure.[69]

Meanwhile, in response to the explosion on AC-5, engineers locked the Atlas valves in the open position. On AC-6, the guidance system performed to virtual perfection on a semioperational flight. The Surveyor model went to the coordinates in space it was intended to reach (simulating travel to the Moon) even without a trajectory correction in midcourse. With AC-7 shifted to a later launch and AC-8 having problems with hydrogen peroxide instead of the usual source of difficulties, liquid hydrogen, the Atlas-Centaur combination was ready for operational use, although there would be one more research-and-development flight sandwiched among the launches of operational spacecraft (AC-9, see table 3.1). Atlas-Centaur performed satisfactorily on all seven Surveyor launches, although two of the spacecraft themselves had problems. The five successful missions provided more than 87,000 photographs and much scientific information valuable for Apollo landings and lunar studies. Among the data useful for Apollo was information on the composition and strength of the lunar soil. Surveyors 1, 2, and 4 each featured a single burn by Centaur, but Surveyors 3 and 5–7 had dual-burn trajectories. On Surveyors 5–7, moreover, the Atlases were all SLV-3Cs with longer tanks than the earlier LV-3Cs. The weight of payload that the Atlas-Centaur combination could place in an orbit 300 nautical miles above Earth rose from 8,500 pounds on LV-3C to 9,100 pounds, while the weight that could escape Earth's gravitational field increased from 2,300 to 2,700 pounds. Unlike the SLV-3As used with Agenas, the SLV-3Cs had no radio-inertial guidance systems, instead relying entirely on the Centaur for guidance, with only an autopilot on board the Atlas. SLV-3Cs flew only seventeen missions but were successful on all

Figure 17. A vent expelling liquid hydrogen on a one-tenth-scale model of the Centaur in the 10"x10" supersonic wind tunnel at Lewis Research Center on September 6, 1963. The test sought to determine how far to expel the venting hydrogen from the body of the upper stage to preclude explosion near the exhaust nozzle. Courtesy of NASA.

of them before being replaced by SLV-3D, used with the advanced Centaur D-1A.[70]

Before these new vehicles went into operation, an SLV-3C and an original Centaur, known simply as Centaur D, launched with a Delta third-stage solid-propellant motor, the Thiokol Star 37E, on the spectacular Pioneer

Figure 18. An Atlas-Centaur launch vehicle lifting a Surveyor spacecraft from Pad 36A at Cape Kennedy on May 30, 1966. The Surveyor gathered information about the lunar surface for future Apollo landings on the Moon. Courtesy of NASA.

10 mission—NASA's first to the outer planets and the first to reach escape velocity from the solar system. In fact, on June 13, 1983, Pioneer 10 became the first man-made object known to pass beyond the solar system. The Delta stage used a propellant composed of ammonium perchlorate, the CTPB binder employed earlier on Minuteman II stage 2, and aluminum fuel. It

was spin stabilized, and engineers mounted it on a spin table attached to the Centaur, which provided the guidance and control to place the Delta stage on the proper trajectory for Jupiter. This mission, launched on March 2, 1972, was a single-burn effort for the second-stage rocket, which performed a retro-maneuver to establish a safe distance between itself and the third stage following separation. Although the three stages performed slightly less well than predicted, they sent Pioneer 10 on a trajectory that required only minor corrections by the spacecraft's hydrazine thrusters for it to carry out its mission successfully, returning large amounts of scientific data and many photos of the distant planets and interplanetary space.[71]

Well before this launch, NASA had decided to upgrade Centaur to the D-1 configuration, with Lewis Research Center responsible for overseeing the $40 million improvement program, the central feature of which was a new guidance and control computer, developed at a cost of about $8 million. The General Precision drum computer used on the Centaur D worked well but could not be reprogrammed easily or quickly. The center selected an airborne Teledyne computer to replace it. With a memory of 16,384 "words" of 24 bits, the new computer had five times the capacity of the previous one, yet was lighter. It provided guidance, control, direction of propellant use, data management, sequencing, and telemetry. With it in place, General Dynamics could simplify the Atlas to the SLV-3D configuration by removing the autopilot, programming, and telemetry units from SLV-3C and having Centaur perform those functions. Lewis engineers had the Central Inertial Guidance Test Facility at Holloman AFB test a strapped-down inertial guidance system for Centaur as early as January 1967, but Centaur D-1 kept the same basic guidance gyros, accelerometers, and gimbal structure used in Centaur D. (Only later would Honeywell convert to a three-axis strapdown system for the Centaur navigation unit.) The new Centaur had two configurations, the D-1A for use with Atlas launch vehicles and the D-1T for use with Titan. The differences between the two involved details of external insulation, payload fairing diameter, battery capacity, and the like.[72]

The first use of Centaur D-1 and SLV-3D was on Pioneer 11, which had the same mission as Pioneer 10 plus making detailed observations of Saturn and its rings. As on Pioneer 10, the third stage employed the Star 37 motor. Pioneer 10's launch in 1972 had to be delayed twice because of winds at the launch site, so Bruce Lundin, by then director of NASA Lewis, told his managers to come up with a better way to handle wind problems. The system Lewis and General Dynamics engineers developed was ADDJUST (Automatic Determination and Dissemination of Just Updated Steering Terms), in which software for the D-1's new computer enabled the guidance and control system to compensate for predicted winds during launch. This was now feasible, since the Teledyne computer would allow entering very recent weather data just before launch. Ready by the launch of Pioneer 11 on April 5, 1973, ADDJUST served on seven Titan-Centaur launches and all Atlas-Centaur launches into the twenty-first century, with almost no launches needing postponement for upper-level winds. The space shuttles adopted a similar system after it proved its worth on Centaur.[73]

The new Atlas SLV-3D–Centaur D-1A–Star 37 combination delivered a virtually perfect launch, and Pioneer 11 used two midcourse corrections to pass around Jupiter and send back the first photos of the giant planet's polar regions. Traveling much closer to Jupiter than Pioneer 10, its sister spacecraft used the planet's gravity as a sort of slingshot to propel it on a course to Saturn, becoming the first man-made object to fly past that outer planet. Pioneer 11 returned much data including discoveries of Saturn's eleventh moon and two new rings.[74]

After successfully launching Intelsat IV F-29–a larger satellite than the Intelsat IIIs launched by thrust-augmented Thor-Deltas—on January 25, 1971, the SLV-3D–Centaur D-1A proceeded to launch Mariner 10 on November 3, 1973. This was the first planetary mission to use a double burn by Centaur and the first spacecraft to use the gravity-assist technique—for although it launched after Pioneer 11, it got to its first planetary destination sooner, reaching Venus on February 5, 1974, and taking 4,165 photos before employing Venus's gravity to reach Mercury in March 1974.[75]

Between 1973 and May 19, 1983, thirty-two SLV-3Ds launched with Centaur D-1A upper stages. The payloads included High Energy Astronomy Observatories 1–3; Intelsat IVs, IVAs, and Vs; Comstars 1–4; and FLTSAT-COMs 1–5. The first Intelsat IVA, with its weight increased and its communications capabilities doubled, used the upgraded Centaur D-1AR. This took advantage of information from Titan-Centaur missions and adopted a computer-controlled vent and pressurization system for the regulation of propellant pressures in the tanks. Other additions included new attitude control thrusters and a supplemental supply line for the hydrogen peroxide that operated the thrusters. An extension piece for the payload fairing accommodated the additional length of Intelsat IVA and Comstar, making the overall length of the Atlas-Centaur almost 135 feet. On December 6, 1980, with the first launch of an Intelsat V, which had more relay capacity and weight, Centaur began to use engines adjusted to increase their thrust (per engine) from the original 15,000 to about 16,500 pounds. Of the total of thirty-two launches teaming SLV-3D with D-1A and D-1AR, only two

Figure 19. An Atlas-Centaur launching the Mariner 10 spacecraft November 3, 1973, on a mission to explore the planets Venus and Mercury. Mariner 10 reached Venus in three months, taking 4,165 photos before employing Venus's gravity as a sort of slingshot to reach Mercury the next year. Courtesy of NASA.

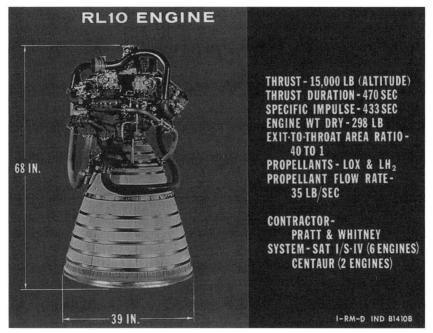

Figure 20. Technical drawing showing characteristics of the 15,000-pound RL10 engine used on Centaur and on the Saturn I (S-IV) upper stage. Courtesy of NASA.

failed—one from an electrical plug on the Atlas that did not disconnect on staging, causing a short circuit and loss of attitude control, and the other from an Atlas fire caused by a hot, high-pressure gas leak. This was a nearly 94 percent success rate, with no failures caused by the Centaur stage.[76]

During the early 1980s, General Dynamics and Pratt & Whitney converted to new versions of both Atlas and Centaur. The Atlas G, with lengthened propellant tanks, was 81 inches longer than SLV-3D. It stood about 137 feet tall when erected with the Centaur and payload fairing attached and developed 438,000 pounds of thrust. Pratt & Whitney made several changes to the Centaur engine, including removal of the boost pumps, for a significant weight savings. Increased pressure in the propellant tanks (to effect the feed by pressure rather than pumps) and modification of the engines to accept lower inlet pressure, together with an increase in the expansion ratio between the nozzle throat and the exit cone from 57:1 to 61:1, left the thrust unchanged at 33,000 pounds. For the attitude-control and propellant-settling engines, hydrogen peroxide gave way to the more stable hydrazine. Altogether, the RL10A-3–3A was a substantially different engine from the RL10A-3–3.[77]

The first Atlas G–Centaur launched on June 9, 1984, to place an Intelsat V in orbit. It did so, but the orbit was not the intended one and was unusable for communications purposes. Investigation indicated that, following separation of the Atlas and Centaur, a four-inch crack was leaking liquid oxygen. The firing of a shaped charge to separate the stages may have dislodged a piece of debris that either caused the crack or enlarged an existing one. The escape of liquid oxygen through the now-four-inch crack produced a lateral force that threw the stage off course and made it tumble. The fix included greater clearance between the shaped charge and the tank blast shield and improved inspection and testing of the area where the crack appeared.

Between March and September 1985, the Atlas G–Centaur launched three Intelsat VA satellites successfully, followed by FLTSATCOM 7 on December 4, 1986. However, on March 26, 1987, the attempt to launch another FLTSATCOM satellite failed. Despite heavy rain, the launch crew went ahead with the countdown, and 48 seconds after liftoff, lightning struck the vehicle. Damage to the flight control computer's memory led to a hard-right yaw command, producing aerodynamic forces that broke the rocket apart. NASA was the recipient of a lot of criticism for its weather forecasting, with a caption on one cartoon declaring that Benjamin Franklin would not have flown his famous kite under such conditions. Four months later, human error, causing a workstand to hit a Centaur tank, produced an explosion with minor injuries. But on September 25, 1989, an Atlas G–Centaur finally did launch the Navy's 5,100–pound FLTSATCOM F-8 satellite into geosynchronous transfer orbit. This was the last in a series of such ultra-high-frequency satellites, part of a worldwide communications system for the Air Force and DoD as well as the Navy.[78]

Meanwhile, forces had been building to commercialize launch-vehicle services. The Air Force was unhappy with NASA's campaign to have all DoD payloads transported on the space shuttles instead of on expendable launch vehicles. There was already competition from the Ariane launch vehicle in Europe, with the prospect of other countries selling launch services to such users as communications-satellite purveyors. On January 28, 1986, the explosion of *Challenger* grounded the remaining shuttles for more than two years. During the second half of 1986, therefore, General Dynamics evaluated the option of building several Atlas-Centaurs with company funds and then selling them to customers together with launch services. Early in 1987 the firm announced that it would sell Atlas-Centaur as a commercial launch vehicle. NASA then signed a commercial contract with the company, although the document provided for much greater oversight by the agency than was customary in such contracts.[79]

General Dynamics decided to designate the commercial vehicles with roman numerals, the first being Atlas I. All would have Centaur upper stages. A key change from Atlas G–Centaur was a payload fairing 14 feet in diameter, an increase of 4 feet. The extra space was needed since manufacturers were designing most new satellites or other spacecraft for the shuttle payload bay, which was roughly 15 feet wide, or for Ariane's 4–meter (13.12–foot) fairing. General Dynamics decided to use an aluminum skin-stringer construction for the fairing because of its low cost. For smaller payloads, the company also offered an 11–foot fairing of similar construction. Because the wider fairings imposed higher aerodynamic loads on the 10–foot-wide launch vehicle, Atlas I had to be strengthened. With the additional drag from the wider fairings plus the added weight of the shored-up structure, the payload capability to GTO decreased from the 5,100 pounds of Atlas G–Centaur D-1A to 4,900 pounds for Atlas I.[80]

On July 25, 1990, the first Atlas I successfully launched the Combined Release and Radiation Effects Satellite into a highly elliptical transfer orbit. Researchers designed this 3,842–pound satellite, a joint NASA–Air Force effort, to study Earth's magnetic field and the magnetospheric and ionospheric plasma by means of chemical releases. Two of the next three launches of the new vehicle, on April 18, 1991, and August 22, 1992, were failures. They take the story beyond the period covered in this book but are worth examining because they illustrate that, even at that late date, engineers could not anticipate all behaviors of rockets and prevent all problems. Both failures were on the RL10 engine, which had previously been virtually flawless. Both involved unsuccessful engine starts. The causes were difficult to diagnose because they stemmed from changes made several years apart. When the boost pumps on the RL10s were eliminated for the Atlas G–Centaur, one feature of the adjustment was an arrangement to chill the engines with liquid helium before liftoff. This was necessary since the lower inlet pressure for the engines meant a lower flow of the liquid hydrogen, which no longer could cool the bootstrap turbines and pumps sufficiently. On the ground, liquid helium flowed through the engine pumps until just before liftoff. Then a check valve closed to keep atmospheric moisture from reaching the turbopumps.

Introduced in 1984, this system was modified a few years later under a new contract that awarded two components of the check valve to different companies to save money. Penny-wise proved to be pound-foolish, because although the two segments satisfied the specifications for tolerance, the clearance was smaller than before. Post-failure testing showed that 30 percent of them stuck in the open position (suggesting that no one really understood how much clearance was required). Apparently this had happened on the two failed launches, causing moisture from the humid air above Cape Canaveral to freeze the pumps during the ascent. As was often the case, the fix was simple once the engineers understood the problem. As one engineer wrote, "These failures demonstrate the need for intense review of failure modes in initial design and for thorough review of seemingly minor changes." This is a lesson that all rocket engineers should long ago have learned. In this case, the cause was elementary, but it illustrated the ways in which the sheer complexity of rockets often led engineers to overlook problems they might have discovered much earlier in a system with fewer components than the roughly 100,000 found on Atlas-Centaur launch vehicles.[81]

Before the first of these two failures, the Centaur had a 95 percent success rate on 76 flights. This included 42 successes in a row for Centaur D-1 and D-1A between 1971 and 1984. The vehicle, as well as its Atlas booster, would continue to evolve into the twenty-first century with the successful launch of an Atlas V featuring a Russian RD-180 engine and a Centaur with a single RL10 engine, signifying both the end of the cold war and the continuing change in the technology. But meanwhile, development of the Centaur had led to the use of liquid-hydrogen technology on both upper stages of the Saturn V launch vehicle and on the space shuttle. Despite a difficult start and continuing challenges, Centaur had made major contributions to U.S. launch vehicle technology.[82]

Atlas E, F, and H, and Their Upper Stages

Partway through the history of Centaur and the various Atlas models used to launch it, the Air Force contracted with General Dynamics, beginning on February 14, 1966, to modify for space launch Atlas Es and Fs that had been in storage since their decommissioning as missiles in 1965. The process began with the newer F models. The primary modification was replacement of the inertial guidance system with the Mod 3G radio-inertial system developed for SLV-3. Rocketdyne also inspected each of the MA-3 engines and fixed or replaced any part that failed to meet specifications. In 1969 the rocket firm started a more extensive program of refurbishment to ensure that the engines in storage would work when called upon.

Following two launch failures in 1980–81, Rocketdyne rebuilt the engines at its plant and performed static tests before installing them on a launch vehicle.[83]

Some 54 Atlas E/F vehicles plus 18 D-models participated from 1963 to 1974 in a BSD effort to improve reentry systems for ballistic missiles. This Advanced Ballistic Re-entry System program analyzed the probable defenses an enemy might erect against ballistic missiles and studied what reentry devices or methods might best penetrate such defenses. Six Atlas Ds and four Fs combined for a series of launches of Orbiting Vehicle One (OV-1) spacecraft, beginning with a failed launch by a D-model on January 21, 1965, and ending with the successful launch of OV-1s 20 and 21 by an F-model on August 6, 1971.

A number of the Atlas launch vehicles carried multiple OV-1 satellites, each of which included an FW-4S solid-propellant rocket motor built by the United Technology Center of United Aircraft. Twelve hydrogen-peroxide thrusters on the propulsion module achieved separation of the OV-1 from the Atlas and provided attitude control during the period when the solid rocket was firing. Controlling the thrusters was a guidance and control system consisting of a programmer with a logic network and a strapdown autopilot with gyros. Although on four of the OV-1 launches the satellite failed to orbit for a number of different reasons, the program for the Air Force's Aerospace Research Support Program succeeded in placing 117 space experiments in orbit to study a variety of phenomena.[84]

An Atlas F successfully launched a radar calibration target and a radiation research payload for the Air Force's Space Test Program on October 2, 1972, using the Burner II solid-propellant upper stage that usually paired with the Thor booster. Another solid-propellant upper stage that operated only once with an Atlas E or F was the Payload Transfer System (PTS), which used the same basic Thiokol Star 37E motor as the Stage Vehicle System (SVS), employed multiple times with Atlas Fs and Es. This motor was also the one used on the missions that launched Pioneers 10 and 11. On July 13, 1974, an Atlas F and the PTS successfully launched Navigation Technology Satellite 1 (NTS-1) to test the first atomic clocks placed in space. These tests confirmed their design and operation and provided information about signal propagation to verify predictions for the Navstar Global Positioning System (GPS). GPS was then in development, destined to become a vital navigational aid, far more accurate than anything that preceded it.[85]

SVS, built by Fairchild Space and Electronics Company in Germantown,

Maryland, used Star 37E motors in two upper stages to place NTS-2 and six Navigation Development System (NDS) spacecraft in orbit between June 23, 1977, and April 26, 1980. The NDS-7 launch failed on December 18, 1981, when the Atlas E launch vehicle went out of control 10 seconds into the mission and crashed to Earth, creating a huge ball of fire. The other seven satellites all supported the development of the GPS. Although one was no longer operating by 1980, the five remaining NDS spacecraft were providing up to six hours per day of accurate positioning data worldwide as a developmental system.[86]

The Air Force used a different upper stage, known as Space Guidance System-II (SGS-II), together with the Atlas E to launch NDS-8 through NDS-11 between July 14, 1983, and October 8, 1985, all four launches being successful. McDonnell Douglas Astronautics Company made the upper stage, using two Thiokol Star 48 motors, also featured on the Payload Assist Module, which the space shuttle and Delta launch vehicle had employed since 1979. (See chapter 2 for a description.) Mounted in tandem, the Star 48 motors yielded roughly 14,700 pounds of thrust for a relatively long burn of 90 seconds beginning some 20 seconds after the upper stage was spun to 95 rpm and then separated from the Atlas E.[87]

The Atlas Es and Fs used other upper stages to launch satellites, including one Agena D. On June 26, 1978, an Atlas F placed the roughly 115–foot Seasat-A oceanographic satellite in orbit. For this mission, the former missile was modified to mate with the Agena upper stage and to carry this particular payload. The other major upper stage used by the Atlas Es and Fs was the Integrated Spacecraft System (ISS), which had a Thiokol Star 37S motor. This was in 1977–78 the latest in the Star 37 series of motors, also used as an upper stage on Thor for launching weather satellites. Little specific information about it is available other than that it had a spherical titanium case and a thrust of approximately 9,800 pounds; its specific impulse was more than 285 lbf-sec/lbm; it had a high propellant mass fraction of 0.925; and it burned for about 44 seconds.

Beginning with a launch of Tiros-N using an Atlas F on October 13, 1978, the ISS served as an upper stage for launching the NOAA-6 through NOAA-11 polar orbiting meteorological satellites plus four Defense Meteorological Satellite Program (DMSP) satellites by September 24, 1988. The only failure in the series was on NOAA-B on May 29, 1980, when one of the boosters on the Atlas F delivered only 75 percent of its programmed thrust, leading to an orbit from which the satellite decayed by May 3, 1981. Until NOAA-7 arrived in its orbit in June 1981 and became operational on

August 24, the National Earth Satellite Service in the Department of Commerce had only one polar-orbiting satellite functioning. After NOAA-7 became operational, it and NOAA-6 provided environmental data about Earth four times a day.[88]

A Thor first stage and a Burner II upper stage had launched the previous DMSP satellites. But on December 20, 1982, the Department of Defense began launching a new block of meteorological satellites, designated 5D-2, that were considerably heavier than the older block—about 1,657 pounds as compared with 1,047. This presumably dictated the use of Atlas E instead of Thor, and the ISS upper stage was necessary because engineers had integrated that stage and the satellite into a single system. This integrated system provided guidance and control for the satellite following separation from the Atlas until insertion into orbit. It also provided second-stage propulsion, attitude control, electrical power, and telemetry for insertion into sun-synchronous near-polar orbits. The final Atlas E–ISS launch of a DMSP during the period covered by this book was on December 1, 1990. This was only partially successful, but the DMSP-10 placed in orbit on that date provided meteorological data to the military until September 26, 1994. Other Atlas E–ISS launch vehicles placed DMSPs 11 through 13 in orbit through 1995.[89]

In February 1983 the Air Force began operating a derivative of SLV-3D known as the Atlas H, which used most of the basic systems on the SLV but employed GE radio-inertial guidance. The particular solid-propellant upper stage used with the Atlas H and previous Atlas Es and Fs to launch the White Cloud Naval Ocean Surveillance System (NOSS) satellites was classified. But the H-models had to be equipped with a new conical adapter to attach the upper stage and payload fairing, which were smaller in diameter than the Atlas. The White Cloud NOSS satellites provided the Department of Defense (primarily the Navy) with the ability to identify naval units by locating transmissions from radios and radars and deducing their affiliation from their operating frequencies. There were four of the satellite clusters for the system launched by Atlas Es or Fs and five by Atlas Hs. Only one of the launches failed, on December 8, 1980, which also marked the first use of an Atlas E as a space-launch vehicle. A malfunction caused the vehicle to deviate from its intended course, and the range safety officer had to destroy it some seven minutes into the launch.[90]

Overall, the Atlas Es and Fs used as launch vehicles had only 4 failures in 41 launches by the end of 1990, yielding a success rate of more than 90 percent. All five launches with the Atlas H were successful.[91]

Conclusions

Conceived as a missile, Atlas turned out to be a versatile launch vehicle, mated with a great variety of upper stages ranging from Able and Agena through Centaur to a succession of solid-propellant designs. Featuring a controversial but "brilliant, innovative, and yet simple" concept, the steel-balloon tank design, both Atlas and Centaur were adaptable and effective. With commercialization, the pair continued to provide launch services beyond the period of this book and into the twenty-first century.[92]

The Centaur proved particularly difficult to develop because of the peculiar properties of liquid hydrogen. But it was also hampered by initial funding arrangements and by having its management shuffled from the Air Force to NASA and, within NASA, from Headquarters to Marshall Space Flight Center to Lewis Research Center. As with other rocket programs, engineers found that the existing fund of knowledge was inadequate to predict everything that might occur in developing and launching an extraordinarily complex machine. Unforeseen problems with rocket technology continued into the 1990s, and engineers had to relearn the lesson that continual and sophisticated testing was the price of success, even if it did not always preclude failure.[93]

Getting Centaur developed required not only adjustment to the unexpected but also a major reorganization within General Dynamics. A previous matrix organization, in which only a small fraction of the engineers working on Centaur actually reported to the program director or project engineer, was ineffective. At the behest of Hans Hueter, Marshall's director of light and medium vehicles, General Dynamics reorganized into a project-type organization for Centaur in which one manager had direct line authority over all engineers working on the upper stage. Coupled with additional funding, the nation's highest priority, and the transfer of technical direction from Marshall to Lewis, this change eventually overcame developmental difficulties. Marshall had been uncomfortable with the risk-taking proclivities of General Dynamics and was preoccupied with Saturn development. Lewis, under Abe Silverstein, proved to have the expertise and the extreme commitment to testing that Centaur needed at that point. As a result, the liquid-hydrogen technology that Lewis, Pratt & Whitney, and Convair/General Dynamics, in particular, had developed became available for use on Saturn upper stages and later on the space shuttle. The other upper stages used with Atlas were derived from various programs and had only to be adapted for use on Atlas's unusual structure. With the partial exception of Agena, those upper stages made fewer contributions to major missions than did Centaur, but taken together, they helped launch a great variety of spacecraft that contributed in various ways to the United States' defense and its space-based infrastructure. This infrastructure provided services including prediction of weather, surveillance, navigation, and communication of voice messages and television pictures. In the case of communications satellites, in particular, the contribution, assisted by other nation's launch vehicles, was not just to the United States but to the entire world.

4

The Scout Family of Space-Launch Vehicles, 1956–1990

The Scout series of launch vehicles was unique in American experience in several ways. It was the first multistage booster to operate exclusively with solid-propellant motors. It remained the smallest multistage vehicle in long-term use for orbital launches. And it was the only launch vehicle developed under the auspices of Langley Research Center, which made many contributions to space efforts but, as the oldest of NASA's component organizations, had a long heritage of aeronautical effort that predated its space-related research. Like the Delta, with which it shared many stages, Scout proved to be both long-lasting and reliable. But in contrast with the Delta, it suffered through a difficult gestation and early childhood.[1]

Since, like Delta, Scout used much technology that had been developed elsewhere, it encountered fewer design and development difficulties than did many other rockets, although there *were* several. But Scout's major problems were primarily matters of systems engineering and quality control. Following a string of early failures, the program underwent a reliability-improvement and recertification process, after which one Scout engineer stated that he and his colleagues had "all underestimated the magnitude of the job" when they undertook the vehicle's development. "The biggest problem we had was denying the existence of problems that we did not understand." Once the project accepted that it had these problems and examined them, it learned from the process and went on to produce a long-lived, reliable small launcher used by NASA, the Department of Defense, and foreign countries. Its payload capability increased almost fourfold by its final flight in 1994. By that time, it had launched a great variety of scientific and applications payloads, Transit navigation satellites, and experiments to help understand the aerodynamics of reentry, among other types of missions. Counting partial successes as failures, by one account Scout had 104 successful missions out of 125, for an overall 83 percent success rate. The 21 failures were mostly in the early years, however, with 15 of them occurring by June 1964. In the 91 missions since that time, only 6 failures or partial failures occurred, for a 93 percent success rate. But accounts differ and do not seem to be fully reconcilable. Thus, these figures give only an approximate tally.[2]

Scout operated beyond the end of the period covered by this history, but even before its last flight, a comparatively new company, Orbital Sciences Corporation, had teamed with Hercules Aerospace to develop a new launch vehicle, Pegasus, whose first flight occurred in 1990. Since Pegasus was in some sense a follow-on to Scout, even though it was not part of the Scout family of launch vehicles, a postscript to this chapter discusses the new multistage rocket.

Conception and Early Development

During 1956, Scout had its origins in the imaginations of a creative group of engineers at Langley's Pilotless Aircraft Research Division (PARD) on remote Wallops Island in the Atlantic Ocean off Virginia's Eastern Shore. This group included Maxime A. Faget, later famous as a spacecraft designer; Joseph G. "Guy" Thibodaux Jr., who promoted the spherical design of some rocket and spacecraft motors beginning in 1955; Robert O. Piland, who put together the first multistage rocket to reach the speed of Mach 10; and William E. Stoney Jr., who became the first head of the group responsible for developing Scout, which he also christened. Wallops, established as a test base for Langley in 1945, had a history of using rockets, either individually or in stages, to gather data on both aircraft models and nose cones of rockets at transonic, supersonic, and then hypersonic speeds. Such data made it possible to design supersonic aircraft and hypersonic missiles at a time when ground facilities were not yet capable of providing comparable information. It was a natural step for engineers working in such a program to conceive a multistage, hypersonic, solid-propellant rocket that could reach orbital speeds.[3]

In 1957, after a five-stage vehicle at Wallops had reached a velocity of Mach 15, PARD engineers including Thibodaux, Faget, Piland, and Stoney began to study in earnest how to extend the speed of solid-propellant combinations still further. The group learned that Aerojet had developed the largest solid-propellant motor then in existence as part of its effort to convert the Jupiter to a solid-propellant missile for use aboard ship. Called the Jupiter Senior, the motor was 30 feet long and 40 inches in diameter and weighed 22,650 pounds, more than three times as much as the Sergeant missile's powerplant. Using a propellant of polyurethane, ammonium perchlorate, and aluminum, the Jupiter Senior motor provided a thrust of up to about 100,000 pounds for 40 seconds in two successful static firings in March–April 1957. It eventually amassed a record of 13 static tests and 32 flights without a failure, and it prepared the way for the Aerojet motors used in Polaris and Minuteman.

About the same time that the PARD engineers learned about Jupiter Senior, they found out that Thiokol had discovered a way to improve the Sergeant motor by shifting from the polysulfide binder used on the missile to a polybutadiene-acrylic acid binder with metallic additives. This offered a possible 20 percent increase in specific impulse. The information on these two motors led Stoney to analyze a four-stage vehicle with the Jupiter Senior as stage 1, the improved Sergeant as the second stage, and two X248 motors from Vanguard as the third and fourth stages. Even after Sputnik, in early 1958 NACA Headquarters informed the PARD team that it would not be receptive to developing a fourth launch vehicle when Vanguard, Jupiter C, and Thor-Able were well along in development or already available.[4]

However, with plans moving forward for what became NASA, in March 1958 NACA Headquarters asked Langley Aeronautical Laboratory, as the future research center was then called, to prepare a program of space technology. As a necessary part of this program, Langley included Scout (which only later became an acronym meaning Solid Controlled Orbital Utility Test System) to investigate human space flight and problems of reentry. The program called for $4 million to fund five vehicles for these purposes. By May 6, 1958, when the vehicle had become part of the space program, further analysis suggested that the third stage needed to be larger than the X248, but by then, plans for America's space efforts were becoming so extensive that the extra costs for such development were hardly a limiting factor. By that time also, Langley had arranged for a contract with Thiokol via Army Ordnance for four improved Sergeant motors. Meanwhile, the Air Force had become interested in Scout, reaching an agreement with NASA that the space agency would develop the vehicle and the military air arm would then consider modifying it for its own purposes, calling the result the Blue Scout.[5]

Langley assigned Stoney as project officer but gave Thibodaux's Rocket Section at PARD the responsibility for the initial five contracts needed to develop the Scout vehicle. The first contract under the new program was with Hercules' Allegany Ballistics Laboratory for four motors each for the third and fourth stages. Since the X248 already existed, the urgent part of this contract—which the Navy's Bureau of Ordnance arranged with ABL on October 23, 19589—was for development of an enlarged X248, designated X254, for use on the Scout's third stage. It used the same cast double-base propellant as the X248 and the same fiberglass-resin case but had five instead of four slots in the internal-burning grain. The propellant length grew from under 40 to over 71 inches, and the average thrust rose from about 2,600 to upwards of 13,000 pounds of force, depending on ambient temperature. PARD dubbed the new third-stage motor Antares I, with X248 becoming Altair I for the fourth stage.[6]

The next contract, with Aerojet, became effective on December 1, 1958, after several months of negotiations. The name of the Aerojet first stage changed from Jupiter Senior to Aerojet Senior, also called Algol I. As developed by December 1959, the motor was 29.8 feet long and 40 inches in diameter. With a steel case and its ammonium perchlorate-polyurethane-aluminum propellant configured with an eight-point gear (a cylinder with eight gear-tooth-shaped, squared-off "points" radiating from it) as its internal cavity, it yielded a specific impulse of only about 214 lbf-sec/lbm and a low mass fraction of 0.838 but an average thrust of more than 100,000 pounds, depending upon the ambient temperature.[7]

Although PARD already had a NACA contract with Thiokol for the improved Sergeant motor, it signed another one with the same firm under NASA auspices on December 5, 1958, for additional motors. The earlier contract proved fortunate, since after initial success in static firings, Thiokol was encountering unspecified difficulties with a new propellant. The Scout second-stage motor, which came to be called Castor I, had the same grain design as the Sergeant, but it used a polybutadiene-acrylic acid-aluminum-ammonium perchlorate propellant (PBAA, first employed as an interim propellant on Minuteman I) and was 20.5 feet long to Sergeant's 16.3 feet with an identical 31–inch diameter. Once Thiokol engineers overcame problems with the propellant, the Castor yielded a specific impulse of almost 275 lbf-sec/lbm to only 186 for Sergeant (although the two measures of performance were not comparable since, for Scout, the Castor was used at altitude with a larger expansion angle than was used at ground level for Sergeant). According to one set of figures supplied by Thiokol, Castor's average thrust was 64,340 pounds compared to Sergeant's 41,200. Because of the delays with Castor, Stoney decided to contract with Aerojet for Jupiter Junior motors as backups, using the same polyurethane propellant as Algol I but the same case as Castor. These Aerojet variants had about the same performance as

Castor. Never actually used for NASA's Scouts, they did find use in some Air Force Blue Scouts.[8]

The fourth contract for the Scout vehicle was for the guidance and control system. On October 20, 1958, eight companies submitted proposals. After prolonged negotiations, on January 12, 1959, NASA awarded the contract to Minneapolis-Honeywell. Like the Vanguard, on which Minneapolis-Honeywell provided the gyro sensors, the Scout used a strapped-down inertial reference package with miniature integrating rate gyros detecting deviations from the programmed path and initiating error signals proportional to the variances in pitch, yaw, and roll. An electronic signal conditioner converted the outputs of the gyros to the appropriate control signals. Located in a transition section between the third and fourth stages, the guidance and control system, including a programmer, permitted control of the pitch angle to meet the requirements of each launch, even though the Scout itself could be launched at an angle as much as 20 degrees from the vertical. Eventually, the yaw axis could also be programmed at a preselected angle to meet the needs of orbital insertions in a process called "yaw torquing."

Control of the aerodynamically stable first stage occurred the old-fashioned way, with jet vanes in the rocket-motor exhaust and aerodynamic control surfaces in the tips of four fins. Most of the control during the first-stage burn came from the jet vanes, made of 85 percent tungsten and 15 percent molybdenum (changed in the 1970s to copper-infiltrated tungsten). The fin-tip surfaces provided control during a coasting period between first-stage burnout and second-stage ignition. Hydraulic servoactuators received signals from the control system and operated the control surfaces. For stages 2 and 3, a variety of hydrogen peroxide–powered motors provided pitch, roll, and yaw control. Compressed nitrogen fed the 90–percent hydrogen peroxide to the motors. The fourth stage remained gyroscopically stable through spin imparted by small motors. Blockhouse command initiated first-stage ignition, with a guidance program timer directing the ignition of the next three stages, the ejection of a fourth-stage heat shield before third-stage ignition, and the spin-up of the fourth stage before ignition of its motor.[9]

Only on April 21, 1959, after it had awarded all of these other contracts, did NASA select an airframe contractor. The initial contract was only for four airframes and a launcher (launch tower), with a cost estimate of about $1 million, but twenty-two contractors, including all major aircraft builders, expressed a desire to bid on it. So many were interested because of potential follow-on contracts with not only NASA but the Air Force. After analyzing the proposals, NASA selected the Chance Vought Corporation of Dallas for a fixed-price contract worth $1,069,300. Through a series of follow-on contracts, Chance Vought (whose name changed through Ling-Temco-Vought and LTV with various suffixes to the Vought Corporation and LTV Aerospace and Defense Company) acquired responsibility not only for the airframe and launchers but for systems management and motor procurement under the overall management of Langley Research Center until January 1991, when responsibility for Scout management transferred to Goddard Space Flight Center, of which the Wallops Flight Facility had become a part in 1982.[10]

Because of "development difficulties" with the X254 (Antares)—which included problems with bonding the propellant to the case, the apparent cause of burn-throughs in testing—the first launch of the Blue Scout was delayed. Since development of the third stage did not reach completion until June 1960, this was undoubtedly also at least one reason for a delay in the launch of the first NASA Scout, which originally was scheduled for the middle of October 1959. By March 9, 1960, Hercules had completed six successful static tests of the modified X254, but in testing at the Arnold Engineering Development Center's high-altitude chamber in Tennessee, the nozzle's fiberglass lining came loose. Engineers corrected this problem by May, yet as late as June 30 the motor still produced high levels of vibration. These were worrisome to the Air Force because they were "detrimental to the operational function of the destruct system," though they apparently did not delay a NASA launch from Wallops. (The Air Force usually launched Blue Scouts from its own facilities at Cape Canaveral.)[11]

The Antares third-stage motor was not the only part of Scout to encounter difficulties. Tests of the reaction control system were not completed until February 1960, and testing of the airframe and heat shields also took longer than anticipated. Under pressure from NASA Headquarters, on March 7, 1960, Langley director Floyd Thompson agreed, however, that an unguided, partial or "Cub" Scout could be launched to collect engineering data. This would not be an official test, so it is not listed in all chronologies of Scout launches. Besides lacking guidance, the vehicle contained live motors only in stages 1 (Algol) and 3 (Antares), with dummies for stages 2 (Castor) and 4 (Altair). Engineers set the controls to produce an intentional spinning motion to stabilize the vehicle. Stage 1 functioned as designed, except that the combination (coupling) of the intentional roll and a "natural bending mode" produced an unanticipated structural failure about 30 miles above the launch site on Wallops Island. This occurred roughly at burnout of the Algol stage, so the Antares motor never ignited. Moreover, the third-stage heat shield tore off as the rocket passed through the difficult transonic region with its shock waves and high dynamic pressure.[12]

Engineers for the project did not need to worry about the structural failure since the full Scout, with guidance, would not need to roll. But to prevent recurrence of the loss of the heat shield, they performed wind tunnel tests that indicated high external pressures on the front of the shield as the vehicle reached Mach 0.90. To correct for these loads, they had holes drilled in the shield to equalize the pressure inside and outside the fiberglass structure.[13]

The four-stage Scout used the four solid-propellant motors as the principal structural elements, with connecting sections designed and produced by Chance Vought. PARD's practice had been to place heavy magnesium sections between stages of multistage rockets. Chance Vought instead made considerable use of either thin aluminum skins with internal stiffening or fiberglass monocoque construction, although there was also some use of steel and magnesium. Besides the troublesome third-stage heat shield, there were also heat shields around the fourth stage and the payload. The last of these was a honeycombed fiberglass structure with a nose cap made of Inconel, an alloy of nickel, chromium, and iron. From jet vanes to this nose cone, the original Scout measured some 71 feet.[14]

The first four-stage Scout with all stages live flew on July 1, 1960, in the first of nine developmental flights labeled ST-1 (Scout Test 1) through ST-9, all launched from Wallops. Engineers programmed the ST-1 vehicle for a probe trajectory with a payload designed at Langley containing an acceleration package and a radiation package. After reaching an expected altitude of 2,020 nautical miles with a maximum velocity of 22,000 ft/sec (about 15,000 mph), the 193–pound payload was to descend unpowered through the atmosphere. In the event, the first-stage motor functioned successfully, but as the vehicle passed through the transonic speed region about 15 seconds after launch, the third-stage heat shield again fell off as in the previous test. It turned out that the holes newly drilled to equalize pressure inside and outside the shield were too numerous, counteracting the intended effect.

Furthermore, there was vibration in the third-stage motor, which had a deleterious effect on the hydrogen-peroxide reaction-control jets. Whether or not this was the cause, at 136 seconds after launch the third stage began rolling, and the reaction-control jets could not immediately counter the rotation. After 210 degrees of rotation, the rolling stopped. But a problem with the tracking radar made it seem in the control room that the vehicle was veering significantly off course. The safety officer waited as long as he dared and then activated a signal to hold the ignition of the fourth stage. When it became apparent that the radar signal had been erroneous, there was no way to countermand the hold-fire signal. Since the fourth stage never had a chance to fire and thus did not fail, and since the instrumentation in the nose made radiation measurements to an altitude of 875 miles, the Scout program deemed this launch a success, although the heat shield and vibration problems suggest that the success was only partial.[15]

On October 4, 1960, ST-2 was much more successful. Although it carried an Air Force Special Weapons Center radiation payload instead of a NASA one, it launched from Wallops, reaching a maximum altitude of 3,500 miles and a range of 5,800 miles. Of the seven remaining flights in the NASA developmental series—all launched between December 4, 1960, and March 29, 1962—three were failures and four were successful. NASA treated the missions as operational, having them carry Explorer spacecraft, ionosphere probes, and one reentry payload. The December 4 launch of ST-3 was the first attempt at an orbital mission with the Scout. The Algol IA first stage again performed properly, but the Castor IA second stage did not ignite, owing to an ignition system defect that had gone undetected during checkout. Consequently, ST-4 on February 16, 1961, became the first entirely solid-propellant launch vehicle and the first rocket from Wallops to achieve orbit. In a payload designated Explorer 9, it carried a 12–foot inflatable sphere designed to measure the density of the atmosphere. Although a radio beacon on the balloon satellite failed to operate, researchers did gather some data.[16]

ST-5 and ST-6 both attempted to launch satellites to determine the effects of micrometeoroids on spaceflight, and both were failures. On June 30, 1961, the third stage of ST-5 failed to ignite. For ST-6, the Scout stages all functioned but the satellite reached an initial perigee of only 61.9 miles, lower than intended, resulting in an unplanned reentry on August 27, 1961, only two days after launch. The problem lay with the so-called "hot" separation system that was common to all three upper stages. Joining each upper stage with the stage below was a frangible diaphragm, which formed a sort of clamp with "the threaded periphery that engages two structural threaded rings at the separation plane," connecting the stages, as one engineer described it. When the motor ignited, blast pressure ruptured the diaphragm, disengaging the rings and allowing the stages to separate. The procedure worked fine when the second stage ignited, and the third. But at 530,000 feet, when the third stage burned out and remained linked to the fourth stage to continue providing guidance and control until the fourth stage spun up and ignited, it worked less well. In this case, the lighter mass of the fourth stage, as compared with the previous stages, permitted a greater

Figure 21. Scout ST-4 launching from Wallops Island, Virginia, on February 16, 1961. The first entirely solid-propellant launch vehicle (and the first rocket from Wallops) to achieve orbit, it carried a 12-foot inflatable sphere designed to measure the density of the atmosphere in a payload designated Explorer 9. Courtesy of NASA.

deflection of the trajectory as a result of the energy released when the diaphragm burst.

The solution the Scout team devised, in still another piece of cut-and-try engineering, was a "cold" separation procedure for the fourth stage only (and, where applicable, the fifth). Following spin-up, explosive bolts and springs effected separation of the fourth from the third stage. The springs were compressed by clamps, and the explosive bolts released the clamps, allowing the springs to exert equal pressure on four positions at 90–degree

Figure 22. Scout ST-5 in launch position on June 30, 1961. Note the slender contour of the launch vehicle and its angle from the vertical, which ranged from 0 to 20 degrees, depending on the destination. Courtesy of NASA.

intervals around the circumference of the stage. Only then did the Altair fourth-stage motor ignite, with the spinning keeping it in the proper attitude to reach the desired orbital path.[17]

Before this new system could be implemented, ST-7 launched a successful probe on the afternoon of October 19, 1961. Like a similar probe launched (in the morning, for purposes of comparison) on ST-9, this payload provided data on electron densities and the composition of the ionosphere. Between these two launches, on March 1, 1962, ST-8 was the first Scout to carry the new cold separation system. More significant, perhaps, this was the first Scout to use a payload kick stage (not a true fifth stage) and the first mission of Scout to gather data on reentry heating, which proved important to the successes of Mercury, Gemini, and Apollo. Little information is available about the NOTS-17 kick stage, but its name indicates it was developed by the Naval Ordnance Test Station. The motor may have been a variant of a 17–inch spherical motor that will be discussed below in conjunction with Blue Scout. ST-8 used Scout's first two stages to ascend to an altitude of 705,000 feet, after which the vehicle began to descend with the nose rotating downward. Then the third and fourth stages plus the kick motor all added to its downward velocity. The goal was 19,000 mph, but the vehicle achieved only about 15,340. Still, the mission is usually listed as a success.[18]

ST-9 was not only the final developmental vehicle but the first Scout launch with Antares IIA replacing Antares IA as the third-stage motor. NASA initiated work on the IIA as well as a new Altair II fourth stage in a development contract with Hercules in September 1960, followed by a production contract for fourteen of each motor, executed through the Navy in June 1961. ABL designed Antares IIA, after which Hercules produced it at its Bacchus Works in Magna, Utah. Described as "an improved version" of Antares I, Antares II, like its predecessor, used a fiberglass-resin case. Unlike the previous third-stage motor, however, Antares II employed a composite modified double-base propellant including ammonium perchlorate, HMX, nitrocellulose, nitroglycerin, and aluminum, among other ingredients. Even with a smaller nozzle expansion ratio, this yielded an increase in average thrust from about 13,000 pounds for the older motor to about 21,000 for Antares IIA.[19]

The successful ST-9 launch on March 29, 1962, must have included a developmental version of the motor known as X259A1, because the A2 unit did not complete its development until June of that year. The Altair II fourth stage, initially procured under the same contract, did not fly until June 28, 1963, as a replacement for Altair I. Much shorter at about 59 inches than the roughly 114–inch-long Antares II, Altair II had a similar cast modified double-base propellant with an identical specific impulse and an average thrust of almost 5,900 pounds, about double Altair I's. Used successfully in various versions on twenty-eight Scout missions, Altair II did exhibit some problems. These did not prevent it from functioning but included low thrust, an inferior igniter, a too-light insulator, and low burst strength of the fiberglass-resin case. The last problem required modification of the fila-ment-winding pattern to increase the strength of the case—another instance of trial-and-error engineering.[20]

With the advent of Antares II, Scout transitioned from its initial X-1 ver-sion to the X-2. The X-1 had a payload capacity, when launched in an easterly direction into a 300–nautical-mile circular orbit, of 131 pounds. With the

Table 4.1. Growth in Scout Orbital Capabilities

Year[a]	Scout Version[b]	Payload Capability[c]
1960	X-1	131 pounds
1962	X-2	168 pounds
1963	X-3	193 pounds
1963	X-4	228 pounds
1965	A-1	268 pounds
1966	B-1	315 pounds
1972	D-1	408 pounds
1974	F-1	425 pounds
1979	G-1	458 pounds

Sources: Leiss, "Scout," 3–4, 437–38; Wilson, "Scout," 449, 459; McDowell, "Scout Launch Vehicle," 100–102, 105–6; Krebs, "Scout-E1," 2, 5. This table differs in many particulars from Bille et al., "Small Launch Vehicles," 210–11.

a. Year of first launch.

b. There apparently was no C version that ever flew. An X-5 version dating from 1968 appears to have had no greater payload capability than the X-4. An E-1 from the same year as the F-1 seems to have been less used; its orbital capability was uncertain but slightly higher than the F-1's.

c. To an orbit of 300 nautical miles above Earth when launched in an easterly direction.

Table 4.2. Timeline

Program	Dates of Operation
Notsnik	July 25–August 28, 1958
Scout	July 1, 1960–August 5, 1994
Blue Scout	September 21, 1960–after 1976
Pegasus	April 5, 1990–present

X-2, the comparable orbital weight rose to 168 pounds. Table 4.1 shows the growth in payload into the same orbit over the course of Scout's history.

Because the story of Scout and related vehicles became quite convoluted and overlapping, the timeline in table 4.2 may help readers to keep related programs straight.

Blue Scout

While NASA was in the process of developing Scout, the Air Force had continued to work with the civilian agency on Blue Scout. Meetings with the Air Force had begun before the creation of NASA. They started as early as June 4, 1958, with Langley representatives going to USAF headquarters. A series of pre-NASA meetings led to a memorandum of understanding between NASA and the Air Force signed by Air Force chief of staff Thomas D. White on October 31, 1958, NASA deputy administrator Hugh L. Dryden having already signed it for the space agency. The agreement called for a coordinated effort in which the air arm would develop specifications to modify the NASA vehicle so it met Air Force needs.[21]

By the end of February 1959, the Air Force had assigned primary responsibility for the development of Blue Scout to its Ballistic Missile Division, with a project office at BMD under Maj. Donald A. Stine. The two organizations set up a NASA–Air Force Scout Coordinating Committee, which first met on May 8, 1959. The Air Force initially agreed to transfer funds to NASA for procurement of ten (later reduced to nine) Blue Scouts, which the civilian agency would acquire through amendments to NASA contracts. Because of the payloads the Air Force expected to launch, Blue Scout required thicker walls and more mounting studs for the third and fourth stages. By September 1960 the Air Force had evolved its designs to include a Blue Scout 1, Blue Scout 2, and Blue Scout Junior.[22]

The first of these to launch was Blue Scout Junior. This flew in a variety of configurations, most but not all of them four-stage. All of them used Castor I (the regular Scout second-stage motor) in place of Algol I in the first stage, and Antares I (the regular Scout third-stage motor) in the second stage. The third stage used an Aerojet motor called Alcor. Not a great deal is known about this propulsion unit. Its unspecified solid propellant yielded a specific impulse of 230 lbf-sec/lbm and a vacuum thrust of almost 8,000 pounds of force. The stage was 4.6 feet long and 1.65 feet in diameter. It owed its 0.9 mass fraction to a nozzle made of "heat-resistant plastic" and a spiral-wound case bonded with steel strips, both of which were light. Unlike the regular Scout, the Blue Scout Junior carried no guidance and control system, the first two stages having fins with the third and higher stages being spin-stabilized.[23]

The motor for the fourth stage of most Blue Scout Juniors was NOTS model 100A, designed by the Naval Ordnance Test Station in the California desert at the behest of the Air Force Special Weapons Center (AFSWC) at Kirtland AFB in Albuquerque, New Mexico. AFSWC had long worked with Wallops on rocket projects there and would be intimately involved with the use of Blue Scout Junior to gather information, but it is not entirely clear how it came to select NOTS to develop the model 100A. Besides contributing to Polaris and other strategic missiles, NOTS had been engaged in developing tactical missiles, in doing propulsion research, and in helping to solve the problem of combustion instability. It had even developed the first known aircraft-launched orbital vehicle, known as Notsnik in a play on the Soviet "Sputnik."[24]

Physicists and engineers at NOTS, many of whom worked on the highly successful Sidewinder tactical missile, had already been thinking about a possible satellite before the launch of Sputnik, but once the beeping Soviet satellite was in orbit, they sought to develop a solid-propellant launcher to place a primitive scanner in orbit. Like Sidewinder, the project began without either a requirement or a funding authorization from the Navy, although the Notsnik team quickly obtained support from the Bureau of Ordnance and the Bureau of Aeronautics as well as ARPA and elsewhere for their satellite-launching effort. The project developed rapidly but also evolved in the process, with the projected satellite becoming a radiation monitoring payload.

Unable to get Sergeant motors from the Army for their launch vehicle, the Notsnik team members developed the idea of launching a small, multistage solid-propellant rocket from a Douglas F4D-1 Skyray jet aircraft. Since the Skyray, which burned liquid fuel, served as the rocket's first stage, Notsnik was not the first all-solid-propellant multistage satellite-launching vehicle, but it did predate Scout by a couple of years. With the Skyray releasing the vehicle in a pull-up maneuver at 41,000 feet and a speed of about Mach 0.9, Notsnik used a 5–inch high-velocity aircraft rocket (HVAR) to push it away from the aircraft. The next two stages each consisted of two pairs of HOTROCs, modified antisubmarine rocket (ASROC) motors, designed at NOTS but furnished with thinner cases and additional propellant for Notsnik. Each pair supplied a maximum of 28,400 pounds of thrust for 4.86 seconds. The following stage was an X241, the original ABL third stage for Vanguard, with a thrust of 2,720 pounds for 36 seconds. The fifth stage (or sixth, if the HVAR is counted) was an "extruded" motor, elliptical in cross section, which added 1,155 pounds of average thrust for 5.7 seconds. Finally, there was a spherical apogee kick motor to place the satellite in orbit with only 172 pounds of thrust for a single second.[25]

Under the official rubric Project Pilot, the slender Notsnik launch vehicle that was less than 14.5 feet long made its first orbital attempt on July 25, 1958, and its sixth on August 28, 1958. The first attempt and the third (on August 16) reportedly reached orbit, although there is considerable uncertainty that they really did. In any event, Notsnik led to the larger Caleb rocket, some 16 feet long and 2 feet in diameter. Caleb never got beyond the prototype stage, but before programs using the rocket ended about 1962, there were at least plans to include the NOTS 100A motor in its third stage.[26]

In the meantime, NOTS had completed development of the 17–inch spherical model 100A rocket motor as the fourth stage of Blue Scout Junior on May 5, 1961. Like Thibodaux and others at PARD, the engineers at NOTS recognized the advantages of spherical motors. One of these was that at high altitudes, spin stabilization was easier to maintain with the spherical shape, since aerodynamic drag was not critical there. Also, the sphere permitted the combustion chamber walls to be only half as thick as with cylindrical shapes, significantly reducing weight and improving the mass fraction. These factors led NOTS to develop a series of such motors in the late 1950s.[27]

The model 100A motor used a propellant composed mainly of ammonium perchlorate, polyurethane, and aluminum configured with a seven-point internal-burning star in the center of the grain and a nozzle recessed inside the sphere. The nozzle was made of graphite, but the case was made of hydrospun and welded steel, yielding a mass fraction of a comparatively low 0.887 despite the spherical shape. With a nozzle expansion ratio of 30:1, the motor achieved a specific impulse in vacuum of about 265 lbf-sec/lbm. It burned for 36 seconds with an average thrust of 895 pounds. A modified version of the NOTS 100A motor used beryllium as the fuel in place of aluminum in a composition called Arcane 35D developed by Atlantic Research Corporation. NOTS had also developed a 100B variant of the 100A with a different method of payload attachment for NASA use. The 100B cases were likewise loaded with a beryllium propellant, but it is not clear that the beryllium-fueled motors were ever actually used. Beryllium offered an improvement in specific impulse, but as one prominent rocket engineer wrote in 1992, the fuel was "a highly toxic powder absorbed by animals and humans when inhaled. The technology with composite propellants using beryllium fuel has been experimentally proven, but its severe toxicity makes its application unlikely."[28]

When the first Blue Scout Junior launched from Cape Canaveral on September 21, 1960, the fourth-stage motor's development was not yet complete, but apparently a NOTS 100A was still used. The primary objective of the flight was to test the vehicle itself. The Air Force reported that its "vehicle development objectives . . . appear to have been accomplished," although the payload, a radiation probe, did not work. The Blue Scout Junior performed normally until 8 seconds before burnout of the fourth stage, at which time the telemetry ceased to function. The Air Force assumed that the fourth-stage burn ended normally except for the lack of data, with the 32.8–pound payload reaching an altitude of 14,400 nautical miles.

With a 29–pound radiation probe as the AFSWC payload, the second Blue Scout Junior launched from Cape Canaveral on November 8, 1960, but this mission was a clear failure. Following second-stage ignition, the telemetry got progressively weaker until its signal was too faint to use. Long-range cameras and radar indicated that an explosion occurred at 62 seconds after launch. The third and fourth stages apparently never ignited. Engineers deduced from available data that a loose pressure tapping in the head of the second-stage motor probably permitted hot gases to leak into the transition section between the second and third stages. This was consistent with a recorded increase in temperature in this area and the decline of the telemetry signal. Apparently the hot gases finally eroded supports for paddles designed to dampen resonant burning, and the paddles blocked the nozzle. A rapid rise in pressure would have led to rupture of the motor case, as the long-range cameras recorded.[29]

In all, there were 25 known Blue Scout Junior launches from Cape Canaveral, Vandenberg (or the Navy's nearby Point Arguello), and Wallops Island, with the last one on November 24, 1970. All were suborbital, and 22 were listed as successful, for an 88 percent success rate, although telemetry continued to be a problem and the payloads did not always work. Besides radiation probes, there were tests of an emergency rocket communications system, ion engine tests, probes of the magnetosphere, a test of a supersonic combustion ramjet (scramjet) engine, research into loss of communications during vehicle reentry into the atmosphere, and an ultraviolet astronomy mission. At least six of the launches used only the first three stages, and the scramjet test from Vandenberg on January 11, 1967, used only one stage. The launches from Wallops apparently all carried NASA payloads. But in 1962 the civilian agency and the Air Force agreed that since Blue Scout Junior differed from the larger Scouts in having no guidance and control system, and since the Air Force used most of the vehicles, the air arm would directly procure the vehicles themselves as well as the motors not used on the larger Scout, working through NASA only for the common motors. Accordingly, in June 1962 the Air Force issued its first direct contract to Chance Vought for seven Blue Scout Juniors.[30]

Blue Scout 1 was a three-stage version of Scout with only the Algol IB, Castor IA, and Antares IA motors. The first launch of Blue Scout 1 occurred at Cape Canaveral on January 7, 1961. It was successful, attaining 960 nautical miles of altitude and a range of 1,025 nautical miles. The payload was a 392–pound collection of eight Air Research and Development Command experiments. A Navy recovery ship did not arrive until 17 hours after a 90–pound recovery capsule had landed in the ocean. Since the beacon and flashing lights had ceased to function by then, the ship was unable to retrieve the capsule, but six of the eight experiments yielded usable data, presumably

Figure 23. An Air Force Blue Scout launch on May 9, 1961, at Cape Canaveral. Official U.S. Air Force photo, courtesy of the 45 Space Wing History Office, Patrick AFB, Fla.

via telemetry. There were subsequent Blue Scout 1 launches on May 9, 1961, and April 12, 1962, with Blue Scout 2 flying missions between those dates.

The May 9 flight from Cape Canaveral veered off course during the second-stage burn, and the range safety officer destroyed it. The cause was "a failure in the electrical power line to the hydrogen peroxide control system." Apparently "a break or disconnect occurred in the wire to the yaw-right control motor." On June 6 a postflight analysis "indicated that the control malfunction was not due to any inherent shortcomings of the control system or any of the subcomponents. As a result," an Air Force document states, "there appears to be no required or recommended action relative to this matter for future flights."[31]

The April 12, 1962, Blue Scout 1 also launched down the Atlantic Missile Range (AMR) with the objective of measuring the effects upon a vehicle's radio transmissions of an ionization sheath that formed when the vehicle reentered the atmosphere. The first stage performed properly, but the Castor second stage failed to ignite because of an electrical problem. "Review actions" led to unspecified "corrective action." This apparently was the final launch of Blue Scout 1.[32]

Blue Scout 2 was a four-stage vehicle employing, at least initially, the Algol IB in stage 1, the Castor IA in stage 2, the Antares IA in stage 3, and the Altair IA in stage 4. Most sources list only three flights, all in 1961: D4 on March 3, D5 on April 12, and D8 on November 1. One source calls D8 "the only orbital attempt by the Blue Scout." But a once-classified General Dynamics publication about booster performance and costs lists the Blue Scout as still being in the inventory in May 1965. And one Internet site lists the May 18, 1967, launch of a Transit satellite as being performed by a Blue Scout 2. A presidential report to Congress prepared by the National Aeronautics and Space Council simply lists this launch as being done by a Scout, and a NASA launch list does not differentiate between Transit and other military launches, on the one hand, and NASA's own uses of Scout on the other.[33]

Whether or not they were counted as Blue Scouts, the Navy continued to procure some Scout vehicles at least as late as fiscal year 1967, and the Air Force until fiscal year 1976. Of the first 92 Scouts, NASA paid for 54, the Navy 19, and the Air Force 14, with the other 5 being funded by the Atomic Energy Commission or European users. On April 1, 1961, now–Lt. Col. Stine and the management staff for Blue Scout transferred from the Ballistic Missile Division to the Space Systems Division. On October 2 the Air Force issued a delivery order to NASA to procure 19 Blue Scout vehicles. Whereas earlier Blue Scout vehicles had been launched by uniformed ("blue suit") Air Force personnel, on January 10, 1970, an agreement between NASA and DoD stated that NASA would contract for Scout launches from Vandenberg AFB for both NASA and DoD. Thus it appears that there was a gradual blurring of the lines between Blue and NASA Scouts, but whichever they were called, they continued to perform launch services for the armed forces as well as the civilian space agency.[34]

Meanwhile, Blue Scout 2 flight D4 was successful in carrying a 172–

pound AFSWC radiation probe with six experiments to an altitude of 1,380 nautical miles and a range of 1,720 nautical miles down the AMR. It met all test objectives and yielded "valuable radiation measurements." The D5 flight launched a 365–pound payload of Air Force geodetic and radiation-measuring experiments on a probe to about 1,000 nautical miles. Although the launch succeeded in meeting only seven of eleven primary test objectives with one other objective partly satisfied, the Air Force counted the mission a success. Since flight D8 would take place before D7, the Air Force decided that it would constitute the last flight in the Blue Scout Development Test Program. D8 was to be an orbital flight to test the Mercury tracking and communications network. It would be the only orbital attempt in this program, but since NASA was expected to have launched at least five orbital Scout vehicles by mid-1961, the Air Force decided on April 28, 1961, that flights D7 and D9 would be part of the applications rather than the development program for Blue Scout.[35]

Using the Air Force's D8 to perform testing for Project Mercury was presumably done because no NASA Scout was available. In any event, the last developmental Blue Scout lifted off normally from Cape Canaveral on November 1, 1961. Soon after launch, however, its flight became so erratic that it began to break apart from the flight loads, leading the range safety officer to signal its destruction about 30 seconds after launch. Postflight analysis revealed that a technician had switched the wires to the pitch and yaw rate gyroscopes. When the control system signaled a pitch change, the Scout responded with a yaw correction, and vice versa. Since the Scout team had used the fourth stage from a backup vehicle to replace a faulty one on the D8 vehicle and the first stage on the backup Scout was itself faulty, there was no second Mercury-Scout flight attempt. Mercury-Atlas 5, scheduled for launch in the middle of November, would provide a test of the tracking network, so the Blue Scout development program ended on a sour note.[36]

Crisis and Resolution

Both Scout and Mercury went on with their operations. But NASA, at least, was not satisfied with progress in its Scout program. The Air Force, despite the many Blue Scout failures, kept Stine as director of its program until his retirement on November 30, 1962. At NASA, however, after the failures of three of its first six Scout launches, and despite the success of the next three, Langley Research Center replaced Bill Stoney as director of its Scout Project Office about the end of March 1962 with Marine Corps Lt.

Col. George Rupp. Rupp had served as a project officer on the Bullpup tactical missile. He remained as Scout director until he retired from the service in June 1963, but his first four months in the new post were hardly a honeymoon in his civilian assignment. From April 26 to August 31, 1962, three of four Scout missions failed.[37]

Actually, it is not entirely clear how many of these four missions were Rupp's direct responsibility and how many were Stine's. NASA had procured Scout vehicle 111 (S-111), which launched from the Pacific Missile Range on April 26, 1962. But it carried a Naval Research Laboratory payload, Solar Radiation (SOLRAD) IVB, to monitor the Sun's ultraviolet and X-ray emissions. In addition, though this was not revealed at the time, the Scout launched an electronic intelligence satellite. Thus it may have been operationally under Stine's aegis. The Scout operated normally until it lost attitude control during the third-stage burn. Examination of evidence following the flight indicated that the vehicle had no hydrogen peroxide for control in the third stage because of a faulty ground system for loading the propellant.[38]

S-112 was not a NASA-procured vehicle, and since it carried an unspecified Air Force satellite, presumably it fell within Stine's area of responsibility. Launched on May 23 or 24, 1962, it failed during operation of the second stage, apparently because an electrical malfunction of the destruct system caused the vehicle to self-destruct. It and the previous failures led to a "comprehensive technical review of the complete Blue Scout Program" by the Air Force, suggesting that S-112 was a Blue Scout despite its not being included in many accounts of that program. On August 22 or 23, the launch of S-117 from Point Arguello by Air Force personnel (with the vehicle listed in a presidential report to Congress as a Blue Scout) succeeded in placing another unspecified Air Force satellite in orbit. But on August 31 another NASA-procured Scout, S-114, failed on what was presumably a NASA reentry test mission. The cause was again electrical, and together with an apparently different electrical failure of S-112, it "prompted the complete refurbishment of the wiring system and the upgrading of the ignition system and heatshield design."[39]

There followed three successful launches in a row, holding out hope that the modifications had resolved existing problems. On December 16, NASA Scout S-115 successfully launched Explorer 16 into orbit to measure the effects of meteoroid puncture on spacecraft flying near Earth. Two Blue Scouts, S-118 on December 18, 1962, and S-126 on February 19, 1963, launched Transit 5A and an unspecified Air Force satellite, respectively. These were followed by two more failures and then three successes.[40]

Figure 24. Eugene D. Schult, one of the original nine members of the Scout Project Office, and head of the Scout project from June 1963. Courtesy of NASA.

It was only on July 20, 1963, with the launch of S-110 that NASA recognized it had a major crisis in its Scout program. By then Rupp had retired and been replaced as director of Langley's Scout Project Office by Eugene D. Schult, a member of the original Scout team. S-110 carried the NOTS-17 motor for a reentry test flight launched from Wallops. Only two and a half seconds into the flight, observers could detect a flame above the fins on stage

1. It spread almost immediately to engulf the bottom of the Algol stage. "It was obvious something terrible had happened," recalled Bud English, one of nine original members of the Scout Project Group. "There had been a burn-through of the first stage nozzle a few seconds after takeoff. The vehicle went through some wild gyrations. It got about 300 feet high and broke into three parts." The rocket's destruct system sent components in various directions within a mile of the launchpad. One flaming piece went through the roof of an assembly building, landing in the front seat of an automobile someone had parked there, "burning that car to a crisp," as engineer Tom Perry of the recovery team remembered.[41]

The first stage for this launch was Aerojet's Algol IIA, which had already flown on eight missions before S-110. It kept the steel case of the firm's Algol I but improved the mass fraction from 0.838 to 0.898. The main ingredients for the propellant were unchanged from Algol I but in a different formulation, with an internal-burning four-point star in place of an eight-point gear. While the steel nozzle still featured a graphite insert, its expansion ratio increased from 4.64:1 to 7.35:1. These changes actually reduced average thrust from about 100,000 to some 87,000 pounds, despite a higher specific impulse, but this thrust was extended over a longer burn time—from about 35 seconds for Algol I to 48 seconds for Algol IIA.[42]

After the failure of Scout vehicle S-110, the recovery team combed the salt marshes but did not find enough nozzle debris to determine conclusively why this nozzle had failed when eight previous ones had not. The producer of the nozzle had changed, however, and by July 30, 1963, the seven-person Scout 110 Review Committee appointed by NASA Headquarters had decided that the failure resulted from flaws in the nozzle structure that the manufacturer had not detected. The committee therefore recommended a redesign of the nozzle. Specific details have not been available, but Algol IIB, as the motor with the new nozzle was called, did not enter the inventory for some time. In the interim, rebuilding and extensive testing of existing nozzles revealed and presumably corrected faults not previously discovered.[43]

There were thirty-six successful flights using Algol IIB until March 5, 1968, when Scout vehicle S-160C launched Explorer 37 with a Solrad 9 payload to measure the Sun's X-ray emissions, including those from solar flares. The satellite went into the wrong orbit because of an Algol IIB nozzle anomaly. The spacecraft operated normally and sent back important data, but after this anomaly the program came up with a lightweight, ablative reinforced plastic nozzle for Algol IIC, which was qualified in 1970.[44]

Meanwhile, the first-stage nozzle on Algol IIA posed no problem on the launch of S-132 on September 27, 1963, from the Pacific Missile Range on an Air Force orbital mission. However, the attitude control system failed during the coast period before fourth-stage ignition, causing this mission also to fail. Engineers evaluating the problem through ground testing found that radiation from the hot nozzle of the Antares IIA third-stage motor was adversely affecting the hydrogen peroxide system. So they increased the insulation for components of the reaction control system and added a heat shield around the nozzle.[45]

These were essentially individual cut-and-try solutions to specific design problems, but the Scout 110 Review Committee was also investigating the larger picture, looking at Scout's subsystems and the history of Scout failures. The findings were in one sense reassuring: no two Scout failures had been caused by the same problem. But the large number of failures, including many recent ones, suggested a need for greater procedural consistency and also for requalifying all Scout vehicles then in storage awaiting launch. As Eugene Schult, then head of the Langley Scout Project Office, later remembered in 1990, "We did things differently at Wallops than at the Western Test Range. The Air Force had its own way of doing things; the contractor had his ways; and we had our ways. It was a problem trying to coordinate them." Or as Bud English put it, "There simply were not good standardized vehicle safeguards and checkout procedures, which were needed to have a successful vehicle."[46]

To improve quality control, a team from NASA, LTV, and the Air Force initiated procedures to bring the manufacturing and launch teams into closer contact. A so-called tiger team followed each vehicle from the manufacturing plant in Dallas to the launch site. One problem with the early Scouts had been pressure to get them ready and launched in a hurry. Now the emphasis changed to reliability. LTV began to procure motors directly instead of having NASA do it. The contractor accepted and stored them in Dallas and then sent complete launch vehicles to the launch sites, instead of having the stages assembled there.[47]

The other half of the corrective action was recertification. All twenty-seven existing Scout vehicles went back to the LTV plant for disassembly and inspection using X-rays and microscopes. Standardization became the order of the day. The recertification and reliability-improvement program lasted fourteen months, though the first recertified Scout, S-122R, with the R indicating that it had been refurbished and recertified, launched on December 19, 1963, from Vandenberg. It placed in orbit Explorer 19, an inflated sphere that measured air density at high latitudes. This was the beginning of a series of 14 launches through August 1965 in which there was only a single failure. The next 12 launches through October 1966 were also successful, so beginning with the first recertified Scout, there was a stretch of 26 launches with one failure, for a 96 percent success rate.[48] The crisis was over.

Further Development

In the interim, Scout had not ceased to develop. Besides the evolution of Algol IIA to IIB in stage 1, Castor II replaced Castor I in the second stage. NASA had contracted with Thiokol to produce the new second-stage motor in 1963, and it replaced Castor I in August 1965 for the S-131R launch of Secor 5 on the tenth of the month from Wallops. This was an Army geodetic satellite, but the mission also served as a test flight for both Castor II and the FW-4S motor that was new in the fourth stage. S-131R also introduced the use of yaw torquing in the control system. This was a technology whereby the launch site's location no longer limited the orbital inclination of a satellite. Engineers could now program the guidance and control system to introduce a dogleg into the launch vehicle's trajectory to provide added flexibility for orbiting spacecraft.[49]

The shift from polybutadiene-acrylic acid as the binder in Castor I to the hydroxy-terminated polybutadiene in Castor II, combined with an increase in propellant weight, paradoxically decreased the average thrust from nearly 54,000 pounds of force for Castor I to only slightly more than 52,000 for Castor II, but this sacrifice increased the burn time from about 27 seconds to about 38 for an overall increase in total thrust from 1,632,000 to 1,953,400 pounds of force. Also used on the third stage of Delta, the FW-4 (with an S designating the Scout version and a D the Delta version) was built by the United Technology Center (UTC, previously United Technology Corporation). Called Altair III, it had a fiberglass case like the Altair II that it replaced but a higher mass fraction of 0.911 as against Altair II's 0.877. Its polybutadiene-acrylic acid-acrylonitrile composite propellant had a slightly higher vacuum specific impulse than the composite modified double-base propellant in Altair II (an increase of 5 lbf-sec/lbm), but the average thrusts of the two motors were roughly comparable. Since the FW-4S burned a few seconds longer, however, it yielded an increase in total impulse from 141,300 to 173,000 pounds of force. The combined result of the two new motors was to increase the payload capability of Scout in a circular 300–nautical-mile orbit launched in an easterly direction from 268 to 315 pounds.[50]

The next major upgrade of Scout began with a $2.5 million contract from

Langley to LTV in 1969 for design, development, and qualification of a new first-stage motor. LTV in turn contracted with UTC, which had produced the FW-4. Called Algol III, this new motor kept the steel case and jet vanes of Aerojet's Algol II but increased its diameter from 40 to 45 inches without necessitating modifications to the rest of the Scout or the launching equipment. This substantially increased the propellant mass without a change in the motor's length and with no substantial change in the propellant mass fraction. The polybutadiene-acrylic acid-acrylonitrile propellant yielded a substantial increase in specific impulse over Algol II when combined with aluminum fuel and ammonium perchlorate (although in different proportions than in the Aerojet motor). With an increase in burn time, the new motor also had a substantially higher total impulse (although the absence of comparable figures precludes giving details). However, with the same upper stages, Scout D-1 with the Algol III first stage could deliver 404 pounds of payload to a 300–nautical-mile circular orbit as compared with 315 for Scout B-1 with the Algol II first stage.[51]

One of the first missions with Algol III as the first stage propulsion unit was the successful S-170CR launch from the San Marco platform off the coast of Kenya by an Italian launch crew. This mission on November 15, 1972, placed the 410–pound Explorer 48 satellite in a nearly circular orbit above the equator to measure the distribution of galactic and extragalactic gamma radiation. The San Marco launches, the first of which occurred on April 26, 1967, filled a need for equatorial launches that could place satellites in orbits not achievable by Scouts from Cape Kennedy/Canaveral, let alone Wallops and Vandenberg, the three U.S. launch sites for Scout. Wanting a satellite effort of its own, Italy reached an agreement with NASA for the launches, of which there were nine from the floating platform, the last occurring on March 25, 1988.[52]

Another new motor for Scout, available in the late 1960s but not used on Scout until 1974 and then only once, was the Alcyone fifth-stage satellite-injection motor. Earlier versions of this motor had been used as retrorockets on the Ranger launches with the Atlas–Agena B combination and also with Vela launches by Atlas–Agena D as well as Titan IIIC. Produced by Hercules at its plant in Magna, Utah, the motor was small—not quite 33 inches long and 19 inches in diameter. It used a composite modified double-base propellant with ammonium perchlorate, nitrocellulose, nitroglycerine, and aluminum among its ingredients. Even with a fiberglass case, it had an unremarkable propellant mass fraction of 0.883, but it provided an average thrust of 5,880 pounds for 8.9 seconds. This first and apparently the only launch of a true five-stage Scout occurred on June 3, 1974, from the Western Test Range. It placed an Explorer 52 in a highly elliptical orbit "to study the plasma properties of the magnetosphere in the vicinity of the magnetic neutral point over the earth's north polar cap."[53]

About 1972 the Thiokol division in Elkton, Maryland, received a contract to develop a new fourth-stage motor that eventually replaced the FW-4S as the Altair IIIA motor. Thiokol successfully fired a demonstration motor in 1972, but in 1973 the motor failed three times in qualification tests. Available sources do not specify the nature of the failures, but they were serious enough that UTC was awarded a backup contract for more FW-4S motors, although this contract was cancelled when the FW-4S, long in the inventory, "exceeded safe operating temperatures" in a "simulated altitude test." In 1974 Thiokol successfully tested four of the new motors in static tests and quality-assurance firings, and the motor first flew on a Scout mission that year.[54]

Exactly why the Scout program shifted to the Thiokol motor, designated Star 20, is not clear. The new motor with a carboxy-terminated polybutadiene binder combined with ammonium perchlorate and aluminum had a specific impulse 2 lbf-sec/lbm higher than the FW-4, and its total impulse was also slightly higher, at 173,500 pounds of force instead of 173,000. The Star 20 at 663 pounds weighed a pound less than the FW-4S, but since its propellant mass fraction was also slightly lower, at 0.908 as against 0.911, the gain in performance was not great. Thiokol did develop a higher-energy propellant for a Star 20A using hydroxy-terminated polybutadiene and HMX in addition to ammonium perchlorate and aluminum. This yielded a specific impulse of more than 290 lbf-sec/lbm. Apparently the Star 20A flew twice, but information on dates and missions seems not to be readily available.[55]

The first mission to employ Altair IIIA on the fourth stage of a Scout was the launch of ANS 1, the first cooperative satellite between the Netherlands and the United States, launched by NASA from the Western Test Range on August 30, 1974, into a polar orbit. Designed to increase scientific knowledge of stellar ultraviolet and X-ray sources, the satellite went into a "highly elliptical rather than a near-circular orbit" because of an Algol (first-stage) malfunction. Despite the different orbit, all of the "experiments were turned on and returning data" as of the end of the year, and the data were useful, although the orbit apparently did cause loss of some observation time over the twenty-month operational lifetime of the satellite.[56]

In 1977–79, under contract to the Vought Corporation, Thiokol produced a new third-stage motor for Scout at its Elkton Division. This was the Star 31 for the Antares IIIA stage, employing a hydroxy-terminated polybutadiene binder combined with ammonium perchlorate and aluminum. With this propellant, the specific impulse rose from about 285 for the composite modified double-base propellant used by Hercules in Antares IIB to more than 295 lbf-sec/lbm. In addition to the higher-performance propellant, Thiokol used a composite case made of Kevlar 49 and epoxy. Introduced commercially in 1972, Kevlar 49 was Du Pont's registered trademark for an aramid (essentially nylon) fiber that combined light weight, high strength, and toughness. Lighter than fiberglass, it yielded a mass fraction of 0.923 compared with Antares IIB's already high 0.916. Burning much longer than Antares IIB, Antares IIIA produced a lower average thrust, but its total impulse in the vacuum of the high altitudes where it operated was 840,000 pounds, compared with 731,000 pounds for Antares IIB. No doubt because of the greater erosive propensities of the Antares IIIA motor, which had a higher chamber pressure than Antares IIB, the newer motor used 4–D carbon-carbon (pyrolitic graphite) for the nozzle throat insert.[57]

The first Antares IIIA served on a Scout launch from the Western Test Range on October 30, 1979. The spacecraft was the 402–pound Magsat, which gathered data on Earth's magnetic field. This successful flight was allegedly the first to use the 4–D carbon-carbon nozzle throat material on any rocket motor. This was also the first Scout to fly in the G-1 configuration with Algol IIIA as the first stage and Castor IIA, Antares IIIA, and Altair IIIA as the second, third, and fourth stages. Scout continued in service until 1994, with all remaining launches using this configuration. The G-1 could place up to 454 pounds of payload in a 300–mile circular orbit. This compared with only 131 pounds for the original Scout back in 1960.[58]

Summary and Conclusions

Operating for nearly three and a half decades, Scout obviously was a successful launch vehicle. Neither its payload capacity nor its reliability matched those of Delta. But clearly it filled a niche in the launch vehicle spectrum or it would not have lasted so long. In the beginning, it had to overcome some growing pains. Many of its motors and other components experienced developmental problems, including Castor I, Antares I, and Altair II as well as the reaction control system, airframe, heat shields, fourth-stage frangible diaphragm, and the nozzles on Algol IIA and IIB. Thus, like other missiles and launch vehicles, Scout suffered from the inability of designers always to foresee problems their handiwork might face, resulting in the familiar trial-and-error engineering. But as in so many other cases, engineers were able to correct the problems once they understood them or at least arrived at fixes that worked.

The last of the Scout launches during the period of this history occurred on May 9, 1990, when a Scout G-1 launched two small satellites for tactical communications, known as Lightsats, or MACSATs M-1 and M-2, from the Western Test Range. While the gravity gradient boom on one of the satellites did not deploy, the other functioned during Operation Desert Storm, but most analysts decided that there was no real need for such small satellites, given the other communications capabilities the military already had. The final Scout launch of all, on May 9, 1994, successfully placed the second Miniature Sensor Technology Integration satellite (MSTI-2) for the Air Force and the Ballistic Missile Defense Organization in orbit. (A Scout had launched MSTI-1 on November 21, 1992.) MSTI had as its primary goal the use of miniaturized sensor and seeker technologies to detect launches of ballistic missiles. MSTI-2 imagery also contributed to studies of volcanic activity in Chile.[59] Thus, to the end of its career, Scout continued to serve the military as well as the civilian space program.

Postscript: Pegasus

As the Scout program neared its end, there appeared to be a continuing need for a small launch vehicle, raising questions about why Scout had been decommissioned. To fill the void, Orbital Sciences Corporation (OSC) in Fairfax, Virginia—a firm founded in the early 1980s by three men associated with the Harvard Business School—began developing the Pegasus launch vehicle in 1987 in what became a joint venture with Hercules Aerospace the following year. Hercules had responsibility for three new solid-propellant rocket motors and a payload fairing, while OSC provided all other mechanical and avionics systems, software (ground and flight), a mechanism for dropping the three joined stages of the launch vehicle from a launch airplane, and overall systems engineering. The two companies evenly split the more than $50 million development cost.

In July 1988 the Defense Advanced Research Projects Agency (soon to become ARPA again) awarded OSC an $8.4 million contract for one Pegasus launch vehicle with fixed-price options for five additional launches of Pegasus. This was the first completely new U.S. space-launch vehicle designed since the 1970s. It was eventually to be launched from a specially modified Lockheed L-1011 "mothership," although NASA's Dryden Flight Research Center launched the first six Pegasus vehicles from the same B-52

aircraft that had served as the launch platform for the X-15 and other research airplanes. Since the launch aircraft carried the three stages to about
38,000–42,000 feet and to speeds of about Mach 0.5 to Mach 0.8, Pegasus
had a considerably higher payload capability than a similarly sized vehicle
launched from the ground.

The first Pegasus vehicle—rolled out in August 1989 and launched from
the B-52 on April 5, 1990—successfully placed two spacecraft in orbit: an
ARPA/Navy experimental communications and data-relay satellite and a
NASA Goddard Space Flight Center bus containing experimental canisters
and a payload-environment instrument package, both of which remained attached to the third stage of Pegasus. The remaining launches from the B-52
followed through 1994, after which Orbital Sciences' L-1011 assumed launch
duties.[60]

Antonio L. Elias at OSC conceived the new launcher as a less expensive
way to place small payloads in orbit than a ground-launched vehicle. Although similar in concept to Notsnik, Pegasus was actually based on a two-
stage antisatellite vehicle, the Vought ASM-135A, fired from an Air Force F-
15 in 1985. Being released by a carrier aircraft, Pegasus could begin operating
under rocket power while already at considerable speed and altitude in an
environment of lower drag than prevailed near the ground. The vehicle had
triangular (clipped delta-shaped) wings that provided some lift at altitude,
while the first solid-propellant stage could operate with a nozzle expanded
to a higher ratio than would have been effective for a ground launch. Pegasus, being launched horizontally, also did not have to be turned as abruptly
in pitch to reach orbit as would a vertically launched vehicle, saving loss of
thrust through vectoring. Dynamic pressures operating on the fuselage were
lower than for a ground-launched vehicle, allowing for lighter construction.
Finally, the carrier aircraft freed the vehicle from limitations imposed by a
fixed launchpad, permitting Pegasus to be launched from almost anywhere
above Earth into any inclination from the equator.[61]

To reduce costs, Orbital Sciences drew upon the experience of its joint
partner, Hercules Aerospace, in solid-motor design and construction, and
it used "commercial off-the-shelf components" where possible. Thus, OSC
derived the flight computer for guidance and control from "a battle tank fire
control computer." Similarly, for the inertial guidance unit, the firm used
technology from a Navy torpedo. It also used a GPS receiver to assist in the
guidance effort. The system employing these devices, located in the third
stage, controlled the vehicle aerodynamically during most of the roughly
73 seconds of first-stage burn by providing commands to operate three all-

moving tail surfaces electromechanically. During the final seconds of first-stage propulsion, small solid-propellant thrusters embedded in the bases of the movable tail fins ignited to provide additional control in the thinner atmosphere the vehicle had then reached. Gimballing (Flexseal) nozzles in the second and third stages provided control in pitch and yaw, while a cold nitrogen-gas reaction-control system provided roll control during the operation of the upper-stage motors and control in all three axes during coast.[62]

Nielson Engineering & Research of Mountain View, California, provided the aerodynamic design of Pegasus exclusively through use of computational tools rather than wind tunnels. Some 94 percent of the external structure consisted of a carbon-fiber, epoxy-matrix composite material with an additional 5 percent being aluminum (in the aft structure to support the fins) and 1 percent titanium. To protect the wings from aerodynamic heating at speeds up to Mach 8, engineers designed "selectively applied additional layers of graphite composite," which charred and ablated. A substance called Korotherm protected the leading edges of the fins, while "less critical areas of the wing and fuselage" used artificial cork to protect the surfaces.[63]

The propellant Hercules used in all three stages of Pegasus consisted of a hydroxy-terminated polybutadiene binder with 88 percent solids loading (meaning that the binder constituted only 12 percent of the propellant). It featured a slow burning rate. In the Orion 50S first-stage motor, the propellant yielded about a 295 lbf-sec/lbm specific impulse with a nozzle expansion ratio of 40:1. Burning for about 72 seconds, it produced an average thrust of 109,419 pounds. At 29.1 feet long and 4.17 feet in diameter, this was by far the largest of the three stages. The Orion 50 second-stage motor, by contrast, had the same diameter but was only 8.92 feet in length. With a 65:1 nozzle expansion ratio, it too yielded a specific impulse of about 295 lbf-sec/lbm. Over a 71–second burn time, it delivered an average thrust of 27,605 pounds of force. The Orion 38 third-stage motor was 4.39 feet long and 3.17 feet wide. It burned for 65 seconds. With a specific impulse of 291 lbf-sec/lbm, it used a nozzle expansion ratio of 60:1 and had an average thrust of 7,772 pounds.[64]

The three motors used graphite-epoxy composite cases. Their nozzles incorporated low-erosion 3–D carbon-carbon throat inserts with carbon-phenolic exit cones. All three motors burned to depletion of the propellant. For stage separation, stage 1 used a linear shaped charge and stages 2 and 3 used springs. Beginning with the second launch, a liquid-propellant Hydrazine Auxiliary Propulsion System was available as a fourth stage. With a thrust of 150 pounds for what one source claimed was 241 seconds, it could lift a payload into a higher orbit or provide more accurate orbital injection than was possible with the three stages alone.[65]

Orbital Sciences and Hercules subsequently completed development of a growth and stretch version of Pegasus, designated Pegasus XL, and OSC with various subcontractors including Hercules and Thiokol developed a four-stage, ground-launched vehicle named Taurus using derivatives of the Pegasus motors plus Thiokol's Castor 120 as an initial stage designated stage 0. Castor 120 was another motor, like Hercules' Orion series, that used HTPB as the binder. Taurus first flew in 1994—although not with a Castor 120, which was not yet available—while from April 5, 1990, until their thirty-seventh flight on March 22, 2006, Pegasus and Pegasus XL accumulated only three failures, for a 92 percent success rate.[66]

Pegasus was not in any sense a part of the Scout launch vehicle family. Developed during the final three years of the period covered in this history, it essentially belongs to the era of launch vehicle history that followed the end of the cold war. But as an Air Force fact sheet on the vehicle concludes, "Pegasus occupies a critical niche in the space launch inventory, filling the void left by Scout."[67] For that reason, Pegasus makes a fitting postscript to the history of Scout.

5

Saturn I through Saturn V, 1958–1975

By far the largest U.S. launch vehicle, Saturn V stood some 363 feet tall with its 80 feet of payload included. This made it taller than the Statue of Liberty—equivalent in height to a 36–story building and taller than a football field is long. Comprised of roughly five million parts, it was a complex mass of propellant tanks, engines, plumbing fixtures, guidance and control devices, structural elements, and more. Its electrical components, for example, included some five thousand transistors and diodes, all of which had to be tested both individually and in conjunction with the rest of the vehicle to ensure that they would work properly when called upon. Ultrasonic inspections determined whether five miles of tubing and an acre of adhesive bonds would stand up to the rigors of launch and travel through both the atmosphere and (for the upper stages) the harsh environment of space. Dye-penetrant and other tests similarly checked 2.5 miles of welding. Immense and powerful, with 7.5 million pounds of thrust provided by the first stage alone, Saturn V would launch a 95,000–pound payload bearing three astronauts to the Moon—not once but six times.[1]

Using clusters of rocket engines in its lower stages to achieve its massive thrust, Saturn V was based on the earlier development of vehicles that ultimately came to be designated Saturn I and Saturn IB. These two interim space boosters were in turn based upon technologies developed for the Redstone, Jupiter, Thor, Atlas, Centaur, and other vehicles and stages. Thus, although the Saturn family was the first group of rockets developed specifically for launching humans into space, it entailed not so much new technologies as a scaling-up of existing or already developing technology and an uprating of engine performance. Even so, the scaling-up and uprating created significant technological hurdles and a need to find empirical solutions to problems they raised. Existing theory and practice were by no means adequate either for the unprecedented scale of the Saturns or for the reliability needed to carry human beings not merely into Earth orbit but all the way to

Figure 25. Schematic drawing showing the comparative sizes and shapes of several U.S. launch vehicles including two Atlas versions, Saturn I, Saturn V, and Nova, which never went into production. Notice how much Saturn V dwarfs the two Atlases and the already huge Saturn I. NASA Historical Reference Collection, Washington, D.C. NASA-USAF photo courtesy of NASA.

lunar orbit whence, six times, a landing craft would carry two astronauts to the surface of the Moon. Thus, even though the Saturns made extensive use of existing technology, their very size and scope led to advances in the state of rocketry's art.[2]

Early Developments and Saturn I

The organization that did the initial research and development of the Saturn family of vehicles was Wernher von Braun's unit at the Army Ballistic Missile Agency (ABMA), which became the nucleus of NASA's Marshall Space Flight Center in mid-1960. Already in April 1957, when DoD projected a need for a very large booster to launch communication and weather satellites as well as space probes, ABMA had begun to study a vehicle with 1.5 million pounds of thrust in its first stage. The preliminary studies for what it called a Super-Jupiter envisioned clustering four engines to achieve that much thrust. In December 1957 ABMA delivered to DoD a proposal for such a space vehicle, using E-1 engines under development by Rocketdyne. But then the Advanced Research Projects Agency (ARPA), created on February 7, 1958, pointed ABMA's studies in a different direction. Even after Sputnik, funding was limited, so ARPA urged the use of existing and proven engines to get the new booster developed as quickly as possible at minimal cost. ABMA shifted to eight uprated Thor-Jupiter engines to provide the 1.5 million pounds of thrust in the first stage, calling the new concept Juno V. This led to ARPA order 14–59 on August 15, 1958, initiating what von Braun and others soon started calling the Saturn program. (ARPA officially sanctioned the name on February 3, 1959, in a memorandum that discarded the Juno V designation.)[3]

Von Braun's group discussed the cluster idea with Rocketdyne, which expressed interest but protested that the available money was inadequate for modifying the Thor-Jupiter engines to increase the thrust to the needed 188,000 pounds apiece. ABMA argued that the North American Aviation division could use leftover hardware and persuaded the firm to sign a contract on September 11, 1958, for uprating the Thor-Jupiter engine. As already seen,[4] the Saturn I first-stage engine that resulted from Rocketdyne's efforts, called the H-1, also benefited from work on an X-1 engine that an Experimental Engines Group at Rocketdyne had begun working on in 1957. Thus the urgings of the von Braun group and internal efforts at Rocketdyne combined to set Saturn I on its developmental path.[5]

Meanwhile, the engineers at ABMA did some scrounging in their stock of leftover components to meet the demands of ARPA's schedule within their limited budget. The schedule called for a full-scale static firing of a 1.5–million-pound cluster by the end of 1959. Instead of a huge new tank for the first stage, which would have required new techniques and equipment, the frugal ABMA engineers found rejected or incomplete Redstone and Jupiter propellant tanks and combined one of the Jupiter tanks (8.83 feet in diameter) with eight from the Redstone (5.89 feet wide) to provide propellant reservoirs for the clustered engines. In such a fashion the Saturn program got started, with funding gradually increased even before the transfer of the von Braun group to NASA.[6]

By the early Saturn years, von Braun and his group had evolved a distinctive approach to developing rockets and managing the process. Having begun their rocket-making activities before a large fund of technical information was available, von Braun and his early German associates had to use trial and error as their basic methodology, although even in the early 1930s they consulted technical and scientific sources bearing on problems they faced. They also conducted or commissioned fundamental research. But their initial equipment for gathering data was primitive, so it was often difficult to find out why rockets failed. Still, they learned what they could from failures and moved on. In the process, they developed several well-known characteristics. First, they were conservative as already seen, making their rockets more capable and sturdier by a significant margin than the expected performance required. Second, they were methodical, testing each component many times to ensure it would work, and doing this in a step-by-step process.[7]

A third characteristic of the von Braun group was a preference for in-house rather than contract development. This has often been labeled the arsenal approach, but even at the Redstone Arsenal the group had to contract out some of the work. German technical education emphasized hands-on experience, something von Braun himself had acquired. And the Germans as well as an increasing number of Americans at Marshall Space Flight Center preferred to design and build their early hardware before turning it over to industry to fabricate according to in-house specifications. Then they liked to carefully supervise the performance of the contractors and to test the products at Marshall even after testing at contractor facilities and elsewhere. In-house design had already become an impossibility with the Jupiter engine and was still less possible with the H-1, but Marshall retained its preference for hands-on design.[8]

As at Peenemünde, von Braun maintainued his role as an overall systems engineer despite other demands on his time as center director at Marshall. At the frequent meetings he chaired, he continued to display an uncanny ability to grasp technical details and explain them in terms everyone could understand. Yet he avoided monopolizing the sessions, helping everyone to feel part of a team. He also fostered communication among his key technical people by use of weekly notes. Before each Monday, he required his project managers and laboratory chiefs to submit one-page summaries of the previous week's developments and problems, forcing them to digest everything their team had done as they gathered and condensed the information. Then von Braun wrote marginal comments and circulated copies back to all the managers. He might suggest a meeting between two individuals to solve a certain problem or himself offer a solution. Reportedly, the roughly thirty-five managers were eager to read these notes, which allowed them to keep up with overall developments, not just problems and issues in their own areas. The notes thus integrated related development efforts and spurred attempts to solve problems across disciplinary and organizational lines.[9]

Meanwhile, ABMA had initially considered using an Atlas or Titan as a second stage for the Saturn, with a minimally modified Centaur as the third

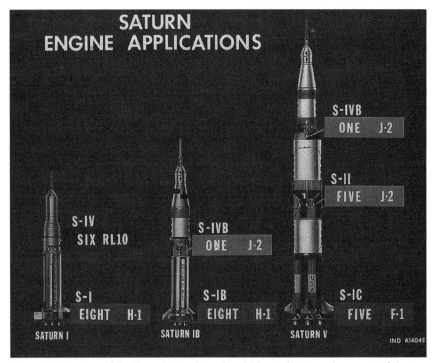

Figure 26. Drawing of the three Saturn launch vehicles indicating placement of their engines. Courtesy of NASA.

stage. However, after the decision to transfer the von Braun group to NASA but before the actual formation of Marshall Space Flight Center, NASA convened a Saturn Vehicle Team chaired by Abe Silverstein and including other representatives from NASA Headquarters, the Air Force, the Office of Defense Research and Engineering, and ABMA (von Braun himself). Silverstein argued that all upper-stage engines should use liquid hydrogen as their fuel, convincing the others, even though von Braun had reservations. But as the former German later told William Mrazek, he was not greatly concerned about the difficulties of the new fuel because many Centaur launches were scheduled before the first Saturn launch with upper stages. The group at Huntsville could profit from what these launches revealed to solve any problems with the Saturn I upper stages.[10]

Saturn I (called Saturn C-1 at first, and rechristened only in February 1963) was originally going to have a third stage, but in the spring of 1961, NASA decided to drop the third stage and to use for the second stage the same Pratt & Whitney RL10 engines being developed for Centaur. Meanwhile, on April 26, 1960, NASA had awarded a contract to the Douglas Aircraft Company to develop the Saturn I second stage, which confusingly was called the S-IV.[11]

The H-1 Engine

As partly related in chapter 2, development of the H-1—a regeneratively cooled engine burning RP-1 and liquid oxygen—had already gotten under way at Rocketdyne's main plant in Canoga Park, California, with early development testing occurring at Santa Susana and fabrication to take place at Rocketdyne's plant in Neosho, Missouri. Because the H-1s would be clustered in two groups of four each, there were two types of engines. An H-1C was used for the four inboard engines, which were incapable of rotating to steer the first stage. The four outermost H-1D engines were gimballed. Both versions used bell-shaped nozzles and regenerative cooling, but the H-1Ds used a collector or aspirator to channel the turbopump exhaust gases, which were rich in unburned RP-1 fuel, and deposit them in the exhaust plume from the engines to prevent the still combustible materials from collecting inside the first stage's boat tail (rear structure) where they might ignite.[12]

Development of the H-1 proceeded with remarkable swiftness. Rocketdyne had built an engine and static-tested it at full power by December 31, 1958, less than four months after signing the contract. However, the engine produced only 165,000 pounds of thrust rather than the 188,000–pound goal, and it still used a 20–gallon pressurized oil tank to lubricate the turbopump gear box, as the Thor-Jupiter engine had. Later versions of the engine, with thrust uprated to 188,000 pounds at sea level, eliminated this by using RP-1 with an additive to lubricate the gearbox. This modification did require a blender that took fuel from the turbopump, mixed it with the additive, and supplied it to the gearbox. The H-1 also featured a simplified starting sequence. Instead of auxiliary start tanks under pressure to supply oxygen and RP-1 to begin operation of the turbopump, a solid-propellant device started the turbines spinning. However, the H-1 kept the hypergolic ignition procedure used in the Atlas MA-3 and the later Thor-Jupiter engines.[13]

Rocketdyne delivered the first 165,000–pound H-1 engine to ABMA on schedule, April 28, 1959. Von Braun and his engineers conducted the first static test on this engine four weeks later, with an eight-engine test following on April 29, 1960. This test lasted only 8 seconds, but on May 17 a second static test of eight engines lasted 24 seconds and generated a thrust of 1.3 million pounds. That fall, the engine passed its preliminary flight rating tests, leading to the first flight test on October 27, 1961.[14]

Meanwhile, Rocketdyne had begun the process of uprating the H-1 to

188,000 pounds of thrust, apparently by adjusting the injectors and increasing the fuel and oxidizer flow rates. Although the uprated engine was ready for its preliminary flight rating tests on September 28, 1962, its uprating created some problems that were not solved by that time. As discussed in chapter 2, these included combustion instability, cracks in the liquid-oxygen dome, and longitudinal cracks in the regenerative cooling tubes. Rocketdyne engineers replaced the flat-faced injectors of the Thor-Jupiter engine with plates featuring six cooled baffles (to fix the combustion instability), developed an improved aluminum-alloy dome more resistant to the stress corrosion that was causing the cracks there, and used stainless steel for the tubular walls of the combustion chamber in place of the previous nickel alloy, which was subject to embrittlement by sulfur in the RP-1 at the higher temperatures the uprated engine produced.[15]

The S-IV Second Stage

Von Braun's prediction that Centaur development would solve the problems with liquid hydrogen seems to have been accurate at least with respect to the Pratt & Whitney RL10 engine. Such was not the case with the S-IV second stage, designed and produced by Douglas, which had no role in Centaur. NASA did take over the contracts for Centaur from the Air Force, and research under NASA contracts could not be proprietary. So NASA arranged for Douglas to have technical discussions with Convair, which built the Centaur. Douglas was able to benefit from Convair's experiences with the vagaries of liquid hydrogen. But according to Ted Smith, a key Douglas engineer, the larger size of the S-IV made Convair's cooperation rather irrelevant in designing the Saturn I second stage. Pratt & Whitney's cooperation in designing the S-IV stage, which held six RL10s rather than the two in Centaur, may therefore have been more important.

Douglas did use a tank design similar to that of Convair, with a common bulkhead between the liquid oxygen and the liquid hydrogen. But Douglas also employed much of its own Thor experience in its use of materials and methods of manufacture. The honeycomb material in the common bulkhead of the propellant tank, for example, was different from Convair's design, drawing upon Douglas's experience with panels in aircraft wings and some earlier missile designs. In any event, Douglas was able to make the larger tanks and S-IV stage in time for the first launch of a Saturn I featuring a live second stage on January 29, 1964. This launch, SA-5, was also the first with 188,000–pound H-1 engines in the first stage, and it succeeded in orbiting the second stage.[16]

Remarkably, this launch took place despite a major accident only five

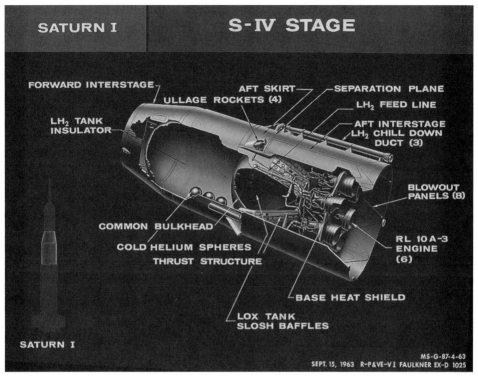

SATURN I

S-IV STAGE

FORWARD INTERSTAGE

ULLAGE ROCKETS (4)

LH₂ TANK
INSULATOR

AFT SKIRT

SEPARATION PLANE

LH₂ FEED LINE

AFT INTERSTAGE
LH₂ CHILL DOWN
DUCT (3)

BLOWOUT
PANELS (8)

COMMON BULKHEAD

COLD HELIUM SPHERES

THRUST STRUCTURE

RL 10 A-3
ENGINE
(6)

BASE HEAT SHIELD

LOX TANK
SLOSH BAFFLES

SATURN I

MS-G-87-4-63
SEPT. 15, 1963 R-P&VE-VI FAULKNER EX-D 1025

Figure 27. Cutaway drawing of the Saturn I S-IV upper stage showing its components. Courtesy of NASA.

days earlier. Douglas engineers and technicians knew that they had to take special precautions with liquid oxygen and liquid hydrogen. The latter was especially insidious. If it leaked and caught on fire in the daylight, the flames were virtually invisible. Infrared TV cameras did not totally solve the problem because of the difficulty of positioning enough of them to cover every cranny where hydrogen gas might hide. So crews in protective clothing carried brooms in front of them. If the broom caught on fire, they knew that hydrogen was leaking and burning.

Despite such precautions, on January 24, 1964, at a countdown to a static test of the S-IV, the stage exploded. Fortunately, the resultant hydrogen fire was short-lived, and a NASA committee with Douglas Aircraft membership determined that the cause was a rupture of a liquid-oxygen tank resulting from the failure of two vent valves to relieve the pressure that built up. The relief valves were incapacitated by solid oxygen, which had frozen because helium gas to pressurize the oxygen tank had come from a sphere submerged in the liquid hydrogen portion of the tank. The pressure got so high because the primary shutoff valve for the helium failed to close when normal operating pressure was reached in the oxygen tank. Testing of the helium shutoff valve showed that it did not work satisfactorily in cold conditions. Since this valve had previously malfunctioned, it should have been replaced before this time. In any event, Saturn project personnel did apparently change it to another design before the launch five days later. The committee "found that no single person, judgment, malfunction or event could be directly blamed for this incident," but "had the test operations personnel had the proper sensitivity to the situation the operation could have been safely secured" before things got out of hand.[17]

Guidance and Control System

Guidance and control of Saturn I (as well as IB and V) came from an instrument unit perched above the uppermost stage of the launch vehicle. The engineers in the von Braun group were able to design this unit in-house to a greater degree than was true of many other elements of the rocket (and its successors in the Saturn family). The basis of the guidance and control system was a stabilized platform that continued the evolution from the LEV-3 of the V-2 through the ST-80 of the Redstone, the ST-90 of the Jupiter, and an ST-120 used on the solid-propellant Pershing missile. The counterpart for the Saturns was the ST-124, which the engineers at ABMA began to work on in 1958, with the first mock-up of the overall instrument unit completed on June 15, 1961, and scheduled to fly on the Saturn-Apollo vehicle SA-5. (Earlier flights with only the first stage active used the ST-90, and even on SA-5 the ST-90 flew alongside the ST-124.) Because the ST-124, along with the rest of the instrument unit, continued to evolve through Saturn IB and Saturn V development, there is comparatively little information about the Saturn I system. We do know that to work with the ST-90 on the earlier flights, it used an adaptation of the IBM computer employed as the guidance computer on Titan II. With the flight of SA-9 in 1965, Saturn I started to use a more developed instrument unit that more closely resembled the one on Saturns IB and V. A notable feature of the improved instrument unit on the later flights of Saturn I was an upgrade of the environmental control unit. Earlier versions had used pressurization with an inert gas to prevent overheating. The improved system used methanol-water coolant that circulated through cold plates and absorbed heat, venting it into the exterior atmosphere or space in the form of boiled-off water. This system was not only lighter than the pressurized design but also half as tall.[18]

Stage S-I and Flight Testing

The S-I first stage was rather slow in being developed—largely, it would seem, because of the late decision about the number of stages that would be placed above it in the launch stack. It was not until May of 1961 that NASA Headquarters agreed to a proposal Marshall had sent up in March to change the design so it could carry either one or two upper stages. Meanwhile, design and testing of stage 1 had long since begun as an in-house effort in Huntsville. But that too could not progress very far until July 27, 1959, when the engineers and technicians completed the last Jupiter airframe for the Army and could begin retooling to support the Saturn project.[19]

On the first four Saturns, the S-I had no fins, with a cluster of four large and four stub fins added for later flights to provide greater aerodynamic stability. The borrowed Redstone and Jupiter propellant tanks had to be lengthened to increase their capacity, but the engineers retained the 5.89–foot Redstone and 8.83–foot Jupiter diameters so technicians could still use the tooling and welding equipment from the earlier projects. With one Jupiter and four Redstone oxygen tanks and four Redstone fuel tanks, all comparatively small, there seemed to be only minor potential for sloshing, but to avoid loss of control as had happened with Jupiter IB, the engineers designed baffles for each tank. As it happened, these proved inadequate on the very first flight, SA-1. The addition of more baffles on SA-3 successfully tamed the sloshing.[20]

The engineers did a better job of forestalling base heating—a concern that had arisen on Polaris AX-3 and AX-4—in advance of flight testing. As with Polaris, they found it almost impossible to predict how the rocket exhaust would be affected by shock waves as the vehicle passed through the transonic region near the speed of sound. Even for rockets with only one engine in the first stage, at different speeds and altitudes, dead air could heat up to extreme levels and interact with fuel-rich exhaust to cause fires or explosions. With eight engines—hence more places for dead air to collect—the S-I stage was especially vulnerable to such phenomena. So Saturn engineers performed cold-flow tests on scale models and arranged for wind tunnel testing at both the NASA Lewis Research Center and the Air Force's Arnold Engineering Development Center. Using data from the tests in conjunction with theoretical studies, von Braun's team arranged the two sets of four H-1 engines so as to limit areas where dead air could collect. The design of the lower skirt directed airflow to the four fixed engines in the middle, with a fire wall covering the base of the stage. There were "flex-

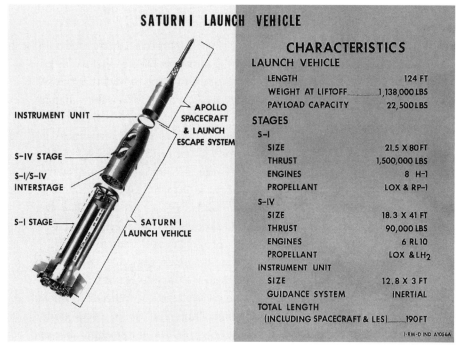

Figure 28. Depiction of the Saturn I launch vehicle and payload, with a list of its characteristics. Courtesy of NASA.

ible engine skirts" around the gimballing outermost engines. The aspirators on these engines, which forced the exhaust from the turbopumps into the engine exhaust, also helped prevent base heating. Static tests subsequently showed that the flexible shield around the outermost engines was insufficiently strong, necessitating redesign. But although clustering had its drawbacks, and wags joked about "cluster's last stand," these arrangements seem to have successfully solved the problems associated with having eight engines in one stage.[21]

Also contributing to a relative absence of problems were other ground tests, especially those done in a dynamic test stand built at Marshall in 1960–61. This facility allowed vibrational testing of the entire vehicle to see if separate oscillatory frequencies from different sources were likely to couple, potentially destroying the vehicle. Engineers did discover that frequencies in the hydraulic system and the gimbals had this potential, so they modified the structure accordingly. And when they found structural problems in the outer liquid-oxygen tanks, they also altered the way propellants flowed to the combustion chambers.[22]

Figure 29. The first Saturn I lifting off from Cape Canaveral on October 27, 1961. The baby of the Saturn family, Saturn I still stood over 190 feet high (considerably more than half a football field), weighed 569 tons on the launchpad, and ascended to an altitude of 85 miles. NASA Historical Reference Collection, Washington, D.C. Courtesy of NASA.

Apart from the sloshing on flight SA-1, the ten flights of Saturn I beginning on October 27, 1961, and ending on July 30, 1965, revealed few problems. There were changes resulting from the flight testing, but all ten flights were counted as successful, a tribute to the thoroughness of von Braun's engineers and their contractors. Flights SA-1 through SA-4 were suborbital and used only dummy upper stages. During SA-4 on March 28, 1963, the Saturn team intentionally shut down an inboard engine after 100 seconds of firing, and the flight successfully continued. Because of the way the propellant lines were designed, the other seven engines could burn longer than normal and compensate for the loss of thrust from the one engine that had ceased to function.

For flights SA-5 through SA-10, the uprated H-1 first-stage engines required lengthened propellant tanks. With a functioning S-IV stage, these flights were able to reach orbital speeds. The S-I stage's new fins provided greater aerodynamic stability, and the second-stage RL10–A3 engines worked satisfactorily. SA-6 unintentionally proved the engine-out capability once again, when an H-1 engine ceased to function 117.3 seconds into the 149–second stage-1 burn. Telemetry showed that the engine's turbopump had stopped supplying propellants. Analysis of the data pointed to stripped gears in the turbopump gearbox. From previous ground tests, Rocketdyne and Marshall personnel knew the gears needed wider teeth, so a redesigned gearbox was already programmed to fly on SA-7 and did not delay flight testing. There were no further problems with H-1 engines in flight. On September 18, 1964, SA-7 placed a 39,000–pound prototype Apollo spacecraft in orbit, leading to a declaration—three flights earlier than planned—that Saturn I was operational.

The last three flights carried Pegasus satellites designed and built by Marshall Space Flight Center. Even in lower orbits, Gemini spacecraft had been hit by meteorites, which posed a potential problem of severe magnitude for Apollo spacecraft and astronauts. Marshall engineers designed the Pegasus spacecraft with wings that collected data on meteoroids. For the last three Saturn I launches, manufacturing of the first stage had shifted to the Chrysler Corporation, with which NASA had contracted on November 17, 1961, to build and test twenty of these. Delivery of SA-8 was delayed, so SA-9 flew the first Pegasus mission on February 16, 1965, followed by SA-8 and then SA-10. Data from the three satellites led to a December 1965 announcement by NASA that the structures of Saturn and Apollo spacecraft were adequate to tolerate the effects of meteorite strikes.[23]

In its final configuration, Saturn I stood some 191.5 feet tall including the payload, with a maximum diameter in the S-I stage of 21.5 feet. Stage 1 delivered 1.5 million pounds of thrust, with stage 2 adding 90,000. On SA-5 it had launched into orbit a payload of 37,700 pounds—reportedly the heaviest weight yet placed in orbit.[24]

Management and All-Up Testing

While Saturn I was undergoing development and flight testing, significant management changes occurred in NASA that profoundly affected the Saturn family of launch vehicles. From November 18, 1959, when NASA assumed technical direction of the Saturn effort, through March 16, 1960, when NASA took over administrative direction of the project and formal transfer took place, to July 1, 1960, when both the Saturn program and the von Braun team of engineers transferred to Marshall Space Flight Center, the administrator of NASA was the capable and forceful but conservative T. Keith Glennan. Glennan had organized NASA, adding JPL and Marshall to the core centers inherited from the NACA. He had supported Project Mercury and the Saturn effort but, in keeping with the fiscal constraints of the Eisenhower administration, he did not believe in a "race to the Moon" with the Soviets. Thus it was a supreme irony that he laid the groundwork for Apollo, which constituted just such a race.[25]

Once John F. Kennedy became president in early 1961 and appointed the still more forceful and energetic but hardly conservative James E. Webb to succeed Glennan, there were bound to be management changes. And Kennedy's famous exhortation on May 25, 1961, "that this nation . . . commit itself to achieving the goal, before this decade is out, of landing a man on the moon and returning him safely to earth" gave an entirely new urgency to the Saturn program. To coordinate not only it but the other aspects of the Apollo program, NASA reorganized in November 1961, and even before that, Webb chose as head of a new Office of Manned Space Flight (OMSF) an engineer with RCA who had been project manager for the Ballistic Missile Early Warning System. D. Brainerd Holmes had finished the huge BMEWS project on time and within budget, so he seemed an ideal person to achieve a similar miracle with Apollo.[26]

Holmes headed one of four new program offices in the reorganized NASA Headquarters, with all program and center directors now reporting to Associate Administrator Robert C. Seamans Jr., who also took over control of NASA's budget. Webb apparently had not fully grasped Holmes's character when appointing him. The second NASA administrator had previously considered Abe Silverstein to head OMSF and rejected him because he wanted too much authority, especially vis-à-vis Seamans. Holmes, however, turned out to be "masterful, abrasive, and determined to get what he needed to carry out his assignment, even at the expense of other programs." Within two weeks of joining NASA, the confrontational Holmes demanded that

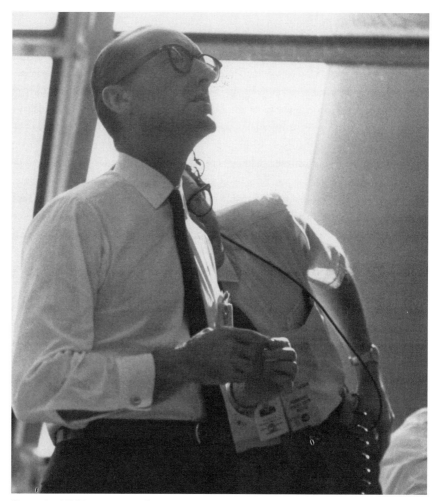

Figure 30. NASA associate administrator George E. Mueller in the Florida Launch Control Center, tracking the Apollo 11 mission. Courtesy of NASA.

he be made independent of Seamans. Webb refused. Less than a year later, in the summer of 1962, Holmes became convinced that Apollo was getting behind schedule and demanded more funding to put it back on track. Again Webb refused. Holmes also sought, in vain, to have center directors with human-spaceflight responsibilities report directly to him rather than Seamans. In frustration, Holmes finally resigned in June 1963.[27]

After consulting with Seamans, NASA deputy administrator Hugh Dryden, and friends in space-related industries, Webb selected another highly regarded engineer but one who turned out to be less confrontational, at least with his bosses, and more "bureaucratically adept." His name was George

E. Mueller. Having earned an M.S. in electrical engineering from Purdue University in 1940, Mueller (pronounced "Miller") began working on microwave experiments at Bell Telephone Laboratories that year. He turned to teaching at Ohio State University in 1946, earned his Ph.D. in physics there in 1951, and became a professor the following year. In 1957 he joined the Ramo-Wooldridge organization as director of the Electronics Laboratories and advanced quickly to become vice president for research and development before formally joining NASA as associate administrator for manned space flight on September 1, 1963.[28]

Perhaps benefiting from a headline that had appeared in the *New York Times* on July 13, 1963, "Lunar Program in Crisis," Mueller was able to get Webb to agree that he could manage Apollo with some freedom. But he really showed his bureaucratic astuteness when, soon after assuming his post, he assigned John H. Disher, who worked on advanced projects, and the assistant director for propulsion in OMSF, Adelbert O. Tischler, to assess how long it would now take to land on the Moon. On September 28 they reported that it was unlikely this could be achieved within Kennedy's decade "with acceptable risk." They believed it would be late 1971 before a landing could be attempted. Mueller took the two men to Seamans's office and had them repeat their findings. Seamans then told Mueller privately to figure out how to get the program back on schedule, exactly the authority and leverage Mueller had evidently sought.[29]

November 1, 1963, saw two major changes that not only offered a way to land on the Moon by the end of the decade but also greatly strengthened Mueller's position so he could achieve that goal. One radical change was all-up testing. In 1971 Mueller claimed that he had been involved with the development of all-up testing at Space Technology Laboratories, although he may or may not have known his organization opposed the idea when Otto Glasser introduced it as the only way he could conceive to cut a year out of the development time of Minuteman at the insistence of Secretary of the Air Force Douglas. In any event, all-up testing had worked for Minuteman and obviously offered a way to speed up the path to the Moon for the Saturn vehicles.[30]

NASA defined all-up testing as meaning that a vehicle was "as complete as practicable for each flight, so that a maximum amount of test information is obtained with a minimum number of flights." This obviously conflicted with the step-by-step procedures of the von Braun group, but on November 1 Mueller sent a priority message to Marshall as well as the Manned Spacecraft Center in Houston and the Launch Operations Center in Florida announcing a deletion of previously planned Saturn I launches with astronauts aboard and directing that the first Saturn IB launch, SA-201, and the first Saturn V flight, AS-501, should use "all live stages" and include "complete spacecraft." In a memorandum marked "For Internal Use Only" and dated October 26, Mueller had written to Webb via Seamans, enclosing a proposed NASA press release about all-up testing and saying, "We have discussed this course of action with MSFC, MSC, and LOC, and the Directors of these Centers concur with this recommendation," referring specifically to eliminating "manned" Saturn I flights but also, by implication, to the all-up testing. The enclosed press release stated that "experience in other missile and space programs" had shown all-up testing to be "the quickest way" to achieve "final mission objectives." The release went on to say that "deletion of the Apollo/Saturn I manned flights saves $50 million and is a step which helps NASA stay within the $5.35 billion authorized for FY [fiscal year] 1964."[31]

If Mueller had already discussed all-up testing with the center directors, this was not apparent at Marshall, where von Braun discussed the message with his staff at a luncheon on November 4, creating a "furor." Many of the staff recalled numerous failed launches in the V-2, Redstone, and Jupiter programs. William A. Mrazek believed all-up testing was insane, and other lab heads and project managers called it "impossible" and a "dangerous idea." Von Braun and his deputy, Eberhard Rees, also had their doubts about it, although in the end they had to agree with Mueller that the planned launches of individual stages would prevent landing on the Moon by the end of the decade. In December, when Mueller and Seamans visited Marshall, Arthur Rudolph, program manager for Saturn V, took Seamans over to scale models of Minuteman and Saturn V, pointing out the difference in size and emphasizing the much greater complexity of the huge liquid-propellant vehicle. When Seamans said, "I see what you mean, Arthur," an encouraged Rudolph repeated his arguments against all-up testing of the huge Saturn V to Mueller. But Mueller, obviously not impressed, reportedly replied, "So what?" All-up testing prevailed over Rudolph's objections and von Braun's doubts.[32]

The second change on November 1 was a reorganization of NASA, placing the field centers under the program offices once again, rather than under Seamans. Mueller obtained authority over Marshall, the Manned Spacecraft Center, and the Launch Operations Center (renamed Kennedy Space Center in December). Meanwhile, on September 1, 1963, Marshall had reorganized to shift from a research-and-development focus to a project-management structure. Beside a Research and Development Operations (R&DO) tier that continued to manage labs "organized by technical discipline," there was an

Industrial Operations (IO) function containing project organizations over-seeing work involving multiple labs. (IO's first head was Robert Young from Aerojet, but he stayed only a year before returning to his parent firm.) For the first time at Marshall, the project managers were equal in rank to the lab directors, with R&DO and IO both reporting directly to von Braun. This arrangement, developed independently from Mueller to facilitate Marshall's extensive dealings with contractors in the Saturn program, ultimately fit well with the management structure Mueller was establishing.[33]

One aspect of the organization in Huntsville that did not fit with Mueller's concepts, however, was Marshall's proclivity for basing technical decisions on their merits instead of schedule or cost, even though project managers were supposed to get jobs done "on time and within budget." A concern with time, budget, and what was now called configuration control, however, had become very important in the Minuteman program and quickly spread to NASA when Mueller arranged for Brig. Gen. Samuel Phillips to join the Apollo program as deputy director and then, after October 1964, director. The slender, handsome Phillips had moved from his post as director of the Minuteman program in August 1963 to become vice commander of the Bal-listic Systems Division. He arrived at NASA Headquarters in early January 1964, and soon arranged for the preparation and issue of a NASA publica-tion numbered 500–1, "Apollo Configuration Management Manual," issued in May and adapted from an Air Force manual.[34]

Phillips expected resistance to configuration management from NASA, and he was not disappointed. Mueller had formed an Apollo Executive Group consisting of the chief executives of firms working on Apollo plus directors of NASA field centers, and in June 1964 Phillips and a subordinate who managed configuration control for him presented the system to the as-sembled dignitaries. Von Braun objected to the premises of the system on the ground that costs for development programs were "very much unknown, and configuration management does not help." He contended further that it was impossible for the chairman of a configuration control board to know enough about all the disciplines involved to decide intelligently about a given issue. Phillips argued that if managers were doing their jobs, such decisions were already being made; configuration management constituted a mere formalization of what should already have been happening. Von Braun ob-jected that the system tended to move decisions higher in the management chain. William M. Allen of Boeing countered that such a move constituted "a fundamental of good management." When von Braun continued to argue the need for flexibility, Mueller explained that configuration management did not mean that engineers had "to define the final configuration in the first instance before [they knew] that the end item [was] going to work." It meant defining the expected design "at each stage of the game" and then letting everyone know when it had to be changed. Center directors like von Braun did not prevail in this argument, but resistance continued in industry as well as NASA centers, with the system not firmly established until about the end of 1966.[35]

Mueller and Phillips introduced other management procedures and infrastructure to achieve control of costs and schedules. Mueller's concept of "alternate paths" ensured that if something went wrong with one element of Apollo, development in other areas would continue and the program would make adjustments to keep on schedule. Phillips converted from a system in which data from the field centers came to Headquarters monthly to one with daily updates. To this end, he quickly contracted for a control room in NASA Headquarters similar to the one used for Minuteman, with data links to field centers. A computerized system stored and retrieved the data, which included information on parts, costs, and failures. Part of this system was a NASA version of the Navy's Program Evaluation and Review Technique (PERT), developed for Polaris, which most prime contractors had to use for reporting cost and scheduling data. In the Phillips document collection is a long April 29, 1964, briefing to NASA on Polaris management by the Navy Special Projects Office, indicating further that the general did not just rely on Air Force procedures in shoring up the civilian space agency's systems management.[36]

Despite von Braun's resistance to configuration management, Phillips recalled in 1988, "I never had a single moment of problem with the Marshall Space Flight Center. Their teamwork, cooperation, enthusiasm, and energy of participation were outstanding." More specifically with regard to von Braun, Phillips said: "Wernher directed his organization very efficiently and participated in management decisions. When a decision had been made, he implemented it—complied, if you will, with directives." No doubt Phillips was seeing through the rose-colored glasses of memory. But his comments reflected the German-American's propensity to argue either until he was convinced that the contrary point of view was correct or until he saw that argument was futile. Then he became a team player.

Phillips had been on the receiving end of V-2s in England during World War II, and he was prepared to dislike the German who had directed their design and development. But the two became friends. He commented that von Braun "had that rare gift of giving his undivided attention" to people with whom he spoke. "He could make a person feel personally important to him and that [his or her] ideas were of great value. . . . My wife says that she always felt like Wernher had been waiting all day just to talk to her." When asked in 1988 about von Braun's contributions to the space program, Phillips observed that a few years before, he probably would have said that American industry and engineering could have landed on the Moon without German input. But "when I think of the Saturn V, which was done so well under Wernher's direction and which was obviously . . . essential to the lunar mission . . . I'm not sure today that we could have built it without the ingenuity of Wernher and his team."[37]

The contributions of Mueller and Phillips were also critical to the ultimate success of Apollo. Phillips was hesitant about characterizing the rather geeky-looking Mueller but did say that "his perceptiveness and ability to make the right decision on important and far-reaching [as well as] complex technical matters" was "pretty unusual." Mueller's biggest shortcoming, according to Phillips, was "dealing with people." Others observed a certain coldness in his behavior toward subordinates. Even John Disher, who had worked with and admired Silverstein, characterized Mueller as the "only bona-fide genius I've ever worked with." But Disher had to observe that while Mueller was "always the epitome of politeness, . . . deep down he [was] just as hard as steel." Similarly, the human space program's director of flight operations, Chris Kraft, who obviously dealt with a great many people and had frequently clashed with Mueller, said, "I've never dealt with a more capable man in terms of his technical ability." Difficult though he undoubtedly was, without Mueller (and Phillips), American astronauts probably would not have gotten to the Moon before the end of the decade.[38]

Saturn IB

Long before George Mueller reported to NASA, engineers in von Braun's group and elsewhere had begun to develop Saturn IB, consisting of a modified Saturn I with its two stages redesigned (as S-IB and S-IVB) to reflect the increasing demands placed upon the intermediate version of the Saturn, plus a further developed instrument unit with a new computer and additional flexibility and reliability. The S-IVB would serve not only as the second stage of Saturn IB but, with further modifications, as the third stage of Saturn V, exemplifying the building-block nature of the development process. Saturn IB, much like Saturn I, originally had another name, C-1B, when NASA announced on July 11, 1962, that it would develop the new two-stage vehicle. Rebaptism as Saturn IB did not occur until February 1963, followed in August of that year by contracts to Chrysler for the S-IB and Douglas for the S-IVB stages.[39]

The S-IB Stage

Chrysler could easily shift from production of the last S-I stages to the S-IB, almost identical in size and shape. Working at the huge government-owned Michoud Assembly Facility near New Orleans, with its 1,828,000 square feet of floor space devoted at that time to manufacturing, Chrysler's engineers did have to modify the upper portion of the stage to accommodate the increase in girth and weight of the S-IVB stage from those of the S-IV. The cluster pattern for the eight H-1 engines did not change, although uprating did increase their thrust in two increments to 200,000 and then 205,000 pounds. To complement this improved performance, the S-IB engineers further added to the payload capacity of the overall Saturn IB by reducing the weight of the stage by some 19,800 pounds. Part of the reduction came from redesigned and smaller aerodynamic fins. Flight experience with Sat-

Figure 31. View of the huge Michoud Assembly Facility near New Orleans, where Chrysler and Boeing had large manufacturing areas for the Saturn IB and Saturn V launch vehicles. Courtesy of NASA.

Figure 32. Saturn IB first stages in final assembly at the Michoud Assembly Facility. Especially in the stage to the viewer's right, the arrangement of the eight H-1 engines is clearly shown, with the four inboard engines not gimballing and the outboard ones capable of gimballing for directional control. Courtesy of NASA.

urn I also revealed that the initial design of the stage had been excessively conservative, and engineers were able to trim propellant tanks, a "spider beam" that provided structural support, and other components as well as to remove "various tubes and brackets no longer required." But production techniques, most tooling, and the sequence of manufacturing did not change significantly.[40]

Uprated H-1 Engines

Just how Rocketdyne increased the thrust of each of the eight H-1 engines from 188,000 to 200,000 pounds for the first five Saturn IBs (SA-201 through SA-205) and then to 205,000 pounds for the remaining vehicles (starting with SA-206) is not entirely clear, but it would appear, as with the uprated Saturn I engines, that the key lay in improving the flow rates of the propellants into the combustion chambers, with resultant increases in chamber pressure.[41]

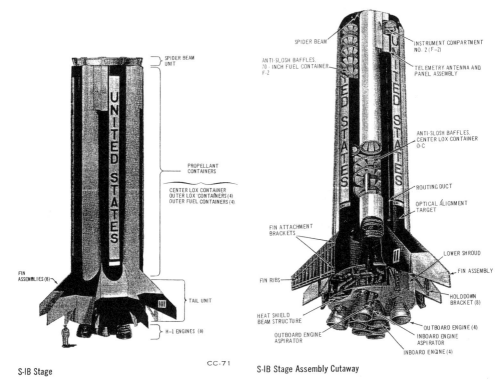

S-IB Stage   CC-71   S-IB Stage Assembly Cutaway

Figure 33. External and cutaway views of the S-IB stage for Saturn IB illustrating many of its details, including the clustering of the H-1 engines, which did not change from the Saturn I configuration. In the view at the left, note the huge size of even this one stage compared with a human figure. Taken from NASA, "Saturn IB News Reference," 3-1. Courtesy of NASA.

J-2 Engine

In its second stage, Saturn IB featured a new and much larger engine, the J-2, with thrust exceeding that of the six RL10s used on Saturn I. This was the liquid-hydrogen/liquid-oxygen engine the Silverstein committee had recommended for the Saturn upper stages on December 15, 1959. When NASA requested proposals from industry to design and build it, five companies competed, most notably Pratt & Whitney, Aerojet, and North American Aviation's Rocketdyne Division. Pratt & Whitney, builder of the RL10, might have seemed the obvious choice, but while NASA's source evaluation board judged all three firms capable of providing a satisfactory engine, Pratt & Whitney's price tag was more than twice the others.' Rocketdyne underbid Aerojet, making an assumption of less testing time, but even if the testing times were equalized, it appeared that Rocketdyne's cost was lower. Thus, on May 31, 1960, NASA administrator Glennan decided to negotiate with Rocketdyne for a contract to design and build the engine. The von Braun group and Rocketdyne then worked together on the design. A final contract signed on September 10, 1960, stipulated that the engine ensure "maximum safety for manned flight." Also stipulated was a conservative design to speed up development.[42]

Rocketdyne began developing the J-2 on September 1, 1960, with a computer simulation to assist with the configuration. Most of the work took place at the division's main facility at Canoga Park in northwestern Los Angeles, with firing and other tests up at the Santa Susana Field Laboratory in the nearby mountains. By early November the Rocketdyne engineers had designed a full-scale injector, and by November 11 they had conducted static tests of it in an experimental engine. Rocketdyne also built a large vacuum chamber to simulate engine firings in space. Although there were various proposals for stages using the engine, by the end of 1961 it was evident that the J-2 would provide power not only to the second stage of Saturn IB but the second and third stages of Saturn V (then still known as Saturn C-5). In the second stage of Saturn V, there would be a cluster of five J-2s, while the S-IVB second stage of Saturn IB and the S-IVB third stage of Saturn V would each have a single J-2. At this time, the thrust of a J-2 was expected to be 200,000 pounds, but the engine would later be uprated.[43]

Rocketdyne's engineers borrowed technology from Pratt & Whitney's RL10. But since the J-2 was more than thirteen times as large as the 15,000–pound-thrust RL10, they initially tried flat-faced copper injectors similar to designs Rocketdyne had been using in its liquid oxygen/RP-1 engines. Heating patterns for liquid hydrogen turned out to be quite different from those for RP-1, and the injectors got so hot the copper burned out. The RL10 had used a porous, concave injector of a mesh design, cooled by a flow of gaseous hydrogen. Rocketdyne would not adopt that design until Marshall insisted that firm representatives pay a visit to Lewis Research Center in 1962 to look at samples. Under pressure, Rocketdyne adopted the RL10 injector design, and problems with burnout ceased. In this instance, in-house experience took advantage of an established design from another firm and benefited from it, even though this occurred only under pressure from the customer, illustrating the sometimes difficult process of technology transfer. In this way Rocketdyne avoided further need for cut-and-try engineering in injector design, which was still, in the words of NASA assistant director for propulsion A. O. Tischler, "more a black art than a science."[44]

In-house Rocketdyne expertise seems to have been more effective in the design of the combustion chamber, consisting of intricately designed cooling tubes made of stainless steel with a chamber jacket made of Inconel, a nickel-chromium alloy capable of withstanding high heat levels. With a computer aiding design by solving a variety of equations having to do with energy, momentum, heat balance, and other factors, designers used the liquid hydrogen to absorb heat from combustion on its way to the injector, "heating up" in the process from −423°F to a gaseous temperature of −260°F. The speed of passage through the cooling tubes varied, depending upon computer calculations of the needs of different locations for cooling.[45]

Because of the low density of hydrogen and the consequent need for a higher-volume flow rate for the hydrogen than for the liquid oxygen (although, by weight, the oxygen flowed faster), Rocketdyne decided to use two different types of turbopumps, mounted on opposite sides of the thrust chamber. For the liquid oxygen, the firm used a conventional centrifugal pump of the type used for both fuel and oxidizer in the RL10. This featured a blade that forced the propellant in a direction perpendicular to the shaft of the pump. It operated at 7,902 rpm and achieved a flow rate of 2,969 gallons per minute. For the liquid hydrogen, an axial-type pump used blades operating like airplane propellers to force the propellant in the direction of the pump's shaft. Operating in seven stages (to one for the liquid-oxygen pump), they ran at 26,032 rpm and sent 8,070 gallons per minute of liquid hydrogen to the combustion chamber. A gas generator provided fuel-rich gas to drive the separate turbines for the two pumps, with the flow first to the one for the hydrogen and then the oxygen pump. The exhaust gas from the turbines passed into the main rocket nozzle not only for disposal but to add slightly to the thrust.[46]

In testing the J-2, engineers experienced problems with: (1) insulation of the cryogenic liquid hydrogen, (2) sealing it to avoid leaks that could produce explosions, and (3) a curious phenomenon known as hydrogen embrittlement in which the hydrogen in gaseous form caused metals to become brittle and break. To prevent the embrittlement, high-strength superalloys had to be coated with copper or gold. Solving problems that occurred in testing often involved trial-and-error methods, with engineers and technicians never knowing if a given "fix" actually solved an existing problem or created a new one. Even exhaustive preflight testing did not always expose potential problems, but engineers, in particular, always sought to find problems on the ground rather than in flight.[47]

Rocketdyne completed the preliminary design for the 200,000−pound-thrust J-2 in April 1961, with the preflight readiness procedures completed in 1964 and the engine qualified in 1965. The engine was gimballed for steering, and it had a restart capability, using helium stored in a tank within the liquid hydrogen tank to operate the pneumatic system. Soon after the 200,000–pound version was qualified, Rocketdyne improved the thrust to 205,000, 225,000, and then 230,000 pounds at altitude with no change in design. The increase came from adjusting the ratio of oxidizer to fuel (con-

SATURN V NEWS REFERENCE

J-2 ENGINE FACT SHEET

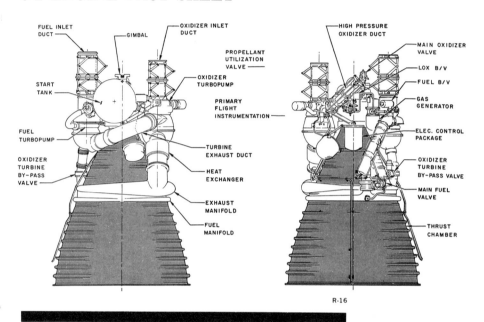

LENGTH	11 ft. 1 in.
WIDTH	6 ft. 8½ in.
NOZZLE EXIT DIAMETER	6 ft. 5 in.
THRUST (altitude)	225,000 lb.
SPECIFIC IMPULSE	424 sec. (427 at 5:1 mixture ratio)
RATED RUN DURATION	500 sec.
FLOWRATE: Oxidizer	449 lb/sec (2,847 gpm)
Fuel	81.7 lb/sec (8,365 gpm)
MIXTURE RATIO	5.5:1 oxidizer to fuel
CHAMBER PRESSURE (Pc)	763 psia
WEIGHT, DRY, FLIGHT CONFIGURATION	3,480 lb.
EXPANSION AREA RATIO	27.5:1
COMBUSTION TEMPERATURE	5,750°F

Note: J-2 engines will be uprated to a maximum of 230,000 pounds of thrust for later vehicles.

Figure 34. Schematic drawing of J-2 engine developed and produced by the Rocketdyne Division of Rockwell International, showing its components and characteristics. (Sources differ on flow rate and other measurements, so these should not be accepted as fully definitive.) Courtesy of NASA.

trolled by the propellant utilization valve, which was electrically operated). The 200,000–pound-thrust engine used a mixture ratio of 5:1, but the later versions could adjust the mixture ratio in flight up to 5.5:1 for maximum thrust and as low as 4.5:1 for a thrust level of only 175,000 pounds. During the last portion of a flight, the valve position shifted to ensure the simultaneous emptying of the liquid oxygen and the liquid hydrogen from the propellant tanks (technically, a single tank with a common bulkhead, but referred to in the plural as if there were separate tanks). The 225,000–pound-thrust engine had replaced the 200,000–pound version in the production line by October 1966, with the 230,000–pound version available by about September 1967. As uprated versions became available, Rocketdyne ceased producing the lower-rated ones.[48]

The S-IVB Stage

Even with six RL10s, the S-IV stage had only been about 39.7 feet tall by 18.5 feet in diameter. To contain the single J-2 and its propellant tank, the S-IVB had to be 58.4 feet long and 21.7 feet wide. On 21 December 1961, NASA selected Douglas to modify its S-IV to accommodate the J-2. At that time, it appears that NASA already envisioned the S-IVB as the third stage of what became the Saturn V. Douglas had already designed the S-IV to have a different structure than the Centaur, with the latter's steel balloon replaced by a self-supporting aluminum structure of a skin-and-stringer construction more in keeping with the "man-rating" planned first for Saturn I and then for Saturn IB, the vehicle that would actually launch the first Apollo astronauts into orbit.

The propellant tank borrowed a wafflelike structure with ribs from the Thor tanks that Douglas had designed. The common bulkhead between the liquid hydrogen at −423°F and the liquid oxygen at −297°F required only minor changes from the smaller one in the S-IV. After conferring with Convair about the external insulation used in Centaur to keep the liquid hydrogen from boiling away rapidly, Douglas engineers had decided on internal insulation for the S-IV. They chose woven fiberglass threads cured with polyurethane foam to form a tile that was shaped and installed inside the fuel tank. This became the insulation for the S-IVB as well.[49]

For steering the S-IVB during the firing of the J-2, Douglas had initially designed a long and slender actuator unit to gimbal the engine, similar to devices the firm had used on aircraft landing gear. Marshall engineers said the rocket stage required stubbier actuators. When this proved to be the case, Douglas subcontracted the work to Moog Servocontrols of East Aurora, New York, which built the actuators to Marshall specifications. The

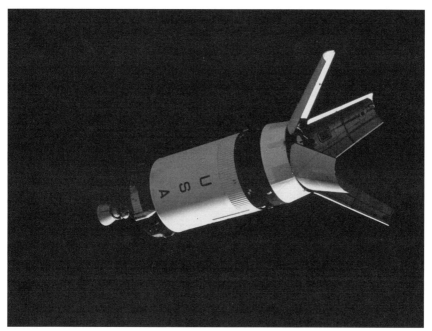

Figure 35. Saturn IVB (Saturn IB second stage) in space during the Apollo 7 mission, with the lunar module adapter's four panels in the open position. Courtesy of NASA.

gimballed engine could adjust the stage's direction in pitch and yaw. For roll control during firing of the J-2, and for attitude control in all three axes during orbital coast, an auxiliary propulsion system provided thrust. Controlled, like the gimballing, by the guidance and control system in the instrument unit, the auxiliary system consisted of three 150–pound-thrust engines built by Thompson-Ramo-Wooldridge of Cleveland. These small engines burned nitrogen tetroxide and monomethyl hydrazine and were located on the aft skirt of the stage.[50]

The Instrument Unit

The instrument unit (IU) used on Saturn IB, like its almost identical twin on the later Saturn V, had a 21.67–foot diameter, stood 3 feet high, and weighed about 4,100 pounds. Its ST-124–M inertial platform used nitrogen-gas-bearing gyros for stabilization and pendulous integrating gyro accelerometers (also gas-bearing devices) to measure rates of change in attitude and velocity. Dry nitrogen supplied from reservoirs in the IU held friction in the gyros to a minimum and ensured high levels of accuracy in the signals indicating vehicle attitude and speed that the ST-124–M system sent to the guidance computer. The Bendix Corporation, which had made the ST-120 inertial platform for the Pershing missile, was the contractor for ST-124 inertial platform and sensing equipment, although the Nortronics Division of Northrop built the rate gyro package. The gyros and accelerometers for the Saturn guidance and control system were less than half as heavy as those for the Jupiter and weighed the same as the ones for the Pershing missile. But where the Pershing's were made of aluminum, beryllium, and monel, those for the Saturn were made entirely of beryllium to provide better thermal and structural stability.[51]

Designed and developed in-house by Marshall Space Flight Center and its predecessors and built by the Eclipse Pioneer Division of Bendix, the Saturn inertial platform system underwent operational tests on the 35,000–foot high-speed test track at Holloman Air Force Base in New Mexico. By the mid-1960s the sleds on the test track—part of the Central Inertial Guidance Test Facility of the Air Force Missile Development Center (AFMDC)— could achieve speeds of more than 1,200 mph and subject the inertial system to intense vibrations and to accelerations upwards of 7.5 Gs, simulating launch and early flight conditions. The sled tests, arranged by the Holloman test facility's Joachim H. Gengelbach, a former Peenemünde V-2 engineer, began in 1962. Early tests in 1962 and 1963 identified such problems as that "the servo power supply could not meet all electrical load requirements under dynamic conditions for all inertial components" and "the gyros had a tendency to oscillate under very high vibration inputs."

With a redesigned accelerometer signal processor, the ST-124 inertial platform resumed testing on the sled track in 1964. Performance was generally satisfactory, and the reports made only minor suggestions for improvement, noting, "The performance of all gas bearings exceeded expectations," and, "The ST-124 performance was much better than of any prior system tested on the AFMDC high-speed test track." To this point, the tests had been on earlier versions of the stabilized platform, but in 1965 Holloman tested the ST-124–M version used on Saturn IB. These tests demonstrated "the functional integrity of the ST124M platform."[52]

Besides the inertial platform, the instrument unit included a high-speed digital computer, an analog control computer, a measuring and telemetry system, a tracking system, an electrical system, and an emergency detection system. The launch vehicle digital computer, or LVDC, controlled sequencing as well as guidance and control in both stages of Saturn IB and all three stages of Saturn V. Made by IBM, it was programmed to control an initial roll and tilt of the launch vehicle and set it on its planned trajectory soon after launch. Thereafter it controlled the vehicle into its designated orbit of Earth and during coasting flight. Containing 40,800 components, the

computer weighed 78.5 pounds and took up 2.10 cubic feet of space in the
instrument unit. This comparatively low weight resulted from the pioneering use of a magnesium-lithium alloy, significantly lighter than either magnesium or aluminum, for the computer's chassis. Employing what was called
"triple modular redundancy," the computer had three critical circuits that
were identical. If one circuit's output was at variance with the other two, all
three circuits "voted," with the two that agreed prevailing. This corrected for
random errors, preventing computer failure and providing better reliability
than use of three independent computers, according to Marshall's Astrionics System Handbook.[53]

Besides three sets of three accelerometers each (nine total for triple redundancy) in the instrument unit, there were nonredundant rate gyros,
fixed to the body of the instrument unit of Saturn IB to sense "angular rates
about the pitch and yaw axes." Structural bending of the launch vehicle necessitated these additional accelerometers. Signals from the inertial platform and the accelerometers went to a data adapter in analog form. The
analog control computer and the digital computer, communicating through
the data adapter, processed the various signals, then determined the adjustments necessary to keep the vehicle not only on course but free from abrupt
changes that would induce undue vehicle bending and propellant sloshing.
They selected and sent the correct control signals for engine gimballing or
for attitude-control vectoring by the small auxiliary engines on the S-IVB
stage.[54]

The measuring and telemetry system used a variety of sensors to gather
information about conditions on the launch vehicle and transmitted these
data to the control center on the ground for analysis. The information gathered ranged from sound levels and temperatures to pressures, flow rates for
fluids, and force levels. During phases of flight when telemetry transmission
was impossible, an onboard tape recorder retained the data for subsequent
transmission. A radio-frequency system permitted ground controllers to
communicate with the guidance and control system during flight for purposes of tracking, command, and analysis of vehicle performance.[55]

Antennas and transponders in the IU supplemented ground-based tracking systems, helping them to following the vehicle from the ground and providing them with in-flight information. A radio command link permitted
ground controllers to update information in the guidance system based on
data gathered on the ground. Extensive verification ensured that ground
controllers transmitted only accurate information. The electrical system provided onboard power from four 28–volt DC batteries to operate equipment
in the other systems. The emergency detection system monitored thrust in

both stages of Saturn IB, the status of the guidance and control system, attitude of the launch vehicle and angular rates of change therein. In case of serious problems requiring instant response, logic circuits commanded a mission abort, while less urgent difficulties produced a visual display in the Apollo spacecraft for an astronaut, if one were on board, to initiate an abort if necessary.[56]

Marshall engineers assumed primary responsibility for building the first four instrument units for Saturn IB. But in February 1964, NASA made IBM the prime contractor for the IU in both Saturn IB and Saturn V. This included building and testing the units, then sending them to Cape Kennedy. Unlike many other Saturn contractors, IBM did not manufacture the IU at its home plant (in Oswego, New York, in this case) but built a research and development complex in Huntsville.[57]

Flights

After completion of its development, Saturn IB stood 224 feet high, with a total liftoff weight of up to 1,300,000 pounds. Its maximum diameter was 21.7 feet, tapering to a mere 2.2 feet for the launch escape tower atop the spacecraft (not used on all launches). The eight H-1 engines in the S-IB stage generated a total thrust of 1,600,000 pounds, while the S-IVB stage provided an additional 200,000 to 225,000 pounds, depending on which version of the J-2 engine it contained.[58]

The first Saturn IB flight, on February 26, 1966, had a complicated mission. It was to test the two-stage launch vehicle plus the Apollo spacecraft systems for "structural integrity, compatibility, communications, [and] separation." A specific concern was whether the Apollo heat shield in the command module could withstand reentry into the atmosphere from the peak of its suborbital trajectory at a speed of 27,300 ft/sec. Also among the mission goals were recovery of the Apollo spacecraft and checkout of ground facilities and equipment at the launch site. This launch marked the first flight tests of an S-IVB stage, a J-2 engine, and a powered Apollo spacecraft. In the event, the first stage performed normally, and the second stage ignited very close to its predicted time after launch, only 0.4 seconds late. J-2 engine cutoff occurred at the desired velocity but 10 seconds past its predicted time because the propellant utilization system had ensured simultaneous exhaustion of both oxygen and hydrogen by slowing the oxygen consumption.

Less propulsion than planned from the service module reduced the speed of the reentry to 26,500 ft/sec—still within test limits. The ablative heat shield exhibited the expected amount of charring, and the spacecraft landed in the Atlantic some 5,400 miles from the launch site and 49 miles from the intended landing point. USS *Boxer* successfully recovered it. The guidance and control system performed its assigned functions, with both S-IB and S-IVB trajectories and terminal velocities being normal. There were no structural problems discovered in either of the stages or the IU, and the quality of data telemetered to the ground was good, with few losses in some 1,200 measurements transmitted. Although parachutes for two data cameras failed to function, crews recovered one camera, which provided good coverage of stage separation and ignition of the second stage. Overall, it was a successful mission.[59]

Delays in development of the Apollo spacecraft for AS-202 caused AS-203 to be the second Saturn IB launch. The "payload" consisted of the S-IVB stage, the IU, a nose cone, and 19,000 pounds of hydrogen in place of the Apollo spacecraft. Thus, while AS-203 exemplified Mueller's alternate-paths methodology, it was a departure from all-up testing. Without an Apollo spacecraft as the payload, the objectives of the mission became evaluation of the S-IVB and IU in orbit and observation of the way liquid hydrogen reacted to orbital conditions. One goal was to gain information about restarting the J-2 in orbit without actually performing a restart. To ignite the engine in orbit once it had been shut down would require the propellants to settle to the back of their tanks under conditions of weightlessness. This required thrust, which venting of gaseous hydrogen (resulting from boil-off of the liquid hydrogen) could provide, supplemented by occasional opening of a liquid oxygen vent valve.

Launched on July 5, 1966, AS-203 performed very much as intended, achieving a planned near-circular orbit of the 58,500–pound payload. This was the heaviest object yet orbited by the United States. The simulated restart of the J-2 occurred as expected. Television portrayal of the interior of the tank showed the fuel settling in the back as anticipated. Also as planned (although the timing had been an unknown), near the beginning of the fifth orbit, the second stage broke apart. As pressure in the fuel tank rose to 39.4 psi compared to only 5 psi in the oxygen tank, the pressure differential had destroyed the tanks' common bulkhead, causing the stage to disintegrate. This event confirmed results of a test a few months before at a Douglas facility, where the bulkhead gave way at about the same pressure differential. Before the anticipated breakup of the S-IVB, the mission had completed all planned experiments successfully.[60]

For AS-202, Saturn IB reverted to a suborbital mission designed to test the Apollo command module's heat shield at a higher heating load than experienced on AS-201. The mission continued to evaluate the launch vehicle and spacecraft for structural integrity, separation, and general functioning. The launch occurred on August 25, 1966, with the 1.6–million-pound-thrust first stage performing satisfactorily, as did the J-2 in the second stage. In an apparent self-contradiction, the J-2 reportedly was the 200,000–pound-thrust version, but it flew at a mixture ratio of 5.5 pounds of oxygen to 1 of hydrogen (characteristic of the uprated J-2s) for the first 350 seconds of the burn, followed by cutback to a ratio of about 4.7:1. Late reduction of the mixture ratio contributed to unexpectedly high performance, leading to second-stage engine cutoff some 13 seconds earlier than foreseen. Neither this nor failure of a liquid-hydrogen recirculation valve in the J-2 detracted from the mission.

The guidance and control system operated satisfactorily, with acoustic and vibration levels remaining within tolerances. There were no structural problems for the launch vehicle, but there were some short circuits in the emergency detection system. When the service module propulsion system ignited 11 seconds after separation, it caused the S-IVB to oscillate and adapter panels in the spacecraft to fall off or fold back. However, the propulsion system in the service module boosted the Apollo spacecraft to an altitude of 706 miles before it began the descent. On the way down, the engine restarted three times, making the reentry path resemble a roller-coaster ride. This trajectory raised the temperatures on the command module's heat shield to 2,700°F. The reentry path for AS-202 was at a less steep angle than for AS-201, increasing the time of reentry, and AS-202's speed was 19,900 mph, up from about 18,000 mph for the earlier mission. Despite the heating load on the shield, the temperature inside the command module reached only 70°F. Some ten hours after launch, USS *Hornet* picked up the command module about 500 miles southeast of Wake Island in the Pacific Ocean, roughly 17,800 miles from Kennedy Space Center. Both the module and the heat shield were in good shape. Obviously, there were some component problems on the mission that needed attention, but overall NASA counted the mission a success.[61]

The next mission, AS-204, was scheduled to launch on February 21, 1967, with astronauts on board, but on January 27 a ground check of the vehicle and spacecraft became a disastrous tragedy. With Virgil I. Grissom, Edward H. White II, and Roger B. Chaffee inside, fire swept the command module. The three astronauts had neither time nor means to escape. Their deaths cast a pall over the entire Apollo program, but NASA administrator Webb determined that NASA "had to find out what happened and fix it and move ahead" with the program.[62]

Investigation of the fire and modifications to the command module considerably delayed the next Saturn IB launch and changed its configuration. When AS-204 finally launched on January 22, 1968, it carried a lunar module instead of a command module, another example of Mueller's alternate-paths approach to keeping Apollo on schedule. The objective of the mission became verification of the lunar module's propulsion systems for both descent to the Moon's surface and ascent back to the command module, which would remain in lunar orbit. This evaluation included the staging and structures of the lunar module and, again, the launch vehicle itself. The mission flew with a redesigned liquid-hydrogen recirculation valve in the J-2 to replace the one that had failed on AS-202. This apparently also was the first launch with a 225,000–pound-thrust J-2 engine. Besides AS-204, the mission bore the out-of-sequence designation Apollo 5. AS-201 and AS-202 had unofficially been called Apollo 1 and 2. But after the fire, the mission scheduled for Grissom, Chaffee, and White became Apollo 1, with the first Saturn V mission designated Apollo 4. Thus, strangely, AS-204 became Apollo 5, with no missions ever officially being called Apollo 2 or 3. Following a successful launch, the S-IVB separated from the S-IB stage, and the second stage carried the payload into a successful orbit at a velocity of 17,515 mph. The new liquid-hydrogen recirculation valve in the J-2 worked properly, as did the guidance and control system. After the lunar module separated from the S-IVB, burns of the descent and ascent propulsion systems in the module were successful, including restart and throttling. AS-204 was a success, preparing the way for the first launch with astronauts aboard, known as AS-205 or *Apollo 7*.[63]

The mission for *Apollo 7* was to test the operation of the redesigned command and service module (CSM—actually, two separate modules but often referred to as if they constituted a single unit) and the performance of the crew, as well as support facilities. Another goal was to demonstrate that the CSM could rendezvous with the S-IVB, in preparation for later rendezvous in lunar orbit with the lunar module ascending from the Moon. The redesigned CSM featured a hatch that was much easier to open, new materials that reduced flammability, and expanded redundancy to lower the possibilities for single-point failure.

Apollo 7 launched on October 11, 1968. The Saturn IB with a 225,000–pound-thrust J-2 in the second stage performed well, with all structural loads within tolerances for the human passengers. More than an hour after the S-IVB placed the 69,345–pound payload in orbit, the second stage dumped its propellants to make it safe for simulated docking with the CSM.

Then the module with the three astronauts separated from the "safed" stage. The rendezvous maneuver was only a simulation because the S-IVB was tumbling, but the astronauts maneuvered the CSM within 70 feet of the stage and remained near it for 25 minutes before moving away. This demonstrated that the spacecraft could rendezvous with the lunar module if it became disabled after lifting off from the Moon. Some seven days after launch, the S-IVB reentered Earth's atmosphere over the Indian Ocean. Meanwhile, the astronauts had performed many other tests in the CSM after the simulated rendezvous. They orbited Earth 163 times before splashing down on October 22 in the Atlantic Ocean, where USS *Essex* picked them up. AS-201 through AS-205 qualified the Apollo spacecraft, as intended, with *Apollo 7* successfully achieving all primary objectives of the flight. This ended the Saturn IB flights for Apollo, although the vehicle was later used in the Skylab and Apollo-Soyuz test projects from 1973 to 1975.[64]

Figure 36. Schematic drawing of the various Saturn engines, with their thrust, fuel, and comparative size. The H-1 and RL10 powered Saturn I, with the H-1 also used on Saturn IB; the F-1 was the first-stage engine for Saturn V, while the J-2 was the second-stage engine on Saturn IB and the second- and third-stage engine for Saturn V. Courtesy of NASA.

Saturn V

Development of some parts of Saturn V began before design of other components that were common to Saturn IB and Saturn V. For example, on January 9, 1959, Rocketdyne won the contract for the huge F-1 engine used on Saturn V but not on IB, while it was not until May of 1960 that Rocketdyne was even selected to negotiate a contract for the J-2 common to both launch vehicles. Configurations were in flux in the early years. Only on July 11, 1962, did NASA officially announce the C-1B (later renamed Saturn IB) as a two-stage vehicle for Earth-orbit missions with astronauts aboard. Yet NASA Headquarters had already formally approved the C-5 on January 25. The point here is that even though Saturn IB was an interim configuration between Saturn I and Saturn V, development of the two vehicles overlapped substantially, with planning for the ultimate Moon rocket occurring even before designers conceived—or at least got approval to develop—the interim configuration.[65]

The F-1 Engine

Development of the F-1 engine by Rocketdyne originated with an Air Force request in 1955. NASA inherited the reports and data from the early development, and when Rocketdyne got the NASA contract to build the engine in 1959, it was, "in effect, a follow-on contract." Since the engine contract preceded a clear conception of the vehicle into which it would fit and the mission it would perform, designers had to operate in something of a cognitive vacuum. They had to make some early assumptions, followed by later reengineering to fit the engines into the actual first stage of Saturn V, which itself still lacked a firm configuration in December 1961 when NASA selected Boeing to build the S-IC stage. Another factor in the design of the F-1 was a decision "made early in the program . . . [to make] the fullest possible use of components and techniques proven in the Saturn I program."

In 1955 the Air Force's Propulsion Laboratory had asked how large a liquid-propellant engine it was possible to build. To answer the question, engineers at the Rocketdyne division of North American Aviation developed a preliminary design for a one-million-pound-thrust engine and reported that they saw no reason why it could not be built. The Air Force said in 1957 to go ahead and build one, following up that request with a contract in mid-1958—apparently without a specific application in mind, only a desire to explore the limits of the technology. During 1957 and 1958, Rocketdyne test-fired such an engine, with much full-scale testing done at Edwards Air Force Base, while Rocketdyne did basic research, development, and production at its plant in Canoga Park. The contractor conducted tests of components at nearby Santa Susana Field Laboratory. At Edwards the organization that became, in 1963, the Rocket Propulsion Laboratory had three test stands, 1–A, 1–B, and 2–A, set aside for the huge engine. The 1959 contract with NASA called for 1.5 million pounds of thrust, and already on April 6, 1961, Rocketdyne was able to static-test at Edwards a prototype engine in which the thrust peaked at 1.64 million pounds.[66]

Burning RP-1 as its fuel with liquid oxygen as the oxidizer, the F-1 did not break new ground in its basic technology. In keeping with NASA guidelines, the powerful engine was to use proven propellants, with the emphasis on reliability, not innovation. But its thrust level required so much scaling up that it still marked a major advance in the state of the art of rocket making. As a Marshall publication said, "An enlargement of this magnitude is in itself an innovation." And as Rocketdyne's William J. Brennan put it, "The giant stride in thrust . . . in itself necessitated many technology advancements and fabrication innovations." For instance, the very size of the combustion chamber, which was 40 inches in diameter (as against 20.56 for the H-1) and had nearly four times the H-1's chamber area (1,257 square inches versus 332), required new techniques to braze together the regenerative cooling tubes. Also because of the engine's size, Rocketdyne adopted a gas-cooled, removable nozzle extension to make it easier to transport.[67]

The entire engine measured 19 feet 8 inches in length and 12 feet 4 inches in diameter. It was bell-shaped and used an expansion ratio of 16:1 with the nozzle extension attached. Its turbopump consisted of a single axial-flow turbine mounted directly on the thrust chamber with separate centrifugal pumps for the oxidizer and fuel that were driven at the same speed by the turbine shaft. This eliminated the need for a gearbox, which had been troublesome on the Thor-Jupiter engine, among others. A fuel-rich gas generator burning the engine propellants powered the turbine. The initial F-1 had the prescribed 1.5 million pounds of thrust, but starting with vehicle 504, Rocketdyne uprated the engine to 1.522 million pounds. This could be done by increasing the chamber pressure through greater output from the turbine, which in turn required strengthening of components, at some expense in engine weight. There were five F-1s clustered in the S-IC stage, four outboard and one in the center. All but the center engine gimballed to provide steering. As with the H-1, there was a hypergolic ignition system.[68]

Perhaps the most intricate design feature of the F-1, and one that certainly caused a great deal of difficulty, was the injection system. As two Rocketdyne engineers wrote in 1989, the injector "might well be considered the heart of a rocket engine, since virtually no other single component has such a major

Figure 37. At the Michoud Assembly Facility, Saturn V first stages with their five F-1 engines visible. Courtesy of NASA.

impact on overall engine performance." The injector not only inserted the propellants into the combustion chamber but mixed them in a proportion adjusted to produce the desired thrust and performance. "As is the case with the design of nearly all complex, high technology hardware," the two engineers went on, "the design of a liquid rocket injector is not an exact science, although it is becoming more so as analytical tools are continuously improved. This is because the basic physics associated with all of the complex combustion processes that are affected by the design of the injector are only partly known." A portion of the problem lay with the atomization of the propellants and the distribution of the fine droplets to ensure proper mixing. Even as late as 1989 they could say: "the atomization process is one of the most complex and least understood phenomena, and reliable information is difficult to obtain." One result of less than optimal injector design was combustion instability, whose causes and mechanisms still in 1989 were "at best, only poorly known and understood."[69] This was even truer in the early 1960s with the F-1 injector.

The team at Rocketdyne knew from experience with H-1 and earlier engines that injector design and combustion instability would be problems. They began with three injector designs, all of them based essentially on that for the H-1. Water-flow tests provided information on the spacing and shape of orifices in the injector, followed by hot-fire tests in 1960 and early 1961. But as Leonard Bostwick, the F-1 engine manager at Marshall, reported, "None of the F-1 injectors exhibited dynamic stability." Designers tried a variety of flat-faced and baffled injectors without success, concluding that it would not be possible simply to scale up the H-1 injector to the size needed for the F-1. Engineers did borrow from the H-1 effort the use of bombs to initiate combustion instability, saving a lot of testing to await its spontaneous appearance. But on June 28, 1962, during an F-1 hot engine test in a test stand built for the purpose at the rocket site at Edwards, combustion instability led to the meltdown of the engine.[70]

Marshall quickly appointed Jerry Thomson, chief of the center's Liquid Fuel Engines Systems Branch, as chair of an ad hoc committee to deal with the problem. A native of Alabama, Thomson had earned a degree in mechanical engineering at Auburn University following service in World War II. Turning over the running of his branch to his deputy, he moved to Canoga Park, where respected propulsion engineer Paul Castenholz and a mechanical engineer named Dan Klute, who also "had a special talent for the half-science, half-art of combustion chamber design," joined him on the committee from positions as Rocketdyne managers. Although Marshall's contingent on the committee was not that large, at Rocketdyne there were some fifty engineers and technicians assigned to a combustion devices team, supplemented by people from universities, NASA, and the Air Force. Using essentially cut-and-try methods, they initially had little success. The instability showed no consistency and started "for reasons we never quite understood," Thomson confessed.[71]

Using high-speed instrumentation and trying perhaps forty or fifty design modifications, eventually the engineers found a combination of baffles, enlarged fuel-injection orifices, and changed impingement angles that worked. By late 1964, even after the explosion of bombs, the combustion chamber would regain its stability. The engineers were never certain that the problem would not recur, but in January 1965 Marshall rated the F-1 injector to be flight ready. Problems were still not over, however. Further testing revealed difficulties with fuel and oxidizer rings containing multiple orifices for the propellants. Steel rings called lands held copper rings for the propellants. Brazed joints held the copper rings in the lands, and these joints were failing. Engineers found a solution in gold-plating the lands to create a better bonding surface. Developed and tested only in the spring and summer of 1965, well after the preliminary flight rating test in December 1964, the new injector rings had to be retrofitted in several engines that had already been delivered.[72]

The injector that resulted from the extravagant reengineering by Thomson, Castenholz, and the others contained 6,300 holes—3,700 for RP-1 and 2,600 for liquid oxygen. Radial and circumferential baffles divided the flat-faced portion of the injector face into thirteen compartments, with the holes arranged so that most of them were in groups of five. Two of the five injected the RP-1 so that the two streams impinged on one another to produce atomization, while the other three inserted liquid oxygen, which formed a fan-shaped spray that mixed with the RP-1 to combust evenly and smoothly. Driven by the 52,900–horsepower turbine, the propellant pumps delivered 15,471 gallons of RP-1 and 24,811 gallons of liquid oxygen per minute to the combustion chamber.[73]

Despite all the work that went into the injector design, according to Roger Bilstein it was the turbopump that "absorbed more design effort and time for fabrication than any other component of the engine." There were eleven failures of the system during the development period. Two of these involved the liquid-oxygen pump's impeller, which required a redesign with stronger components. The other nine failures were explosions. The causes varied. The high acceleration of the shaft on the turbopump was one problem. Others included friction between moving parts and metal fatigue. All eleven failures occasioned a redesign or a change in manufacturing procedures. For instance, Rocketdyne made the turbine manifold out of a nickel-based alloy manufactured by General Electric called René 41, which was a recent addition to the materials used for rocket engines. Unfamiliarity with the welding techniques it required led to cracks near the welds. These necessitated time-consuming research and training to bring welders up to speed on the proper procedures for this alloy, which could withstand not only high temperatures but the extremes in temperature variance entailed by the use of cryogenic liquid oxygen. The resultant turbopump not only provided the speed and high volumes needed for a 1.5–million-pound-thrust engine but did so with a minimal number of parts and high ultimate reliability. Nevertheless, these virtues came at the price of much testing and frustration.[74]

The S-IC Stage

Following Boeing's selection on December 15, 1961, to negotiate becoming the prime contractor for the S-IC first stage, with a preliminary contract signed in February 1962, the firm had to face the increased size of the stage compared with the S-IB. The S-IB had been large enough—21.4 feet in diameter and 80.3 feet in height—but Saturn V's first stage was 33 by 138 feet with a dry weight of 300,000 pounds, compared with 93,000 for the S-IB. The propellant tanks to contain 203,000 gallons of RP-1 and 331,000 gallons of liquid oxygen required tooling of unprecedented size and capabilities, with welds of unexampled length in a launch vehicle. While NASA encouraged use of "proven technology," the sheer size and the need to ensure the safety of human astronauts atop such a large rocket made that virtually impossible. Exacting quality assurance standards meant that welders, among others, had to be taught new methods that could only be developed empirically, since existing techniques would not accommodate the scaling-up the task demanded. Marshall personnel assisted Boeing in improvising the new procedures both for welding and for X-ray inspection. Von Braun's people, for example, devised an electromagnetic hammer that delicately smoothed out bulges produced by welding.[75]

As of mid-December 1961, when Boeing entered the picture, the configuration of the S-IC was still not firm. The original plan had been for it to contain only four engines. Milton Rosen—who was NASA's deputy director for launch vehicle programs from January to November 1961 and became director for launch vehicles and propulsion until April 1963—chaired a committee sometime in late 1961 that consisted of members of his own staff and people from Marshall, including William Mrazek, to discuss the configuration. The Marshall engineers had presented drawings showing four engines with two stout crossbeams supporting them. Rosen argued for a fifth engine at the junction of the crossbeams. Not only would this increase the thrust of the booster; it would eliminate a potentially significant problem. In the space at the middle of the four engines, exhaust gases could accumulate and possibly explode. Base heating could also be a problem there. The fifth engine would obviate this by forcing the heat away from the area with its own thrust. Mrazek was hard to convince, but meeting for two weeks in a Huntsville motel, "including one stretch of 5 almost 24–hour days" according to Rosen, the former technical director of Project Vanguard persuaded Mrazek—and, through him, von Braun—to adopt a fifth engine. On December 21, 1961, a NASA management council approved the decision. The extra thrust it provided became providential as the weight of the Apollo spacecraft increased.[76]

Von Braun's engineers worked closely with their counterparts from Boeing in completing the plans for the S-IC. By mid-1962 there were nearly 500 Boeing technical people working at Marshall and another 600 or so in a refurbished cotton mill in Huntsville. While these people focused on design, some 450 Boeing employees were at the Michoud facilities near New Orleans preparing for the fabrication of the stage in the roughly 1.2 million square feet of manufacturing floor space used at this time by Boeing. (Another 800,000 square feet were still being used by Chrysler for the S-IB.) Having to work much more closely with Marshall than the other two stage contractors, both on the West Coast, Boeing managers sometimes chafed under the meticulous supervision. But the Boeing manager at Michoud, at least, later conceded that Marshall had helped the contractor solve some problems before they arose. Marshall's welding experience, for instance, had contributed to avoidance of difficulties in production, although the intense involvement of von Braun's engineers complicated matters when cost overruns occurred for which the NASA employees bore at least partial responsibility.[77]

Using tooling that Boeing produced, Marshall built some early static test versions plus the first two S-ICs for flight testing. It then sent the equipment to Michoud for Boeing to use. Marshall's technical people involved themselves in the development of the tooling as well as manufacturing concepts, testing such things as welding equipment and jigs for assembly before approving their use at Michoud. Each S-IC was slightly different to accommodate its intended Apollo mission, so tooling had to be flexible as well as effective. Boeing made about 90 percent of the components for the first stage of Saturn V in its Wichita plant, shipping them to either Huntsville or Michoud for processing and assembly. For example, it took dozens of pieces of aluminum, milled down to about one-fifth of their original weight and thickness, to put together the propellant tanks. Shipped to Marshall or Michoud, they underwent welding there to form the tanks.

Because of the immense size of the pieces, they could not easily be moved through a stationary welding device as was normal practice. Instead, a jig held the aluminum pieces in place and the welding tool moved along the seam of the weld. To avoid distortion of the tanks from the heat of the welding, designers arranged for welding areas with temperatures kept below 77°F and humidities below 50 percent. Each S-IC had about 6.2 miles of welding, with every fraction of an inch inspected for defects. In view of their huge size, the propellant tanks contained numerous slosh baffles to prevent the fuel or oxidizer from moving about excessively during launch and flight. Once the tanks were formed, they had to undergo hydrostatic testing to 105 percent of the pressure they would face on a mission. There were countless other tests, but the most extreme ones were probably the static firings of all five F-1 engines. Test stands for these events became available at both Marshall and the Mississippi Test Facility.

Figure 38. A Saturn V stage-1 fuel tank at Marshall Space Flight Center, December 1, 1964. Note the size of the tank compared with the humans in the photo. Courtesy of NASA.

At the latter, the stand had been built after 1961 on the mud of a swamp along the Pearl River near the Louisiana border and the Gulf of Mexico. Mosquito-ridden and snake-infested, this area was home to alligators, wild pigs, and panthers. Construction workers faced 110 bites a minute from salt marsh mosquitoes, against which nets, gloves, repellent perfume, and long-sleeved shirts gave virtually no protection. Spraying special chemicals from two C-123 aircraft did reduce the number of bites to 10 per minute, but working conditions were still challenging. Nevertheless, the stand was ready for use in March 1967, more than a year after the first static test at Marshall. Thereafter the 410–foot S-IC test stand, the tallest structure in Mississippi, became the focus of testing for the F-1s.[78]

The S-IC suffered a series of welding problems in 1963 and was some 6 weeks behind schedule through much of 1964, reaching 19 weeks in October because of delays in production and delivery of parts for the thrust structure, which conveyed the force of the thrust to the rest of the vehicle from the engines (that is, linked the engine and the rest of the vehicle). But by 1965 with the first static tests, the Boeing stage seemed to be in good shape. On April 16, 1965, all five F-1 engines of the S-IC-T test vehicle fired for 6.5 seconds at Marshall. This was followed on August 5 by the first full-

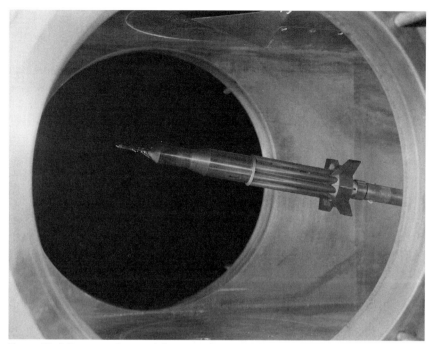

Figure 39. A Saturn rocket model in a wind tunnel at NASA Langley Research Center in 1962. Wind tunnel tests to ensure the aerodynamic integrity of the basic Apollo-Saturn launch configuration were only one part of the elaborate and extensive ground testing of the Saturn launch vehicles. Courtesy of NASA.

duration static test of the first stage. By December 16 of that year, all the S-IC-T static test firings at Marshall had concluded—a total of fifteen hot-fire tests, three of which were for the full duration (not specified, but presumably the 150 seconds of the stage's operation during actual flight). The S-IC-T then resumed testing at the Mississippi Test Facility in March 1967.[79]

The S-II Stage

Meanwhile, the S-II second stage was giving much more trouble than the first stage. NASA had selected North American Aviation to build the S-II on September 11, 1961, several months before Boeing's selection for the lower stage. The division of North American that won the S-II contract was not Rocketdyne, the builder of the F-1 and the J-2 engines. It was the Space and Information Systems Division (previously the Missile Division), headed by Harrison A. Storms Jr., who had managed the X-15 project. An able and articulate engineer, Storms was somewhat mercurial as well as charismatic. His nickname, "Stormy," apparently reflected his personality as much as his name. (Some claimed that "while other men fiddle, Harrison storms.") His subordinates proudly called themselves Storm Troopers, but he could be abrasive, embodying what test pilot and engineer Scott Crossfield called "the wire brush school of management."[80]

When Storms's division began bidding on the S-II contract, the configuration of the stage was very much in flux. Early in 1961 when Webb authorized Marshall to initiate contractor selection, thirty aerospace firms attended a preproposal conference. At that time NASA envisioned a stage with only four J-2 engines instead of five. Its stated height was 74 feet, not the 81 feet 7 inches it would become, and the projected width was 21 feet 6 inches rather than the eventual 33 feet. It still seemed imposingly large, but it was "the precision it would require [that] gave everybody the jitters—like building a locomotive to the tolerance of a Swiss watch," as Storms's biographer put it. This sort of concern whittled the number of firms submitting information on their capabilities and experience from thirty to seven. A source evaluation board eliminated three contractors, leaving Aerojet, Convair, Douglas, and North American to learn that they were now bidding on a stage enlarged to at least a diameter of 26 feet 9 inches—still well short of the final figure, as it turned out. Also missing was information about the configuration of the stages above the S-II, but the Apollo spacecraft proved to be about twice as heavy as originally thought. The Marshall procurement officer did emphasize that an important ingredient in NASA's selection would be "efficient management."[81]

Once Storms's division won the contract, with the stage to be built at a manufacturing plant at Seal Beach on the California coast some 15 miles south of NAA's older plant at Downey, it did not take too long for NASA to arrive at the decision, announced January 10, 1962, that the S-II would hold five J-2 engines. Designers decided to go with a single tank for the liquid hydrogen and liquid oxygen with a common bulkhead between them, as Douglas had done with the much smaller S-IVB. (The S-II contained 260,000 gallons of liquid hydrogen and 83,000 gallons of liquid oxygen, to 63,000 and 20,000 respectively in the S-IVB.) As with the Douglas stage, common parlance referred to each compartment as if it were a separate tank. Obviously, the common bulkhead was much larger in the 33–foot-diameter second stage than in the 21.75–foot-wide S-IVB third stage, so the welding required unusual precision to ensure there was no leakage. The bulkhead consisted of the top of the liquid-oxygen tank, a sheet of honeycombed phenolic insulation bonded to the metal beneath it, and the bottom of the liquid-hydrogen tank. Careful fitting and bonding, verified by ultrasonography, ensured complete adherence and the absence of gaps. Not only did fit have to be perfect, but there were also complex curvatures and a shift in thickness from a maximum of about five inches in the center to somewhat less at the periphery.[82]

Unlike Douglas but like Convair on the Centaur, North American decided to use external insulation, which increased the strength of the tank , as will be seen. Initially, Storms's engineers tried insulation panels, but the bonding failed repeatedly during testing. Using trial-and-error engineering, designers turned to spraying insulation directly onto the tank, allowing it to cure, and then cutting it to the proper dimensions. Once the tanks were formed and cleaned, North American installed slosh baffles inside them.[83]

The decision to put the insulation on the outside of the liquid-hydrogen tank followed from the choice of material for the S-II stage: an aluminum alloy designated 2014 T6. Used for a long time, for example on the Ford Trimotor, the alloy had the unusual characteristic of getting stronger as it got colder. At −400°F, it was 50 percent stronger than at room temperature. So with the insulation on the outside, this material provided a real advantage with the −423° liquid hydrogen. Both the oxidizer and fuel tank walls could be 30 percent thinner than with another material. Unfortunately, aluminum 2014 T6 was also difficult to weld, and with its 33–foot diameter, the S-II presented almost 104 feet of circumference. On the first try at welding two cylinders to one another, welders were about four-fifths around the circular perimeter when the remaining portion of the metal "ballooned out of shape from the heat buildup." The Storm Troopers had to resort to automated and increasingly powerful welding equipment to do the job. Each ring to be welded had to be held in place by a huge precision jig with about 15,000 adjustment screws around the 104–foot circumference, each a fraction of an inch from the next. A mammoth turntable rotated the seam through fixed weld heads with microscopic precision. A huge clean room allowed the humidity to be kept at 30 percent. In all of this, Marshall's experience with welding, including for the S-IC stage, helped Stormy's people appreciably.[84]

Despite such help, there was considerable friction between Storms and his division, on the one hand, and Marshall on the other—especially Eberhard Rees, von Braun's deputy director for technical matters. North American fell behind schedule and had increasing technical and other problems. Marshall officials began to complain about the contractor's management shortcomings, including a failure to integrate elements ranging from engineering and budgeting to manufacturing, testing, and quality control. At the same time, Storms's division was the victim of its own delays on the Apollo spacecraft (see below). As the weight of Apollo payloads kept increasing, the launch vehicle stages had to be reduced in weight to compensate. The logical place to do so was the S-IVB stage, since a pound of reduction there had the same effect as 4 or 5 pounds taken off the S-II (or 14 pounds from the S-IB), basically because the lower stages had to lift the upper ones as well as themselves. But the S-IVB, used on Saturn IB, was already in production. So designers had to turn to reductions in the thickness and strength of the structural members in the S-II.[85]

By mid-1964 the S-II insulation was still a problem. Then in October, burst tests showed that weld strength was lower than expected. This was followed on the twenty-eighth of the month by the rupture of the aft bulkhead of an S-II under hydrostatic testing, even though the pressure had been below specifications. As the date for 1967 launch of the first Saturn V approached, von Braun proposed eliminating a test vehicle to get the program back on schedule. Phillips agreed. Instead of separate dynamic and structural test vehicles, the structural stage would do double duty. On September 29, 1965, this combined structural and dynamic test vehicle, S-II-S/D, underwent hydraulic testing at Seal Beach. While the propellant tanks filled with water, the stage was simultaneously subjected to vibration, twisting, and bending to simulate flight loads. Even though the thinned structure was substantially less strong than it would have been at the low temperatures imparted by liquid hydrogen, Marshall had insisted on testing to 1.5 times the expected flight loads. At what was subsequently determined to be 1.44 times the load limit, the stage broke apart with a thunderous noise as 50 tons of water cascaded through the test site. Not only had the welds failed. The program was short another test vehicle. Storms's people looked at the effect on the cost of the program and concluded that to complete the program after the failure would raise the expense of the contract from the initial $581 million to roughly $1 billion.[86]

When Lee Atwood, president of North American, flew to Huntsville on October 14, 1965, Air Force Brig. Gen. Edmund O'Connor, who had replaced Robert Young as Marshall's director of industrial operations, told von Braun, "The S-II program is out of control. . . . management of the project at both the program level and the division level . . . has not been effective." Von Braun told Atwood that the S-II needed a more forceful manager than William F. Parker, the quiet but technically knowledgeable man Storms had appointed as head of the program in 1961. Von Braun apparently got Atwood's agreement to replace Parker and put a senior manager in charge of monitoring delays and manufacturing difficulties.[87]

In a controversial decision, Storms's division had also won the contract for the Apollo command and service module. As it happened, the day after Atwood's visit to Huntsville, Rees hopped on an airplane for Houston, where

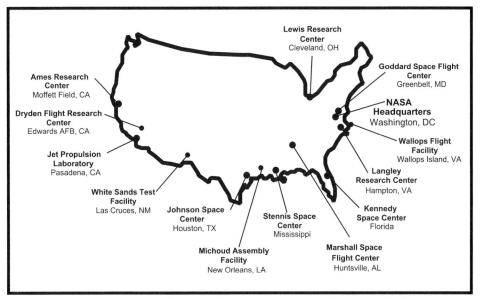

Figure 40. Map of NASA installations in the early 1990s. Some of the names had changed since the mid-1960s, but the locations had not. Courtesy of NASA.

he met with other Apollo managers including Phillips. The Manned Spacecraft Center (later Johnson Space Center) in Houston was managing the programs for the Apollo spacecraft, and Houston manager Joseph Shea had complaints similar to those of Rees and his people about Storms's control of costs and schedules. Phillips decided to head an ad hoc tiger team to visit North American and investigate its problems. The team included people from Marshall and Houston.[88]

The group descended upon North American on November 22, 1965, and in a meeting on December 19, Phillips presented the findings in a briefing and accompanying notes. George Mueller had already conveyed to Lee Atwood his concerns about the S-II and spacecraft programs at Storms's Space and Information Systems Division. In a letter to Atwood dated December 19 he reiterated, "Phillips' report has not only corroborated my concern, but has convinced me beyond doubt that the situation at S&ID requires positive and substantive actions immediately in order to meet the national objectives of the Apollo Program." After pointing to numerous delays and cost overruns on both the S-II and the spacecraft, Mueller wrote, "It is hard for me to understand how a company with the background and demonstrated competence of NAA could have spent 4½ years and more than half a billion dollars on the S-II project and not yet have fired a stage with flight systems in operation." He said Sam Phillips was convinced that the division could do a better job with fewer people and suggested transferring to another division elements like Information Systems that did not contribute directly to the spacecraft and S-II projects.[89]

A memorandum from Phillips to Mueller the day before had been even more scathing: "My people and I have completely lost confidence in NAA's competence as an organization to do the job we have given them." He made numerous specific recommendations for management changes, including "that Harrison Storms be removed as President of S&ID" since "his leadership has failed to produce results which could have and should have been produced." After assuring Phillips and Mueller he would do what he could to correct problems, Atwood visited Downey where the records were kept and was reportedly impressed by the design work. He did not replace Storms, but Stormy himself had already placed his aide, retired Air Force Maj. Gen. Robert E. Greer, in a position to oversee the S-II. In January 1966, Greer added the titles of vice president and program manager, keeping Bill Parker as his deputy. Greer agreed in a later interview that there were serious problems with S-II management. He revamped the management control center to ensure more oversight and incorporated additional management meetings that the Storm Troopers called Black Saturdays, implicitly comparing them with Schriever's meetings at the Western Development Division. The difference was that Greer, who had served at WDD's successor Ballistic Missile Division, held them daily at first, then several times a week, not monthly. With Greer's systems management and Parker's knowledge of the S-II, there was hope for success.[90]

But setbacks continued. On May 28, 1966, in a pressure test at the Mississippi Test Facility, another S-II stage exploded. Human error—a failure to reconnect pressure relief switches after previous tests—was to blame, but inspection revealed tiny cracks in the liquid-hydrogen cylinders that also turned up on other cylinders already fabricated or in production. Modification and repair occasioned more delays. Still, it took the Apollo fire in the command module in January 1967 and extreme pressure from Webb to induce Atwood to separate Information Systems from the Space Division (as it became), to move Storms to a staff position, and to appoint recent Martin Marietta president William B. Bergen as head of the Space Division—actually a demotion for which he volunteered from a position in which he had been Storms's boss. Bergen's appointment may have been more important for the redesign of the command module than for the S-II, and certainly Storms and North American were not solely to blame for the problems with either the stage or the spacecraft. But by late 1967 the problems with both were largely solved or on the way to solution.[91]

The S-IVB Stage

Unlike the S-II, the S-IVB was not a totally new stage used only on Saturn V. It had clearly profited greatly from its development and testing for use as the Saturn IB second stage. But the new version was different in several respects. As the third stage on Saturn V, the S-IVB required control mechanisms to restart the engine in orbit for the burn that would send the Apollo spacecraft on a trajectory to orbit around the Moon. The aft skirt for the S-IVB, which now had to flare out to match the greater girth of the S-II, was heavier than the one for the S-IVB as second stage on Saturn IB. The forward skirt was heavier as well, to accommodate a heavier payload. The auxiliary propulsion and other small engine systems weighed more for the third stage of Saturn V than the comparable second stage on the IB because the lunar missions needed increased attitude control and venting. Finally, the propulsion system was heavier for the Saturn V third stage because of the need to restart. All these additions pushed up the dry weight by some 11,000 pounds. While the first burn of the single J-2 engine would last only about 165 seconds to get the third stage and payload to orbital speed, the second burn would last upwards of 310 seconds and would accelerate the stage and spacecraft to about 24,500 miles per hour for the trip to the Moon.[92]

On the aft skirt assembly, mounted 180 degrees apart, were two auxiliary propulsion modules, each containing three 150–pound-thrust attitude control engines and one 70–pound-thrust ullage control engine. Built by TRW, the attitude control engines burned a hypergolic combination of nitrogen tetroxide and monomethyl hydrazine. They used ablative cooling and provided roll control during J-2 firing and control in pitch, yaw, and roll during coast periods. The ullage control engine, similar to those for attitude control, fired before the coast phase to prevent propellants from moving excessively from the aft end of their tanks. It fired again just before engine restart. There were also two ullage control motors—distinct from the ullage control *engines*—mounted 180 degrees apart between the auxiliary propulsion modules. These motors fired after separation from the S-II stage just before ignition of the third-stage J-2. The two motors, Thiokol TX-280s, each delivered about 3,390 pounds of thrust.[93]

Despite the relatively modest changes in the S-IVB for Saturn V, development was not problem-free. In acceptance testing of the third stage at Douglas's Sacramento test area on January 20, 1967, the entire stage exploded. Investigation finally revealed that a helium storage sphere had been welded with pure titanium rather than an alloy. Excessive testing weakened the sphere and it exploded, cutting propellant lines and allowing the propellants to mix, ignite, and themselves explode. This destroyed the stage and adjacent structures. The human error led to revised welding specifications and procedures. Despite the late date of this mishap, the S-IVB was ready for the first Saturn V mission on November 9, 1967, in which it performed its demanding mission, including restart, without notable problems.[94]

The Saturn V Instrument Unit

Like the S-IVB, the instrument unit had basically the same design for Saturn IB and Saturn V, although it was about 400 pounds heavier at roughly 4,500 pounds in the Saturn V version. The guidance and flight control systems within the IU navigated (determined "vehicle position and velocity"), guided (ascertained "attitude correction signals"), and controlled the vehicle (calculated and issued "control commands to the engine actuators.") As the Saturn V ascended, the ST-124–M inertial platform sensed and measured the acceleration and attitude of the vehicle and sent those data to the launch vehicle digital computer via the launch vehicle data adapter. The digital computer integrated the measurements with the time expired since launch to determine whether the vehicle had reached the desired position, precalculated and stored in the computer's memory. In case of deviation, it computed corrections and sent them to the analog flight control computer, which combined them with data from the rate gyros. (Sensors also included a pair of accelerometers fixed to the body of the S-II stage rather than the IU, as was the case for Saturn IB.) The system then issued commands to change the direction of thrust via engine gimbals and/or six small thrusters on the S-IVB to alter attitude in roll during J-2 burning or to correct the attitude during the orbital coast period.

The guidance and control system also directed engine cutoff and stage separations. Once the vehicle reached Earth orbit, data transmitted from the ground via a radio command system could update the launch vehicle digital computer. An astronaut in the Apollo spacecraft could also change the attitude of the vehicle during Earth orbit. Once the astronauts completed alignment and checkout of their own guidance systems in the Apollo spacecraft, on missions to the Moon the guidance and control system reignited the S-IVB and injected the vehicle into its lunar trajectory. For these missions, once the vehicle had escaped Earth orbit, the J-2 engine shut down for the second time. The launch vehicle's guidance and control system maintained the vehicle's attitude while the command and service module (CSM) mated with the lunar module in "a rather intricate mating maneuver" in which,

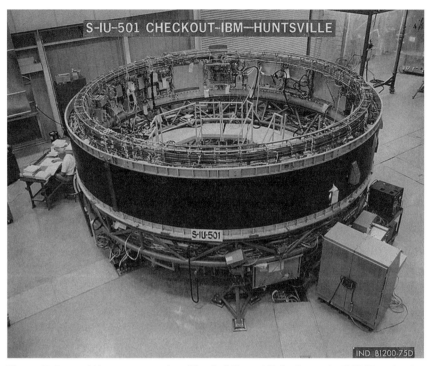

S-IU-501 CHECKOUT—IBM—HUNTSVILLE

S-IU-501

IND B1200-75D

Figure 41. Saturn V instrument unit at Marshall Space Flight Center in 1967. Note its size compared with the technician to the left. Courtesy of NASA.

simply put, the CSM separated from the S-IVB, which was still holding the lunar module in an adapter section. Next, the CSM mated with the LM. Then, the S-IVB and IU separated from the Apollo spacecraft, which continued with the mission using its own equipment. At this point, Saturn V had completed its mission.[95]

Flights of Saturn V

AS-501, the first Saturn V vehicle, was assembled on the launchpad at Kennedy Space Center in June 1967 when word arrived of the discovery of flawed welding in an S-II. It turned out that similar flaws were present in the second stage of Saturn V on the pad. Although these were repaired, other problems kept cropping up to delay the first launch from August until November 9, 1967. But on that day the all-up launch of the unmanned Apollo 4 mission went nearly without a flaw. After going into Earth orbit and completing almost two revolutions, the S-IVB reignited and lifted the instrument unit, the CSM, and a prototype lunar module to a peak altitude of 11,240 miles. The third stage then separated and the service-module propulsion system accelerated the CSM to a speed of 36,537 ft/sec (24,912 mph) before separation of the command module from the service module—comparable to that from lunar reentry. The command module landed in the Pacific Ocean only 9 miles from its aiming point and was recovered by USS *Bennington*.

The instrument unit on this flight included a stiffened external structure to reduce vibrations that had affected the inertial platform on Saturn IB flights. The heavier structure did its job. Guidance and control were essentially as expected—"nominal" in the parlance of spaceflight. There were a few deviations from predicted parameters, including some low-level longitudinal oscillations (known as the pogo effect) in the first-stage engines, but after its difficult parturition, the infant Saturn V seemed quite healthy. The flight demonstrated the structural integrity of the launch vehicle and its compatibility with the Apollo spacecraft. Moreover, the command module and its heat shield had withstood the temperatures associated with reentry into Earth's atmosphere at the speeds they would encounter in a lunar reentry.[96]

Euphoria from this success suffered a dose of reality, however, on April 4, 1968, when AS-502 (Apollo 6) launched. As with AS-501, the vehicle did not carry astronauts on board, but it was considered "an all-important dress rehearsal for the first manned flight," planned for AS-503. The launch went well, but toward the end of the first-stage burn the pogo effect became much more severe than on AS-501, reaching 5 hertz (cycles per second), which exceeded the design specifications and was considered alarming. Despite the oscillations, the vehicle continued on its upward course. Stage-2 separation occurred and all five J-2 engines ignited. Then, at 319 seconds after launch, there was a sudden 5,000–pound decrease in thrust, followed by a cutoff signal to the number-2 engine. This signal shut down not only engine number 2 but number 3 as well (about a second apart, bracketing 414 seconds into the mission). It turned out that signal wires to the two engines had been interchanged. This loss of the power from two engines was a severe test for the instrument unit, but it adjusted the trajectory and the time of firing (by about a minute) for the remaining three engines to achieve (in fact, exceed) the planned altitude for separation of the third stage.[97]

When the IU shut down the three functioning engines in the S-II and separated it from the S-IVB, that stage's lone J-2 ignited and placed itself, the IU, and the payload in an elongated parking orbit. To do this, the IU directed it to burn 29.2 seconds longer than planned to further compensate for the two J-2s that had cut off in stage 2. The achievement of this orbit demonstrated "the unusual flexibility designed into the Saturn V." However, although the vehicle performed adequately during orbital coast, the J-2 failed to restart and propel the spacecraft into a simulated lunar trajectory. After repeated attempts to get the J-2 to restart, mission controllers separated the command and service module from the S-IVB, used burns of the service module propulsion system to position the command module for reentry tests, and performed these tests to verify the design of the heat shield, with reentry occurring "a little short of lunar space velocity," followed by recovery. Although this is sometimes counted a successful mission, and both Phillips and von Braun said a crew could have returned safely, von Braun also said, "With three engines out, we just cannot go to the Moon." And in fact, as restart of the S-IVB's J-2 was a primary objective of the mission and was not attained, the mission was technically a failure.[98]

As Phillips briefed the Senate Committee on Aeronautical and Space Sciences on April 22, 1968, some eighteen days after the flight, pogo was not a new phenomenon, having afflicted Titan II and come "into general attention in the early days of the Gemini program." Aware of the possibility of pogo, von Braun's engineers had tested and analyzed Saturn V before the AS-501 flight and found "an acceptable margin of stability to indicate" it would not develop. The AS-501 flight "tended to confirm these analyses." All of the five F-1 engines had "small pulsations," but each engine experienced them "at slightly different points in time." Thus there was not a problem. But on AS-502, the five 1.5–million-pound engines "came into a phase relationship" where "the engine pulsation was additive."[99]

All engines developed a simultaneous vibration of 5.5 hertz. The entire vehicle itself developed a bending frequency that increased, as it consumed propellants, to 5.25 hertz about 125 seconds into the flight. The engine vibrations traveled longitudinally up the vehicle structure, with their peak occurring at the top where the spacecraft was—and where the astronauts would be, on a flight carrying them. By themselves, the engine vibrations would not have posed a problem, but they coupled with the vehicle's oscillations (bending frequency), which moved in a lateral direction. When the two types of vibrations intersected, with both at about the same frequency, their effects were combined and multiplied. In the draft of an article he wrote for the *New York Times*, Phillips characterized the "complicated coupling" as "analogous to the annoying feedback squeal you encounter when the microphone and loud speaker of a public address system are coupled." This coupling could have interfered with an astronaut's performance of his duties, a significant issue.[100]

To analyze and correct this problem, NASA created a pogo task force that included people from Marshall, other NASA organizations, contractors, and universities. The solution the task force recommended was to de-tune the five engines, changing the frequencies of at least two of the five so that they would no longer produce vibrations at the same time. Engineers did this by inserting liquid helium into a cavity formed in a liquid oxygen prevalve in which a casting bulged out and encased an oxidizer feed pipe. The bulging portion was only half filled with the liquid oxygen during engine operation. The helium would absorb pressure surges in oxidizer flow and reduce the frequency of the oscillations to 2 hertz, which was lower than the frequency of the structural oscillations. Engineers eventually applied the solution to all four outboard engines. Technical people contributing to this solution came from Marshall, Boeing, Martin, TRW, Aerospace Corporation, and North American's Rocketdyne Division.[101] This incidence of pogo showed how difficult it was for rocket designers to predict when and how such a phenomenon might occur, even while aware of and actively testing for it.

Meanwhile, a separate team including engineers from Marshall and Rocketdyne attacked the unknown problem that had caused the J-2 engine failures—not the interchanged signal wires, but whatever had made the number-2 engine in stage 2 fail and the single engine in stage 3 refuse to restart. The team, which included Jerry Thomson from the F-1 combustion-instability effort, examined the telemetry data from the flight and concluded that the problem had to be a rupture in a fuel line. But why had it broken? Raising pressures, vibrations, and flow rates on test stands did not produce a failure but led engineers to suspect a problem with the bellows section in the fuel line. Intended to allow the line to bend around various obstructions, this area had a wire-braid shielding on the outside. On the test stand it did not break under the abnormal strains to which it was subjected, although artificially severing the line did produce measurements that duplicated those from the flight. Finally Rocketdyne personnel tried testing the lines in a vacuum chamber simulating actual conditions in space. Eight lines tested there at rates of flow and pressures no greater than those during normal operation led to failures in the bellows section in all eight lines within 100 seconds. Motion pictures of the tests quickly revealed that in the absence of atmospheric moisture, frost did not form inside the wire braiding as it had in regular ground tests during cryogenic liquid hydrogen flow. The frost had kept the bellows from vibrating to the point of failure, but in a vacuum, without the frost, a destructive resonance occurred. The bellows section was replaced with a stronger design that still incorporated the necessary bends. Testing of the pogo fix on the F-1 and the fuel-line redesign on the J-2 at the Mississippi Test Facility in August 1968 indicated that both worked.[102]

Additionally, the crossed wires that had resulted in the unintentional and unnecessary shutdown of J-2 engine number 3 in the S-II on AS-502 stemmed from a modification. Documentation rules for the modification did not require that the wires be marked with a reference designator. Checks of the wiring for proper resistance did occur at the Mississippi Test Facility, where a full-duration static firing took place. Officials then decided to defer a further test to Kennedy Space Center, where it never took place. To fix this oversight, procedural changes included reference designators for the wires and reviews of all tests deferred from one facility to another. As Mueller said in the Senate hearing, NASA flew the Saturn Vs without astronauts aboard "to be sure we understand how they work," and it had learned more from AS-502 than it could possibly have done "from a successful repeat of [AS-]501." Among the serendipitous lessons, the space agency learned that it was safe to "proceed with two engines out" on an S-II stage.[103]

Despite the problems with AS-502, following the successful *Apollo 7* mission launched by a Saturn IB, NASA decided to make AS-503 (*Apollo 8*) a mission with astronauts aboard undertaking a circumlunar voyage using an actual command and service module plus a lunar-module test article as payload, development of the actual lunar module having experienced delays. Checked and checked again, AS-503 proved to have "many things which needed to be corrected and improved," in the words of Dieter Grau, Marshall's chief of quality and reliability operations. Since the primary objectives of the mission were to demonstrate the performance of the spacecraft, crew, and mission support team, including lunar-orbit-rendezvous procedures, this was not fundamentally a test of Saturn V. But given the problems on Apollo 6, NASA and its contractors would be paying close attention to the launch vehicle's functioning.[104]

The *Apollo 8* launch occurred on December 21, 1968. The S-IB stage carried five 1.5–million-pound-thrust F-1 engines, with the five J-2s in the S-II having 225,000 pounds of thrust each and the one in the S-IVB being a 230,000–pound-thrust engine. All three stages performed without problems. Telemetry revealed no pogo problems and no J-2 failures. After achieving parking orbit, the astronauts checked out the systems, clearing the way for reignition of the S-IVB's J-2 during the second orbit and injection into a lunar trajectory at almost 24,000 mph. Once the spacecraft had separated, the S-IVB used venting of propellants plus auxiliary propulsion motors to place itself in solar orbit. On Christmas Eve the three astronauts entered lunar orbit. They completed ten circlings of the Moon, followed by a burn on Christmas Day to return to Earth, splashing into the Pacific on December 27. For the first time, humans had escaped the confines of Earth and returned from orbiting the Moon. With minor exceptions, the spacecraft operated as designed and the mission achieved all of its primary

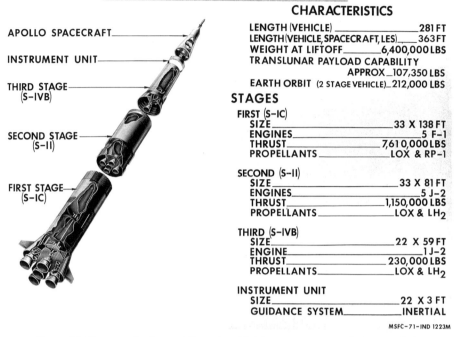

SATURN V LAUNCH VEHICLE

CHARACTERISTICS

APOLLO SPACECRAFT

INSTRUMENT UNIT

THIRD STAGE
(S–IVB)

SECOND STAGE
(S–II)

FIRST STAGE
(S–IC)

LENGTH (VEHICLE) _____ 281 FT
LENGTH (VEHICLE, SPACECRAFT, LES) ____ 363 FT
WEIGHT AT LIFTOFF _____ 6,400,000 LBS
TRANSLUNAR PAYLOAD CAPABILITY
                          APPROX _ 107,350 LBS
EARTH ORBIT (2 STAGE VEHICLE) _ 212,000 LBS

STAGES

FIRST (S–IC)
  SIZE _____ 33 X 138 FT
  ENGINES _____ 5 F–1
  THRUST _____ 7,610,000 LBS
  PROPELLANTS _____ LOX & RP–1

SECOND (S–II)
  SIZE _____ 33 X 81 FT
  ENGINES _____ 5 J–2
  THRUST _____ 1,150,000 LBS
  PROPELLANTS _____ LOX & LH$_2$

THIRD (S–IVB)
  SIZE _____ 22 X 59 FT
  ENGINE _____ 1 J–2
  THRUST _____ 230,000 LBS
  PROPELLANTS _____ LOX & LH$_2$

INSTRUMENT UNIT
  SIZE _____ 22 X 3 FT
  GUIDANCE SYSTEM _____ INERTIAL

MSFC–71–IND 1223M

Figure 42. Diagram of a Saturn V launch vehicle's components and characteristics. Courtesy of NASA.

objectives. The success extended to verification of all the modifications to the launch vehicle since AS-502, with all launch vehicle objectives for the mission achieved.[105]

AS-504 for *Apollo 9* was the first Saturn V to use five 1.522–pound-thrust engines in stage 1 and six 230,000–pound-thrust J-2 engines in the upper stages. It was also the first Apollo-Saturn vehicle and spacecraft combination to be complete, with a virtually final design of the lunar module aboard. It was a test in Earth orbit of the performance of the crew and space vehicles in essential simulation of the activities they would perform in a lunar landing. After the insertion of the S-IVB into Earth orbit following the successful functioning of the first two stages and the instrument unit, on March 3, 1969, the astronauts separated the command and service module from the third stage and docked with the lunar module. On the third day of the mission, two of the astronauts entered the lunar module, fired its descent propulsion system, and then returned to the command and service module. On day 5, they went back inside the lunar module and separated from the command and service module. They jettisoned the descent stage and, using the ascent stage propulsion system, docked with the command and service module after an almost six-and-a-half-hour separation. Staying in orbit four more days, the crew reentered Earth's atmosphere during their 152nd orbit and landed in the Atlantic on March 13, 1969.[106]

Once the spacecraft had separated from it, the S-IVB restarted, burned for 62 seconds, and inserted itself into an elliptical coasting orbit for cooldown of the engines before a second restart. This placed it in an escape trajectory into a solar orbit. During the third burn, a planned dump of liquid hydrogen failed because the helium pressure was no longer sufficient for pneumatic control over the engine valves. There was also rough combustion combined with control oscillations. The launch vehicle thus achieved only nine of eleven primary objectives of the mission, although the other two were partially achieved and the overall mission was considered successful.[107]

The Saturn V for *Apollo 10* (AS-505) and all subsequent Apollo missions through *Apollo 17*, the final lunar landing, used F-1 and J-2 engines with the same thrust ratings as AS-504. *Apollo 10* was in essence a repeat of *Apollo 9* except that the spacecraft maneuvers took place in a cislunar and lunar environment. Launched on May 18, 1969, the Saturn V performed without a hitch, achieving all of its objectives. The mission demonstrated that the Apollo program was ready for a lunar landing, which occurred during the *Apollo 11* mission between July 16 and July 24, 1969. NASA had achieved President Kennedy's goal before the decade ended.[108]

Since this is not an operational history or a history of spacecraft, there is no need to follow the remaining Apollo missions in detail, a task admirably fulfilled elsewhere. Through *Apollo 17* there were six landings on the Moon by twelve astronauts. There were comparatively minor adjustments in the launch vehicles—"in timing, sequences, propellant flow rates, mission parameters, trajectories," to use Roger Bilstein's succinct summation. On all missions there were malfunctions and anomalies that required fine tuning. One of them occurred on the ill-fated *Apollo 13* mission, although it did not contribute to the explosion in an oxygen tank in the service module that prevented the crew from landing on the Moon and nearly cost them their lives before they heroically returned to Earth after six days in space. During the stage-2 burn, pogo-type oscillations in the S-II center engine caused it to shut down 2 minutes and 12 seconds early. But the instrument unit compensated so that by S-IVB cutoff, some 44 seconds late, the stage and its payload were very close to their intended speed and altitude, and the parking orbit was practically the same as planned. It was only long after a successful injection into a lunar trajectory that the oxygen tank exploded.[109]

Besides redesign of the oxygen tank in the service module before *Apollo 14*, NASA and its contractors had to modify the J-2. Evaluation of the *Apollo 13* flight showed that oscillations in the S-II's feed system for liquid oxygen had resulted in a drop in pressure in the center engine's plumbing severe enough to cause cavitation in the liquid-oxygen pump. The bubbles that formed in the liquid oxygen reduced pump efficiency, hence thrust from the engine, and led to automatic engine shutdown. Although the oscillations remained local, and even engine shutdown did not hamper the mission, engineers at the Space Division of North American Rockwell (as the firm had become following a merger with Rockwell Standard) nevertheless developed two modifications to correct the problem.

Figure 43. The Saturn V used on Apollo 16 to launch astronauts John Young, Thomas Mattingly, and Charles Duke on their voyage to the Moon. Shown at Kennedy Space Center, Florida, against the local land- and waterscape, the launch vehicle heads skyward following liftoff on April 16, 1972. NASA Historical Reference Collection, Washington, D.C. Courtesy of NASA.

One involved an accumulator. Like the fix for pogo in the F-1, it served as a shock absorber. It was a "compartment or cavity located in the liquid oxygen line feeding the center engine." Filled with gaseous helium, it served to dampen or cushion the pressures in the liquid oxygen line. This changed the frequency of any oscillation in the line so that it differed from that of the engines as a whole and the thrust structure. The change prevented frequency coupling, which had caused the problem in *Apollo 13*. As a backup to the accumulator, engineers installed a "G switch" on the center engine's mounting beam. This consisted of three acceleration switches that tripped in the presence of excessive low-frequency vibration and shut off the center engine. In a third and apparently unrelated modification, an example of the continued fine-tuning of Saturn V, the propellant utilization valve that controlled engine thrust changed from a motor-driven model to one that was pneumatically actuated. This bypassed electronic circuitry in the interest of simplification and greater reliability, with actuation coming directly from the instrument unit of the Saturn V. Following these modifications, the J-2 and Saturn V were remarkably successful on *Apollo 14* through *Apollo 17*.[110]

Summary and Conclusions

With Saturns I and IB as interim steps, Saturn V was the culmination of the rocket development work von Braun's engineers had been carrying on since the early 1930s in Germany. In this period, the specific engineers working under the charismatic German-American had evolved extensively, although a core of German-Americans remained central to the effort, aided by a great many Americans. There had been a continual modification of technologies in structures, engines, and guidance/control from the V-2 through the Redstone, Jupiter, and Pershing missiles to the three Saturn launch vehicles.

Not all of the technologies used on Saturn V came from von Braun's engineers, of course. Far from it. Abe Silverstein was responsible for the use of liquid hydrogen in the upper stages, something that von Braun had resisted even though his mentor, Hermann Oberth, had recommended it. But once the decision was made, von Braun supported it, and his engineers were partly responsible for its success. The use of five engines on the first stage of Saturn V resulted from Milton Rosen's advocacy. Many other technologies in Saturn V derived from those developed on other programs in which von Braun's team had not participated or for which they were only partly responsible. This is notably true of much liquid-hydrogen technology, which stemmed from contributions by Lewis Research Center, Convair, Douglas, Pratt & Whitney, Rocketdyne, even Storms's division of North American.

But Marshall engineers worked closely with the contractors for the S-II, and S-IVB in overcoming difficulties and made real contributions of their own.

This was also true in the development of the H-1, F-1, and J-2 engines. Rocketdyne had started its illustrious career in engine development by examining a V-2 and had worked with von Braun and his engineers on the Redstone engine, a process that continued through Jupiter and the Saturn engines. But a great many of the innovations that led to the F-1 and J-2 had come more or less independently from Rocketdyne engineers, and even on the major combustion-instability and injector problems for the F-1, Rocketdyne's contributions seem to have been at least as great as those from Marshall engineers. Nevertheless, as Phillips concluded toward the end of his life with regard to Saturn V, "I'm not sure today that we could have built it without the ingenuity of Wernher and his team."

The teamwork, as Phillips apparently meant by his use of "we," was not restricted to Americans and Germans at Marshall but included other NASA centers, industry, universities, and the American military in the overall effort to create the Apollo launch vehicles. Air Force facilities and engineers at Edwards Air Force Base, Holloman Air Force Base, and the Arnold Engineering Development Center made key contributions to various facets of Saturn development. The Air Force's Phillips and NASA's George Mueller (by way of industry and Ohio State) also had important roles. Without their management innovations and oversight, based on both Air Force and Navy developments, it is unlikely that Kennedy's dream would have come true, at least "before the decade [was] out."

So it was a huge, cooperative venture—with many more contributors than can be enumerated here—that led to the unprecedented success of the Saturn launch vehicles. Including military contributions before NASA's creation, one run-out of the costs of Saturn R&D funding put the total at $9.323 billion dollars including post-Apollo (presumably Skylab and Apollo-Soyuz) expenditures—an enormous total in mostly 1960s dollars and their buying power compared to those in the 21st century. In terms of 2006 dollars, the expenditure amounted to roughly $58.6 billion.[111] But for the cost controls and configuration management of Mueller and Phillips, the expenses would undoubtedly have been higher, had Congress and the administrations of Presidents Kennedy and Johnson allowed it. On the other hand, if it had not been for Kennedy's cold-war concerns to demonstrate American prowess to the world, such huge sums would not have been available in the first place.

To return to the contributions of Marshall Space Flight Center, one key element was the essential conservatism of the von Braun team's approach to rocket engineering. At odds with Mueller's all-up testing and configuration

Figure 44. Mission officials relaxing after the successful Apollo 11 Saturn V launch. Shown in the Launch Control Center, the four in the middle foreground of the photograph are, from left to right, Charles W. Mathews, deputy associate administrator for manned space flight; Dr. Wernher von Braun, director of Marshall Space Flight Center; Dr. George E. Mueller, associate administrator for manned space flight; and Lt. Gen. Samuel C. Phillips, director of the Apollo program. Courtesy of NASA.

control, by itself this approach might have led to failure rather than success. But time and again, it provided extra margins of structural strength that could be sacrificed when necessary to accommodate weight growth in the Apollo spacecraft or other unforeseen requirements.

For the most part, the Saturn launch vehicles relied upon established technology. But the very size of Saturn V in particular required unprecedented scaling up that inevitably necessitated new procedures and materials, even new technologies. These and the nature of rocket technology in turn led to a great deal of trial-and-error engineering. Despite this and the many unexpected difficulties that caused such empirical measures, Saturn development, though far from trouble-free, was rapid and highly successful. Whatever we may think today about its costs, it was a triumph of human technical skill, management, and imagination.

6

Titan Space-Launch Vehicles, 1961–1990

While NASA was just getting started with the massive development effort for the Saturn launch vehicles, the Air Force began work on what became the Titan family of launch vehicles, beginning with the Titan IIIs and ending with Titan IVBs. Essentially, most of these vehicles consisted of upgraded Titan II cores with a series of upper stages. Most of the vehicles also used a pair of huge, segmented strap-on solid-propellant motors to supplement the thrust of the Titan II core vehicle. And after September 1988 a limited number of actual Titan IIs, refurbished and equipped with technology and hardware from the Titan III program, joined the other members of the Titan family of launch vehicles. Beginning in June 1989, Titan IV with a stretched core and with seven (instead of Titan III's five or five and a half) segments in its solid rocket motors became the newest member of the Titan family.[1]

Since the initial Titan launch vehicle in this family designed specifically to lift payloads into space, the Titan III, was competitive with the early Saturns in orbital performance, the Department of Defense had to justify a separate development effort for the heavy booster. It did so partly in terms of cost. DoD anticipated a large number of military missions for such a booster during the 1960s, and the expected cost per launch for Titan III was considerably lower than for Saturn vehicles. Other points of justification included the more rapid launch capability afforded by the Titans' lack of dependence on cryogenic propellants (a factor for some proposed missions), greater flexibility resulting from the building-block concept on which the various Titan IIIs were based, fewer logistics and training problems for the Air Force because of previous experience with the Titan II missile, and the fact that part of the cost and development effort would be devoted to large solid-propellant motors, which seemed promising in the wake of Polaris and Minuteman.[2]

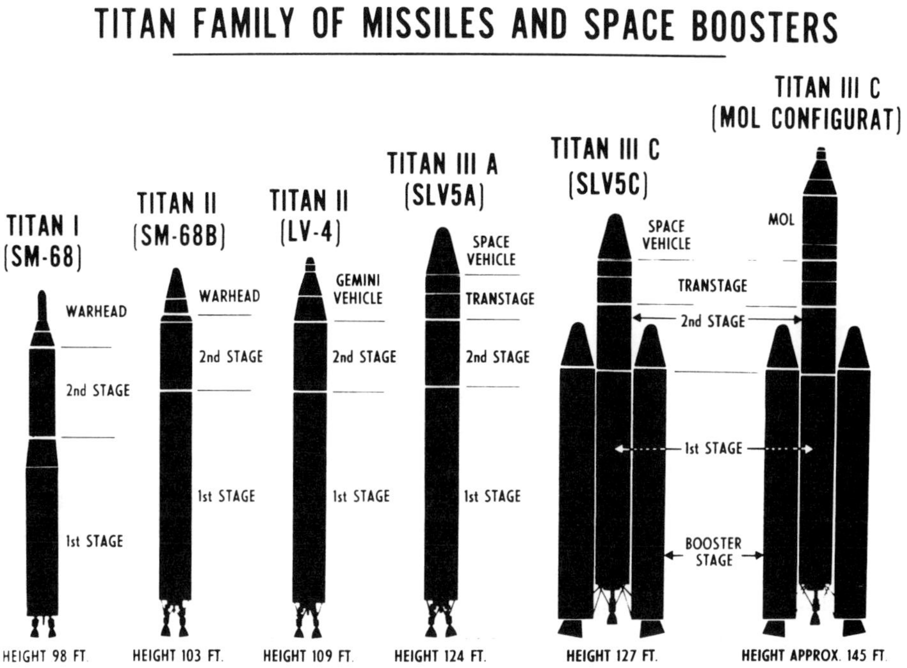

TITAN FAMILY OF MISSILES AND SPACE BOOSTERS

Figure 45. An early depiction of the Titan family of missiles and space-launch vehicles. MOL stood for Manned Orbiting Laboratory, a program that never reached fruition. Official U.S. Air Force photo, courtesy of the 45 Space Wing History Office, Patrick AFB, Fla.

Inception and Early Development of Titan IIIA and C

The development of Titan III was complicated and heavily influenced by changes in management procedures at the DoD level in the new administration of President John F. Kennedy, who assumed office on January 20, 1961. Kennedy had campaigned for the presidency partly on an alleged (but in fact mythical) missile gap between the United States and its cold-war adversary, the Soviet Union. As president he was concerned to ensure that the country at least matched Soviet accomplishments in the space arena. His secretary of defense, Robert S. McNamara, in turn introduced many new management procedures at DoD that greatly affected Titan III development. Some of these stemmed from changes that General Schriever had begun to implement in the Air Force as a result of his experience in missile development.[3]

Essentially, Schriever's reforms entailed the use of systems management, with new proposals for missiles and rockets as well as other types of technology having to be submitted in program packages that included such issues as cost, logistics, management, schedules, operational details, training, and security. With McNamara's approval, the Air Force adopted these reforms on March 14, 1961, with the procurement activities of Air Materiel Command (AMC) shifted to an expanded Air Research and Development Command called Air Force Systems Command. A truncated AMC in effect became Air Force Logistics Command.[4]

All of this changed Air Force methods of missile and rocket development and procurement. But McNamara's more basic shift in management procedures, called planning, programming, and budgeting (PPB), came from another source. Charles Hitch and Roland McKean had written a report for the Rand Corporation (a sort of think tank) entitled *The Economics of Defense in the Nuclear Age*, which McNamara read with interest. Instead of separate budgets for each service, which the individual services divided up with limited controls from above, Hitch and McKean proposed a system that would look at the overall missions and decide, on the basis of projected costs and benefits, what systems made the most sense from the standpoint of the Department of Defense as a whole. This appealed to McNamara because it was similar to management accounting that he had learned about while teaching at Harvard before World War II and had helped to apply in reforming the Ford Motor Company. Hitch made a presentation to McNamara in the early spring of 1961, proposing to convert budgeting for all strategic nuclear forces in a year's time to program budgeting. McNamara bought the idea enthusiastically but told Hitch, whom McNamara appointed as DoD comptroller, to "do it for the entire defense program," not just for strategic nuclear forces, and "in less than a year." Thus began PPB.[5]

These developments shaped the Air Force's proposal for a new combination of elements to provide a space-launch vehicle with enough thrust to carry a wide range of military payloads. On May 15, 1961, Director of Defense Research and Engineering John H. Rubel provided some guidelines for such a launch vehicle. Although PPB was still in its infancy, Rubel indicated the desirability of specifying the design of a new vehicle before system development began. He also suggested that the vehicle should be able to carry a 10,000–pound spacecraft into an Earth orbit at 300 nautical miles above the planet or a 1,500–pound payload to escape velocity.[6]

One possibility for satisfying these levels of performance was use of Titan II with solid motors "strapped" to it. This concept made solid-motor development the key new technology for what became the Titan family of launch vehicles. To decide how to proceed, DoD and NASA agreed through the Aeronautics and Astronautics Coordinating Board to create a Large Launch

Vehicle Planning Group in July 1961. Headed by Dr. Nicholas E. Golovin, who had worked for NASA before joining private industry, and including Air Force and DoD participants, this group was charged with determining the best combination of launch vehicles for a lunar landing, human scientific missions, and advanced military requirements. By September 1961, DoD had agreed to the concept of combining a suitably modified Titan II with strap-on solid motors to satisfy military requirements, and the following month the Large Launch Vehicle Planning Group recommended approval of Titan III, as the vehicle had come to be designated. It would feature solid motors 120 inches in diameter and serve DoD *and* NASA needs "in the pay-load range of 5,000 to 30,000 pounds, low-Earth orbit equivalent."[7]

In conjunction with these developments, the Air Force's Space Systems Division within the new Systems Command had prepared ("in eleven days including two week ends") a report entitled "Titan III, Standardized Space-Launch Vehicle" and sent it to Washington on October 5, 1961. This called for a full-scale test flight of Titan III, including its solid motors, by January 1964—an "extremely compressed" schedule of development. The plan arrived in Washington as the PPB system was still being developed. A key element in the new system was Phase I, later called program definition. Rubel, who was a key player in developing PPB, defined this as "establishing, with considerably greater confidence [than before], the feasibility of accomplishing what is claimed, and establishing organizational and procedural mechanisms for better insuring that we achieve the desired results in accordance with plan." This would enable an accurate forecast of costs, while adaptation of the Navy's Program Evaluation and Review Technique (PERT) from the Polaris program would ensure, again in Rubel's words, "that time-phased compatibility exists with all major program elements." Already on October 13, 1961, Rubel instructed the Air Force to move rapidly so that a Phase I effort "may lead to the development of a family of launch vehicles based on Titan III."[8]

Although Space Systems Division was later to complain about "daily redirection" of the Titan III program from the office of the director of defense research and engineering, initially the development of the launch vehicle got off to a quick start. On November 27, 1961, Col. Joseph S. Bleymaier became system program director within Space Systems Division, with the system program office coming into existence on December 15. Designated on November 31 as Space Booster Building Block Program 624A, Titan III received another endorsement on December 5 from Golovin's committee saying that it was critical for space vehicle requirements after 1965. On December 11, SSD received its first funds to begin a Phase I contractual effort. Already on October 25 the Air Force had requested that the Martin Company, responsible for Titan II, provide a proposal for a Phase I study of Titan III. Martin—or rather the Martin Marietta Corporation, formed when Martin merged with American Marietta on October 10, 1961—responded on January 2, 1962, and won a cost-plus-fixed-fee contract on February 19, 1962. On August 20 of that year, DoD announced plans to develop Titan III as a launch vehicle under Space Systems Division's management, with Martin Marietta as prime (systems integration) contractor. To begin the developmental phase of the Titan III program, Space Systems Division awarded Martin Marietta a contract on December 1 for the vehicle's airframe plus systems engineering, integration, and testing.

Solid Rocket Motors

A second series of contracts, highly significant in their requirements for development of new technology, was for the large solid-propellant rocket motors to boost the Titan III. Final requests for proposals from contractors on this effort went out from Space Systems Division on January 22, 1962. Responses came from Aerojet General, Thiokol, Lockheed Propulsion Company, Hercules Powder Company, Rocketdyne Division of North American Aviation, Atlantic Research Corporation, and a new firm named United Technology Corporation (UTC). On May 9, 1962, the Air Force selected the new firm to develop the solid rocket motors, leading to a Phase I study worth $2.8 million.[9]

How did a recently founded corporation win a major contract against six established and experienced rocket firms? The full answer is buried in unavailable source-selection-board documents, if they even still exist, but one factor was undoubtedly the individual experience and connections of leading managers and engineers at UTC. The president of the new firm—which had initially been called United Research Corporation in 1958, and which had first located in Los Angeles and then moved to new facilities at Morgan Hill in Sunnyvale, California, near Palo Alto—was Donald L. Putt. He had retired from the Air Force as a lieutenant general, after serving as deputy chief of staff for development. Having helped to assemble the team that developed Minuteman, he also had ties to the Jet Propulsion Laboratory through Caltech, where he had earned a master's degree in aeronautical engineering under Theodore von Kármán. Another executive at UTC was Barnet R. Adelman. With even stronger ties to JPL and Minuteman, Adelman had played a major role in founding the new firm and was its vice president, general manager, and director of operations. In 1962 he succeeded Putt as president of United Technology Corporation (later renamed United Technology Center, keeping the acronym UTC).

Putt and Adelman were successful in recruiting a large number of key engineers in the field of solid-propellant rocketry from throughout industry and government. One of them was David Altman, who had worked significantly in propellants and combustion at JPL, then transferred to the Ford Motor Company's Aeroneutronic Systems before becoming director of UTC's research division. Altman recalled that in the mid-1950s he and Adelman had already discussed segmented motors, which became the basis of Titan III's large solid-rocket motors. Other major figures on UTC's team of engineers had worked for Aerojet, Atlantic Research, Thiokol, and the Naval Ordnance Test Station at Inyokern, California, plus Ramo-Wooldridge's Space Technology Laboratories. In March of 1960 the slender, mustachioed Adelman said that UTC had recruited ten of the forty or so "really top-grade men" in solid-propellant technology in the United States.

Not long after the founding of UTC, United Aircraft Corporation purchased a one-third interest in the rocket firm, later becoming its sole owner. When United Aircraft changed its name to United Technologies Corporation in 1975, its solid-propellant division became Chemical Systems Division. (To avoid confusion between United Technology Corporation and the parent firm, United Technolog*ies* Corporation, for the remainder of this chapter the rocket division will be called UTC or CSD, depending on the date, and the parent corporation's name will be spelled out.)[10]

Although Adelman was an early proponent of segmented motors, Aerojet seems to have gotten a head start in testing them. In 1957, before UTC was even founded, Aerojet perceived that the Air Force might soon need huge solid-propellant motors and decided to try slicing a Regulus II—a missile with a 20–inch diameter—into three portions and joining them back together with bolted flange joints. Aerojet successfully fired the segmented motor using the original Regulus propellant and nozzle. The result was performance identical to that of the Regulus's unsegmented motor, proving that segmentation did not hamper performance.

The reason for this approach was the realization that large solid motors would soon become too large for transport on anything but barges. Probably encouraged by Aerojet's initial success, about April 1959 the Power Plant Laboratory of the Air Force's Wright Air Development Center in Dayton sent out requests for proposals for a contract later worth $495,000 to test segmented motors. Not surprisingly, Aerojet won the contract, which it carried out after the Power Plant Laboratory moved to Edwards AFB in California. Aerojet tested 100–inch-diameter motors in the ultimately successful attempt to achieve 20 million pound-seconds of total impulse, although the initial goal was a somewhat more modest 230,000 pounds for 80 seconds. Aerojet first fired a Minuteman first stage (65 inches in diameter), cut in half and reconnected with a lock-ring (or lock-strip) joint configured similarly to subsequent joints used in these tests. It featured inner and outer lock strips held together by a lock-strip key in the center where they joined, and it included an O-ring, which linked the two strips to preclude leaking of expanding gases. This motor provided 160,000 pounds force (lbf) for 60 seconds on May 5, 1961.[11]

Aerojet next tested a variety of 100–inch, segmented motors, beginning with one designated TW-1 on June 3, 1961, and concluding with an FW-4 motor on October 13, 1962. Besides segmentation, the motors tested nozzle materials and thrust vector control systems using fluids that reacted with the rocket's exhaust gases to change the direction of the thrust. With five center segments, a length of 77 feet, and a weight of 876,000 pounds, the intermediate FW-3 tested on June 9, 1962, was the largest solid rocket motor yet built. There were various problems with the Aerojet tests, including spalling (breaking off of fragments) and even ejection of the nozzle throat. But though not all aspects of the tests of these 100–inch-diameter motors were successful, they demonstrated that a large segmented design was feasible, permitting transportation of such huge rocket motors. The lock-strip joint also passed its tests. However, failures of the nozzle throat inserts and thrust vector control systems showed the need for further development in those areas. In other tests of segmented motors, Lockheed and UTC had also encountered problems with nozzles.[12]

Despite Aerojet's priority in actually testing segmented motors, Adelman had obtained a patent for a segmented-joint design to be used in such large solid motors. Assisted by Tom Polter from the nearby Stanford Research Institute in Palo Alto, UTC began development of a specific propellant for large, segmented boosters, starting with the proven PBAN used in stage 1 of Minuteman, with which Adelman would have been familiar from his work on that missile. Using its own funds, UTC successfully tested a P-1 solid-propellant motor on December 15, 1960. The P-1, with a diameter of 87 inches, yielded more than 200,000 lbf for about 75 seconds. The P-1–2, which followed on February 9, 1961, had two middle segments plus a forward and aft closure. With a case composed of low-carbon steel plate and a fixed nozzle using nitrogen tetroxide to provide thrust vector control, it yielded 79 seconds of thrust at 399,000 lbf, providing a preview of the basic type of thrust vector control used on Titan IIIC, which also used nitrogen tetroxide.

Initially, UTC designed the cases with a taper of 1.22 degrees. The firm believed that this tapering prevented erosive burning of the grain in the long internal cavities in the middle of the propellant by providing a larger area as the rapidly expanding gases passed toward the expanded rear of the motor. Later, however, UTC's engineers decided to eliminate taper in its cases. Erosive burning could apparently be prevented by tapering the internal burning cavity alone.[13]

All of the testing done by UTC to this point had preceded its contract with the Air Force for Titan III solid rocket motors awarded in May 1962. But much testing remained to be done, with many attendant problems to be solved. Developmental testing of 120–inch-diameter motors—the size that would actually be used on Titan IIIC—began on February 23, 1963, at UTC's Coyote Canyon test site in the Diablo Mountain Range, 25 miles southeast of Sunnyvale. In the first of four "subscale" tests of 120–inch motors, the motor in question contained only one center segment and achieved a peak thrust of a million pounds. It was about the same size as the liquid-propellant Atlas booster that sent the Mercury astronauts into orbit, yet it produced about two and a half times as much thrust.[14]

This advantage over liquid-propellant engines resulted from several factors. Although liquid propellants typically had higher specific impulses, solids had densities 50 to 70 percent greater than most liquids. They also required no pumping or other complicated plumbing, which added to the weight of liquid engines and thus reduced their thrust-to-weight ratios. Finally, because the sizes and shapes of internal burning cavities in solid-propellant motors could be adjusted to provide huge increments of thrust almost instantly, solid-rocket stages were much better able to lift launch vehicles quickly through the area of high gravity near Earth, with smaller losses in velocity due to gravity, than were liquid stages.[15]

Three more tests with subscale 120–inch rockets followed the one on February 23. They included a sled test at the Naval Ordnance Test Station, China Lake, near Inyokern, California, in June 1964 to see if there was significant hazard in using the 120–inch segments. A one-segment motor was ignited and sent down the sled track, where it smashed into a concrete abutment at 435 mph. Just one of a series of hazard classification tests, it showed that the propellant would not detonate on impact. Other subscale tests included subjecting four motors to altitude testing at the Arnold Engineering Development Center in Tennessee to obtain data on liquid thrust vector control.[16]

In the midst of these subscale tests, already on July 20, 1963, UTC had fired the first full-scale, five-segment motor at Coyote Canyon. According to

Titan historian Robert Piper, this "was, quite properly, considered to be one of the most crucial tests in the development of the Titan III system." Bley-maier, at this time a brigadier general, wrote to Putt that the test was "truly an outstanding and significant event in the life of the Titan III Program." In but one of the 273 test factors measured in this firing, the motor yielded 1.2 million pounds of thrust over a motor-burning time of roughly 112 seconds. The test proved that the thrust vector control system using nitrogen tetroxide worked effectively, yielding an adequate 5 degrees of vector deflection. This system used twenty-four injectant valves arranged in four groups of six. Electrical commands from the guidance and control system caused the valves to spray the fluid into the 5,000–degree exhaust gases, creating a shock wave that deflected the hot gases and thereby would have changed the Titan IIIC's course in a flight test. This was not, however, an operationally configured motor, so testing of such motors would be needed to evaluate their performance in the more difficult environment they would experience in actual flight.[17]

The PBAN propellant UTC developed for the solid rocket motors of Titan IIIC—called stage 0 to enable the lower portion of the liquid-propellant Titan II to remain stage 1—differed from other PBAN propellants in having greater toughness lent by the incorporation of methyl nadic anhydride to the basic polybutadiene-acrylic acid-acrylonitrile-ammonium perchlorate-aluminum propellant. The motors had a mass fraction of 85.3 percent and a configuration in which an eight-point star in the forward closure transitioned, through the five segments and aft closure, into a nearly cylindrical opening. This opening did have a slight taper in the aft direction to help prevent the expanding gases from causing erosive burning of the propellant. The design yielded a propellant specific impulse of slightly more than 265 lbf-sec/lbm.[18]

As the earlier testing of large segmented motors by Aerojet had suggested, it would be difficult to ensure that uncooled nozzles could withstand 5,000–degree exhaust gases flowing supersonically through them for extended periods up to 120 seconds. The nonrotating nozzle was canted 6 degrees from the axis of the motor to direct the undeflected thrust toward the Titan III's center of gravity. Inside it, UTC engineers had emplaced a throat section consisting of three bulk graphite rings. This apparently worked for the subscale motors, but when tried in full-scale tests, the bulk graphite cracked from the thermal stresses. When a nozzle failed during a static test in August 1964, engineers switched to a graphite cloth bonded with a phenolic resin. Although the new design necessitated new tooling and parts, purchase of new materials, and development of new procedures, UTC made the change in six weeks and successfully tested it in September 1964. A major gamble for UTC, the tape-wrapped carbon-phenolic throat marked a significant advance in large solid-rocket technology, working flaw-lessly on the Titans and paving the road for the giant solid-rocket boosters on the space shuttle, which used the same basic technology for its nozzle throats.[19]

Another huge challenge for UTC engineers was to link the segments of the solid rocket motors so that they could withstand 850 psi of pressure as they launched and flew through the atmosphere. The lock-strip joint Aerojet had developed for its segmented motors was a possible solution, but in co-operation with sister Pratt & Whitney Aircraft Division of United Aircraft, UTC engineers developed a different type called a clevis joint, which they tested as early as 1960. It featured "male" and "female" elements that encir-cled the ends of each cylindrical segment of the motor, with the "male" clevis portion facing upward (when the vehicle was on the launchpad) and fitting into the downward-facing, slotted "female" portion in a tongue-and-groove arrangement. Some 237 hand-placed pins and three fixed pins fit through both portions and held them together. An O-ring in a slot around one inside surface of the groove circled the "male" part and formed a seal between them to keep hot gases from escaping. In addition, insulation between the propel-lant and the case protected the O-rings from heating, with putty carefully applied to ensure reliability. Heating of the O-ring could occur only if there was a debonding of the propellant and insulation from the case, allowing a high-pressure gas blow-by because of improper sealing of the O-ring. This could occur at low external temperatures during launch, so UTC engineers used heating strips to prevent this and improve reliability.[20]

Testing of these and other design features of the solid rocket motors involved nine full-scale, 120–inch developmental tests plus five full-scale preflight readiness tests. These tests demonstrated the interchangeability of segments using the clevis joint. Also tested was the motor case, which itself presented some problems. The cases initially consisted of a rolled and welded low-alloy, high-strength steel known as Ladish D6AC, and UTC had chosen two subcontractors, Curtis-Wright and Westinghouse, to provide them. Experience with earlier missiles, including Minuteman, had produced a large database on welding and metal processing. But the Titan 120–inch cases posed new problems because both the thickness of the metal and the size of the segments were greater, and both contractors experienced signifi-cant problems in their welding operations. Nevertheless, the motor cases remained on schedule. Subsequently, with the assistance of both UTC and Westinghouse, the Ladish Company eliminated the need for welding by developing a new process called rolled-ring forging, also known as roll extrusion case forging, which offered additional savings in weight as well as improved reliability for the motor cases.[21]

All of these developments reached fruition on June 18, 1965, when two solid rocket motors, both 84.65 feet long, provided a peak thrust of 2,647,000 pounds before being jettisoned about two minutes after liftoff of the first Titan IIIC, turning over lifting duties to the liquid-propellant center core section of the first Air Force launch vehicle specifically designed to become a large space booster.[22] The launch occurred less than two months after the fifth preflight readiness test and only a little over three years after UTC had won the Phase I contract for the solid rocket boosters from the Air Force.[23]

Stages 1 and 2

Meanwhile, between February 7, 1962, and March 20, 1963, a series of Phase I and Phase II contracts with Aerojet provided for further development of the Titan II and Titan II–Gemini stage-1 and stage-2 engines for use on Titan III. Essentially, however, engineers had already solved most of the problems with the engines during the two earlier programs. Performance of the Titan III stage-1 and stage-2 engines was the same as for the Titan II and Gemini stages, although the designations changed. All three engines bore the designations LR87 for stage 1 and LR91 for stage 2, but the Titan II engines were called LR87–5 and LR91–5 respectively, and the Titan II–Gemini engines were -7s, while the initial Titan III engines carried the -9 suffix. Since there was a Titan IIIA that did not include the solid rocket motors, some of the Titan III first-stage engines would fire at ground level, whereas those used on Titan IIIC would start at altitude after the solid rocket motors lifted the vehicle to about 100,000 feet.

Altitude ignition for the Titan IIIC stage-1 engines required a redesigned solid-propellant start cartridge to get the turbines in the turbopump spinning so they could feed the propellants to the combustion chambers. But since Aerojet had already designed the stage-2 engine to start at altitude, this redesign for stage 1 presented no problems. Additional insulation was also needed around the engine compartment to protect against heat the solid rocket motors would radiate. Finally, both stages 1 and 2 required structural strengthening and other modifications for attaching and bearing the loads associated with the solid rocket motors and a third stage known as Transtage.[24]

Transtage

Already on November 27, 1963, Space Systems Division accepted the first Titan III engine from Aerojet for testing of the Titan IIIA configuration. Seven months later, on June 26, 1964, SSD accepted the first flight-test version of Titan IIIA to evaluate in flight the center core sections of Titan III. Meanwhile, development had gotten started on the Transtage or third stage of the center core. The Air Force decided on a pressure-fed engine that would use the same oxidizer and fuel—nitrogen tetroxide and an equal mixture of hydrazine and UDMH (unsymmetrical dimethyl hydrazine)—as stages 1 and 2. Its two gimballed thrust chambers would each produce 8,000 pounds of thrust and would be capable of up to three starts over a six-hour period. Aerojet won this contract in addition to those for the first two stages, with a Phase I agreement signed in early 1962 and a Phase II (development) award issued on January 14, 1963.[25]

Aerojet designed the Transtage engine, designated AJ10–138, at about the same time as a larger engine for the Apollo service module. The two engines used basically the same design, featuring the same propellants, ablatively cooled thrust chambers, and a radiatively cooled nozzle assembly. Since the Apollo service module's engine bore the designation AJ10–137, its development apparently began earlier, but it also lasted longer. Thus, although Aerojet designed and built them both, and more information is available about the development of the spacecraft engine, it is not clear that any of the latter's early problems and solutions are relevant to the Transtage engine, which was less than half as powerful and roughly only half the length and diameter of its Apollo sibling.[26]

It seems that these two engines were not the only ones with ablatively cooled combustion chambers in this period, because an important NASA publication on liquid-propellant rocket engines issued in 1967 stated that such "thrust chambers have many advantages for upper-stage applications. They are designed to meet accumulated duration requirements varying from a few seconds to many minutes." The ablative cooling, it said, resulted from the pyrolysis, or chemical change by the action of heat, of "resins contained in the chamber wall material." Although construction could vary, in one unspecified example the ablative liner was "fabricated from a phenolic-resin-impregnated high-silica fabric which is wrapped in tape form on a mandrel." Then "a wrap of . . . phenolic-impregnated asbestos is placed on the outer (far) surface of the ablative liner as an insulator." Also, a "high-strength outer shell is composed of layers of unidirectional glass cloth for longitudinal strength and of circumferential-wound glass filaments for hoop strength. The glass wrap is bonded with epoxy resin."[27]

This appears to have been the precise design of the Transtage combustion chamber, the material for which consisted of "fiberglass, asbestos, and resin impregnated silica fibers. Ablative lining." It is not as clear that the Apollo service module's combustion chamber used the same materials, since they consisted of "rubberized, phenolic refrasil inner liner, glass cloth/roving outer wrap," but the ablative principle was at least similar.[28]

Already on July 23, 1963, Aerojet successfully operated a Transtage engine for 4 minutes, 44 seconds, considered "a long duration firing." During that static test, the engine started and stopped three times, demonstrating its restart capability. However, a sterner test of this crucial capability—which would allow the vehicle to place multiple satellites in different orbits on a single launch or to position a single satellite in a final orbit, such as a geostationary one, without a need for a separate kick motor—would occur in the simulated altitude test chamber at Arnold Engineering Development Center. In August 1963, tests there confirmed suspicions from the July 23 test that the combustion chamber would burn through before completing a firing of (undefined) full duration. In addition, gimballing of the engine in a cold environment revealed a malfunction of a bipropellant valve that fed propellants to the combustion chamber, plus a weakness in the nozzle extension, which was made of aluminide-coated columbium and was radiatively cooled with an expansion ratio of 40:1. How Aerojet solved these problems is not readily discoverable, with the official history of Titan III merely stating that "by the close of 1963, an extensive redesign and testing program was underway to eliminate these difficulties so the contractor could make his first delivery of flight engine hardware—due in mid-December 1963."[29]

One Aerojet source, while not commenting on these particular difficulties, does refer to "the error of trying to develop in a production atmosphere" and explains that management and the shops paid scant attention to the small Transtage engine's development at a time when Titan I was starting into production. But presumably the speed required in Transtage's development was also a factor. Obviously, engineers had not expected the difficulties and had to adjust their designs in some fashion to correct them. In any event, engine deliveries began not in mid-December, as initially planned, but in April 1964.

Meanwhile, according to Chandler C. Ross, who had joined Aerojet as an engineer in 1943 and risen into the ranks of management, another "technical problem of some magnitude" had arisen. In Aerojet's own testing of the engine, it could not properly evaluate the nozzle extension with its large 40:1 expansion ratio because the firm could not simulate the high altitude for which the expansion was designed. While the thin atmosphere at high altitudes required the large expansion, it was not appropriate for roughly sea-level testing, where its use would have caused flow separation of the gases produced in the combustion chamber, leading to "all sorts of troubles." So Aerojet tested the engine without an expansion section, letting the gases exhaust from an "opening the same diameter as the nozzle throat." Because "Transtage required exact information on its propulsion performance," Aerojet engineers had extrapolated data from the zero-expansion tests to the altitude that the Arnold Engineering Development Center would simulate. But when actual figures from the simulated altitude tests became available, they were 2.5 percent lower than Aerojet's extrapolations.

This might seem a small discrepancy to the casual reader, but engineers need exact performance information to project orbital injection accurately. The discrepancy had to be investigated. The explanation proved to be simple, but it illustrated that even within the same firm, all data did not always get communicated. Thus, engineers did not have their procedures "down to a science" but sometimes operated with an incomplete understanding of the phenomena they were testing. Here Ross recalled "working on a solid rocket problem a few years previously, where the chemists were using very small rockets to test research fuels which had nozzles with no expansion section." When asked whether they had to apply a correction to their data, the chemists responded that they always decreased the calculated performance by 2.5 percent because flow through the nozzle was restricted. Once aware of the problem, Aerojet engineers found several references to this correction in the literature, explaining the discrepancy. But the engineers who had done the extrapolation had obviously failed to find those references in their own research. Space Systems Division did insist that Aerojet now raise performance by the amount of the discrepancy, but Ross does not say how the firm did so.[30]

Another Aerojet engineer and manager, Ray Stiff, recalled that after engine deliveries began in April 1964, the Air Force began to impose new requirements. Because Transtage needed to be able to perform a 6.5–hour coast while in orbit and then be capable of "a variety of firing, coast, and refire combinations," there were "unique insulation requirements" to protect the propellants from freezing in the extreme cold of space, especially when shaded from the Sun. But this insulation retained heat from combustion, which transferred from the ablative chamber to the injector with presumed dire consequences for continued performance. Stiff does not mention how the Aerojet engineers solved this problem, stating only that the engine's injector was "baffled for assurance of stable combustion."[31]

Other sources reveal that the injector used an "all-aluminum flat faced design" with a "concave spherical face" and "multiple-orifice impinging patterns." The baffle was fuel cooled, so perhaps an adjustment in this feature solved the heating problem. According to an Aerojet history written by former employees and managers, "The injector design has undergone two performance upgrade programs which resulted in the very high specific impulse value of 320 lbf sec/lbm, and the design has been carried over into later versions of the Delta," although the history does not specify what versions those were.[32]

In any event, the two initial Transtage engines each yielded 8,000 pounds (lbf) of thrust with a specific impulse of more than 300 lbf-sec/lbm. Pressurized by cold helium gas, both of the hypergolic propellants were stored in tanks of a titanium alloy that the prime contractor, Martin, machined in its Baltimore Division. The titanium forgings Martin machined came from the Ladish Company of Cudahy, Wisconsin. Although titanium was difficult to machine, it was gaining increasing use for liquid-propellant tanks. With a fuel tank about 4 feet in diameter by 13.5 feet long and an oxidizer tank measuring about 5 by 11 feet, Transtage's propellant containers were hardly huge but were reportedly some of the largest yet produced from titanium. Each overall engine was 6.8 feet long with its diameter ranging from 25.2 to 48.2 inches. Its rated burning time was a robust 500 seconds and its total weight was a mere 238 pounds.[33]

Guidance and Control

The guidance and control system for versions of the Titan III that used a Transtage was located in a control module located on the forward end of the Transtage but separable from it in what was termed a "Siamese twin" relationship. That is, the two were conjoined but could be separated if the mission required it. The proper guidance-and-control system for Titan III became a vexed management issue. From October 1961 to February 1962 the Aerospace Corporation had thoroughly studied the requirements of the vehicle with respect to accuracy and reliability, concluding that there was no system in existence capable of satisfying Titan III's needs. By July 1962, negotiations were in progress with Space Technology Laboratories and the Arma Corporation for a joint venture to develop a new guidance system with high reliability. Then Rubel's Office of Defense Research and Engineering decided that the Titan II all-inertial guidance and control system would be adequate for most of Titan III's needs, forcing Space Systems Division to suspend negotiations with the joint-venture contractors and hold all procurement actions until they were approved by DoD. To the great disappointment of the Titan III program office, it became clear by August 10 that AC Spark Plug's

Titan II guidance system would have to suffice for initial Titan IIIs with some modifications. On that date Rubel recommended to McNamara that he approve full-scale development of Titan III with the modified Titan II system, which the secretary of defense authorized on August 16.[34]

The Air Force then issued a contract for AC Spark Plug to do Phase I development of the modified system between October 3, 1962, and February 4, 1963. There was a separate contract with MIT to determine what modifications were necessary. A follow-on contract authorized AC Spark Plug to do the Phase II development under a cost-plus-incentive-fee contract, which was the preferred instrument under the PPB system, used for all major Titan III components. The resulting guidance-and-control system was similar to the one for Titan II, employing an inertial platform with a four-gimbal arrangement and three gyroscopes plus some axial rate gyros. The principal change was that the data from the system would be fed to a new IBM computer with greater capacity so it could deal with the needs of orbital missions instead of just ballistic guidance and control. The computer would compare changes in the launch vehicle's attitude and accelerations, as signaled by the inertial guidance and rate gyro systems, with the programmed flight path. It would then send corrective signals as needed to the thrust vectoring system in stage 0 and the gimbal systems for the three liquid-propellant stages.

Once the third stage and control module with payload(s) had reached orbit after shedding stages 0, 1, and 2, the guidance and control system would direct up to three burns of the Transtage. It would also send corrective signals to the auxiliary propulsion system on the control module itself for control adjustments during coast periods between firings of the Transtage engines. This auxiliary system was a Rocketdyne SE-9, consisting of four clusters of small thrusters burning monomethyl hydrazine and nitrogen tetroxide. Before burns of the Transtage engines, axial thrusters from the SE-9 also fired to settle the propellants in the third stage's titanium tanks to the rear for proper propellant flow in the virtual absence of gravity.[35]

Testing Titan IIIA

The first of five planned tests of Titan IIIA—without solid rocket motors, all from Cape Kennedy—occurred on September 1, 1964. Stages 1 and 2 performed well, but the failure of a helium pressure valve in the Transtage caused propellant tank pressure to fall so low that the engines ceased firing some 15 seconds before the end of a projected 406–second burn. As a result, the stage and its 3,750–pound dummy payload did not achieve low orbit but fell back into the ocean.[36]

Rated a 95 percent success, this launch was followed by a second attempt with another 3,750–pound dummy payload on December 10. This time the Transtage carried a redundant set of helium pressure valves. The insurance paid off with a totally successful launch into a highly accurate low Earth orbit. During the first orbit, the still unmodified Titan II guidance and control system used on this mission commanded an end-over-end rotation in the pitch axis "to unwind the inertial platform gimbals." Accounts differ on whether the Transtage then succeeded in separating from the dummy payload, but both stayed in orbit for three days.[37]

The third Titan IIIA launch, on February 11, 1965, involved three burns of the Transtage engines. The longest burn placed the stage, with a communications satellite called LES-1 (Lincoln Experimental Satellite) and a dummy satellite, in a low circular orbit. Following a 90–minute coast, the second burn raised the apogee to 1,737 miles. After another orbit and a half of coasting, the third burn roughly circularized the orbit, and the Transtage ejected the two payloads. The LES-1 was supposed to use its kick motor to raise the apogee to about 13,000 miles, but the solid-propellant motor failed to fire. This was a payload failure, so the launch itself was a success, having fulfilled the main objective of firing the Transtage engines the programmed three times.[38]

The fourth launch of a Titan IIIA was even more ambitious, with four separate ignitions of the Transtage engines. On May 6, 1965, the Titan IIIA performed successfully, with an LES-2 experimental communications satellite and a hollow aluminum radar calibration sphere called LCS-1 both ejecting as planned. The 75–pound LCS-1 stayed at an apogee of 1,729 miles, while the solid-propellant motor worked this time, carrying the LES-2 to an apogee of 9,364 miles. After ejection of the satellites, the Transtage coasted for three hours to demonstrate the rest of its 6.5–hour designed battery life and then ignited its engines for the fourth time. So successful was this flight that the Air Force cancelled the planned fifth test flight of Titan IIIA and converted the core vehicle to a Titan IIIC configuration.[39]

Titan IIIC Tests

The first test of a Titan IIIC was the June 18, 1965, launch already mentioned in connection with development of the solid rocket motor. Engineers arranged the first-stage ignition to occur during the tail-off of the solid rocket motors' thrust. Ten seconds later, the system jettisoned the huge solids, and the core stages all performed satisfactorily. The Transtage demonstrated coasting flight after entering its orbit, and on the fourth orbit it released its 21,400–pound dummy payload into an orbit with a 118–mile apogee and a

104–mile perigee—described as a low-Earth, circular orbit—reportedly the heaviest payload yet orbited by the United States.[40]

The second Titan IIIC launch was not as successful. Carrying an LCS-2 radar calibration sphere and an OV2–1 radiation sensor, it lifted off successfully on October 15, 1965, and entered an elliptical orbit 615 by 324 miles. During the second-stage burn, however, both the stage-2 oxidizer and the Transtage fuel began leaking. At Transtage separation, the firing of the separation rockets ignited the leaking propellants, causing a minor explosion. Then, at the end of the second burn of the Transtage, the bipropellant valve in one of the two engines, perhaps damaged in the explosion, became stuck in the open position. With one engine continuing to burn while the other did not, the spacecraft began tumbling and exploded, placing in orbit a record amount of trackable debris—nearly five hundred pieces. Although the launch went well until then, the satellite pieces were useless and the mission a failure.[41]

Engineers modified the bipropellant valves, and a launch on December 21 was somewhat more successful. As on the previous mission, the liftoff and boost phases of the flight went without problems. But after the vehicle reached its first, circular orbit 105 miles above Earth, the Transtage attitude control system's oxidizer valve stuck in the open position, probably as a result of propellant contamination. Despite this, the Transtage completed a successful second burn into a geosynchronous transfer orbit. The long coast that followed this burn exhausted the attitude-control oxidizer, causing the stage to tumble. Two experimental communications satellites, LES-3 and LES-4, and an Oscar IV amateur radio communications satellite went into improper orbits but did function. An OV2–3 satellite with a variety of instruments to measure solar and geomagnetic activity, cosmic rays, and other phenomena remained attached to the Transtage and did not perform its intended functions at all.[42]

This was the third failure out of seven Transtage launches, all caused by defective plumbing. Nevertheless, the Air Force believed that it understood the causes and corrections for the problems well enough that it committed the next launch on June 16, 1966, to launching seven 100–pound Initial Defense Communication Satellite Program (IDCSP) satellites. The gamble paid off. The Titan IIIC launched successfully and placed the seven satellites, plus an experimental 104–pound spacecraft to test gravity-gradient stabilization at high altitudes, in near-synchronous orbits dispersed around the equator. Spin-stabilized and solar-powered, the seven IDCSP satellites could each relay 600 voice or 6,000 teletype channels. Intended as parts of an experimental system, they quickly proved themselves and became the first step in a three-phase program to provide "survivable" strategic and tactical communications.[43]

A follow-on launch of a Titan IIIC with eight IDCSP satellites on August 26, 1966, was less fortunate. Some 78 seconds after launch, with the solid rocket motors still firing, a structural failure occurred in the payload fairing. The malfunction detection system duly opened thrust-termination

Figure 46. Launch of a Titan IIIC from Complex 40, Cape Kennedy, on September 4, 1970. Courtesy of NASA.

ports and set off self-destruct explosives. This was the final failure in the four Titan IIIA and fourteen Titan IIIC developmental launches from September 1, 1964, to April 8, 1970, of which fourteen were successes with a variety of payloads.[44]

Titan 23C

In the interim, during mid-1967 the Air Force negotiated new agreements with the existing Titan IIIC contractors to introduce improvements in their components, calling the new version Titan 23C—a designation not widely used, as the vehicle was often treated as an upgraded Titan IIIC. As examples of these agreements, in May 1967 UTC received a contract for thirty-two more solid rocket motors, and in September 1967 Aerojet got one to improve the stage-1 and stage-2 engines. Plans for the earlier Titan IIICs had included missions carrying humans for the Dyna-Soar orbital spaceplane, but McNamara cancelled that program in 1963. Consequently, the new solid rocket motors no longer included features designed solely to protect humans, such as thrust-termination ports. The thrust vector control system for the new motors was simpler and lighter, with a reduction in tank length. This was possible because UTC changed from a pressure-regulated system requiring constant pressure to a "blowdown" system in which the pressure in the system gradually decreased with use from 1,050 to 600 psi. The new system featured electromechanical instead of hydraulic valves.[45]

The improved engines called for in the Aerojet contract were originally intended for use on a Titan IIIM that would support the Manned Orbiting Laboratory. But Secretary of Defense Melvin R. Laird cancelled this "military space station" in June 1969, and Titan IIIM never flew, although its technologies became features in other Titan configurations. This was notably true of the uprated stage-1 and stage-2 engines, designated LR87–AJ-11 and LR91–AJ-11 respectively. Both were used on the 23C and other Titans. To increase thrust, Aerojet engineers redesigned the stage-1 combustion chamber to function at a higher internal pressure. They also increased the nozzle expansion ratio by using an ablatively cooled nozzle skirt consisting of a laminated silica-phenolic liner over a honeycomb structure with a fiberglass outer cover. Adding the ablative skirt, with an increased expansion ratio of 15:1, reduced the area that had to be regeneratively cooled. Engineers also shortened the path followed by the cooling propellant and changed the cross-section of the coolant tubing, improving the burnout margin of the engine. They redesigned the injector in the attempt to assure stable combustion. A revamped turbopump, including a stiffened high-speed shaft and changes in the lubrication and ball-bearing systems, had greater capability to support the uprated engine. The result, according to Aerojet engineers, was an increase in the thrust of the stage-1 engine from 457,000 pounds at vacuum pressure for the LR87–AJ-9 to 529,000 pounds for the -11 version, while the specific impulse improved from 275 lbf-sec/lbm to 302.[46]

Aerojet engineers redesigned the stage-2 combustion chamber to boost performance without sacrificing stable combustion. A changed spray pattern in the injector achieved more uniform film cooling of the chamber wall, while wider openings for injection of fuel increased the flow to areas of the chamber "exposed to oxidizer rich zones of combustion," reducing the danger of burnthroughs. At the same time, an increase in the overall fuel-to-oxidizer ratio improved the specific impulse of the LR91–AJ-11. Higher injection velocity and better propellant vaporization also added to specific impulse. Engineers reduced the number of tubes in the engine and wrapped the combustion zone with wire to increase its thermal and structural safety margins. The result, at least on the version of the LR91–AJ-11 later used on Titan IV, was a slight rise in specific impulse and thrust—the latter from 100,000 to 105,000 pounds of force.[47]

The initial Titan 23C launches used a new Univac 1824 computer selected for Titan IIIM combined with the remaining components of the old Titan IIIC guidance and control system, but these were now becoming obsolescent. So in October 1970 a successor to AC Spark Plug—called Delco Electronics Division after September 1, 19709—won a contract to replace the older system completely. The new "universal space guidance system" apparently included a Magic 352 computer, with a memory size of 16,384 words expandable to 32,768 words of 24 bits apiece, paired with a Carousel 5B inertial reference (measurement) unit with four gimbals, three gas-bearing gyroscopes, and three accelerometers. The Carousel 5B was a variant of the Boeing 747 guidance system, which was designated Carousel 4. Lack of comparable information makes it impossible to evaluate the differences between the original Titan IIIC's guidance and control system and Titan 23C's and thus to assess any improvements in capability. In the related attitude-control system for the Transtage control module, a simpler and cheaper monopropellant system replaced the Rocketdyne SE-9 bipropellant system. Six pairs of Rocket Research Corporation MR-3A thrusters now furnished the attitude control during coast phases of Transtage operations.[48]

The upgrades for Titan 23C did not all fly on the first missions. First to fly were the new attitude control system for Transtage, along with the Univac digital computer and the LR87–AJ-11 stage-1 engine. The improved thrust-vector control system flew on the second mission, on May 5, 1971, while

Figure 47. Artist's conception of ignition of the Transtage following separation from the Titan III stage 2. Official U.S. Air Force photo, courtesy of the 45 Space Wing History Office, Patrick AFB, Fla.

the new universal space guidance system did not fly until the sixth mission on December 13, 1973. The first launch, on November 6, 1970, carried the initial Defense Support Program (DSP) satellite, which by infrared detection could provide early warning of Soviet ICBMs and other space and missile threats. The intention was to place the satellite in geosynchronous orbit over the Indian Ocean. With the greater thrust of the LR87−AJ-11 first-stage engine, the Titan could now place the second stage in low Earth orbit, eliminating the need for a short Transtage burn to achieve that interim goal. Because someone had failed to align the guidance platform correctly before launch, the Transtage did not remain in the correct attitude during its burns, placing the satellite in an elliptical rather than a synchronous orbit. The apogee was approximately correct at 22,299 miles, but the perigee was way low at 16,186 miles of altitude. This was certainly a launch error, and the launch is listed in some places as a failure. But other people regarded it as a fortunate mistake because the satellite transmitted valuable data on American and Soviet launches as it orbited the globe in its erroneous elliptical pattern. Consequently, an Air Force source counted this as "the 500th satellite to be placed in orbit successfully by a vehicle launched from Cape Canaveral."[49]

The Titan 23C launches from this November 6, 1970, date to March 6, 1982, totaled 22. Counting the first flight as a failure, there were 19 successful missions, with 10 Defense Support Program satellites and 10 Defense Satellite Communications System (DSCS) II satellites, among others, placed in orbit. The two failed launches, on May 20, 1975, and March 25, 1978, each carried two DSCS II satellites that did not reach orbit. In all, between the original IIIC and 23C versions, there were 30 successful and 6 unsuccessful launches, for a success rate of over 83 percent. Four of the failures were due to Transtage problems, without which the overall vehicle would have had a much more successful career.[50]

Titan IIIB

While the IIIA and IIIC were under development, an Air Force program management directive on January 8, 1965, approved Program 624B to acquire what became the Titan IIIB with an Agena D replacing the Transtage in the core stack. The Titan IIIB did not use solid rocket boosters but did employ the -11 stage-1 and stage-2 engines. With the Agena D's 5,800 pounds of thrust compared with Transtage's roughly 1,600, the Titan IIIB could launch a 7,920–pound payload to a 115–mile (100–nautical-mile) Earth orbit, 660 pounds better than the IIIA. Although the predominant guidance system for the B-version was Western Electric Company's radio guidance system, which was compatible with the Vandenberg facilities from which the vehicle was launched, there was also a Titan IIIB version equipped with inertial guidance.[51]

Since the stage-1 engine fired at sea level rather than altitude, the 15:1 nozzles from Titan IIIC were overexpanded for the B-model, so Aerojet engineers reduced the expansion ratio to 12:1. Since Titan IIIB was mostly used to launch photoreconnaissance and related intelligence and communications satellites that were highly classified, the Air Force released little information about the vehicle. So it is not clear when this modification occurred, though it was probably in 1968 after two years of operation. At some point, certainly by 1976, a stretched version of the first stage converted the vehicle to a 24B configuration. And in 1971 a Titan 33B version first operated featuring an Ascent Agena, so called because it became purely a launch stage instead of being attached to the payload while it was in orbit to provide power and attitude control. This was at least one of the configurations of Titan IIIB with inertial guidance, providing guidance and control to the other core stages and eliminating the need for the radio guidance. Between June 29, 1966, and February 12, 1987, various versions of the IIIB with Agena D third stages—including 23B and 34B, which had stretched versions of stage 1 to provide additional propellant capacity—launched some 68 times with only 4 known failures. One of them involved a second stage and 3 involved Agena failures. But this still yielded a 94 percent success rate, much higher than that of Titan IIIC.[52]

Titan IIID

On November 15, 1967, the Titan III Systems Program Office began designing, developing, and ultimately producing the Titan IIID, which essentially added Titan IIIC's solid rocket motors to Titan IIIB. Perhaps more accurately, Titan IIID can be considered a IIIC without the Transtage. By this time, Air Force Systems Command had inactivated Ballistic and Space Systems Divisions (BSD and SSD) and reunited the two divisions in the Space and Missile Systems Organization (SAMSO), headquartered in the former SSD location at Los Angeles Air Force Station. Launched from Vandenberg, like the Titan IIIBs, the D-models carried many photoreconnaissance payloads that were too heavy for the B-models. The Titan IIID could carry a reported 24,200 pounds of payload to a 115–mile Earth orbit, compared with only 7,920 for the B-model. It used a Western Electric Company radio guidance system on stage 2.[53]

Payload details were classified, but there is speculation that the Titan IIID may also have used the Agena D attached to the payload. In any event, it seems clear that the D-model launched 22 heavy imaging satellites between June 15, 1971, and November 17, 1982. Some missions reportedly carried secondary payloads with their own propulsion for achieving higher orbits. It appears that all 22 launches were successful, giving the Titan IIID an enviable—and highly unusual—perfect launch record.[54]

Titan IIIE–Centaur

While the Air Force was developing and using various configurations of Titan III in the mid-1960s, NASA's planners saw no need for the space agency to adopt its own version of the family of launch vehicles. When severe budget restrictions forced the civilian agency to delay development of the reusable launch vehicle that became the space shuttle, however, NASA decided that it needed a more powerful vehicle than the Atlas-Centaur to launch the interplanetary spacecraft then being planned. Consequently, on June 26, 1967, the space agency contracted with Martin Marietta to study the use of a Titan-Centaur combination for missions to Mars and the outer planets in our solar system. When this began to look feasible, in March 1969 NASA Headquarters assigned management of the vehicle to Lewis Research Center, with follow-on contracts going to Martin Marietta (via the Air Force) and General Dynamics/Convair (directly) to study and then develop what became Titan IIIE and to adapt the Centaur D-1 for use therewith.[55]

While the Titan IIIE lower stages were basically the same as the IIID's and thus by themselves not difficult to design, there were challenges in making them compatible with the Centaur D-1. The Centaur had its own guidance and control system and required a new shroud assembly to give it environmental protection while on the ground and during the early portion of the launch. This Centaur Standard Shroud had a diameter four feet greater than the Titan core. Another technical problem involved temperature extremes. Since Centaur used both liquid oxygen and liquid hydrogen, its engines required prechilling. Atlas, the previous launch vehicle for Centaur, used cryogenic liquid oxygen as one of its own propellants, so the temperature differential between it and the second stage was less great than between the Titan second stage and Centaur. All of this was complicated by the fact that General Dynamics, the contractor for both Atlas and Centaur, was a competitor of Martin Marietta, the prime contractor for Titan, and "there was no love lost" between the two firms. Fortunately, rivalry and lack of cooperation at the corporate level did not extend to relations between individual engineers from the two corporations, easing the problems somewhat.

One instance of corporate noncooperation involved the ADDJUST program for the Centaur guidance-and-control computer that allowed it to compensate for winds during liftoff and thus avoid launch cancellations because of heavy breezes. Martin Marietta resisted using this program but ultimately gave in, with the happy result that no Titan-Centaur launches had to be cancelled because of wind. In several areas, the Teledyne guidance and control system had to be adjusted in the Titan (D-1T) version of Centaur,

TITAN CENTAUR

STAGES

"0"-SOLID MOTOR STRAP-ONS
1ST STAGE-N_2H_4-UDMH/N_2O_4 (TITAN)
2ND STAGE-N_2H_4-UDMH/N_2O_4 (TITAN)
3RD STAGE-LH_2/LOX (CENTAUR)

MISSION CAPABILITY

PLANETARY PROBE TO MARS 7500 LBS.
(3,402 Kg)

USE

PLANETARY PROBES
SCIENTIFIC SATELLITES

INITIATED

MID 1969

1ST LAUNCHING

1ST QTR 1974

LAUNCH RATE CAPABILITY

4/YR.

LAUNCH SITE

ETR-PAD 41

Figure 48. A Titan IIIE with the D-1T version of Centaur, including the two five-segment solid rocket motors (stage 0), plus the two core stages topped by the Centaur. Note how the launch vehicle features a bulge from the Titan to the Centaur diameter. NASA Historical Reference Collection, Washington, D.C. Courtesy of NASA.

with more reserve memory and a greater number of functions to accommodate the greater complexity of Titan IIIE, with its two stage-0 solid rocket motors and two core stages that had to be guided and controlled in addition to Centaur itself, as compared with Atlas's stage and a half.

To cope with the Centaur shroud's greater width, engineers simply tapered it where the two rockets joined. This created an odd-looking bulge at the forward end of the launch vehicle but no practical difficulty. More complex was the issue of temperature incompatibility. The insulation had to provide adequate protection without adding excessive weight, which necessitated a carefully calculated balance in the design effort.[56]

The Titan IIIE and Centaur D-1T were eventually ready for a proof flight on February 11, 1974. Carrying a Viking Dynamic Simulator (to determine the flight loads that the first Viking mission to Mars would encounter on its launch) and a Space Plasma High Voltage Interaction Experiment (to provide data for better design of high-voltage systems to operate in space), the vehicle launched from Cape Canaveral and functioned successfully through separation of the Centaur. At that point the upper stage failed to start and could not be coaxed by the guidance and control system to execute a restart. Range Safety had to destroy the rocket and payload. Initial investigation showed that a liquid-oxygen boost pump had failed to supply propellant to the combustion chamber, but it could not pinpoint the cause beyond speculation that a foreign object might have jammed the mechanism. The only concrete result of the investigation was development of a new procedure to show, before launch, whether the boost pumps were functioning properly.

Some four years later the cause of the failure finally came to light. At General Dynamics the employee responsible for anchoring to the wall of each propellant tank a probe that indicated propellant usage had retired before the Titan E-Centaur proof flight and evidently failed to convey to his successor that he had departed from specifications in his work (and had not gotten them changed to reflect his own practice, either). He had always, when securing the clip that attached the probe to the tank wall, used a longer rivet than called for in the drawings. His successor had followed the specs, and in retrospect it became apparent that, with the shorter rivet, the clip had fallen off and jammed the boost pump. This was but one of many instances in the history of rocket development where such a small detail could cause a major failure, showing the immense complexity of such vehicles and the need for meticulous documentation to cover treatment of every single part accurately.

In this case, even though the mission was not a full success, it had achieved enough for NASA to declare a second proof flight unnecessary. The first one had shown that the vehicle was structurally capable of launching a Viking, that Centaur would separate from the Titan as designed, and that the Centaur shroud was structurally sound, providing the thermal protection needed for Centaur to function once separation occurred.[57]

Experience ratified this decision. Between December 10, 1974, and September 5, 1977, the Titan IIIE–Centaur launched two Helios solar probes, two Viking missions to Mars, and two Voyager missions to Jupiter and Saturn, all of them highly successful. Voyager 2 also explored Uranus and Neptune. Besides the "normal" Titan and Centaur stages, the Helios missions used Thiokol Star 37E solid motors, developed for the Delta 2914, as additional stages. All of these missions returned valuable data—and, in the cases of the missions to the planets, huge numbers of photographs—to solar physicists and planetary scientists. The only major problems with the launch vehicles for these missions were the discovery of a bonding problem in the Centaur computer before Helios 1 and a propellant problem for the Titan III core on Voyager 1. Before the Helios 1 mission, it became apparent that some microelectronic modular assemblies in the Teledyne computer were failing. Teledyne engineers had used a tin paste to bond circuitry to a ceramic substrate. In tests this had come loose only rarely, but even the low rate of 1 in 10,000 was a matter of concern because Centaur had 2,400 modules and, when a bond broke and ended an electrical connection, the computer would fail. After testing a number of substitutes for the tin paste, Teledyne engineers found that epoxy held fast, helping to ensure that the Helios mission was successful. The propellant problem for Voyager 1 stemmed from a hardware failure, which left 1,200 pounds of propellant unused and caused the Titan core to underperform. Fortunately, the Centaur computer recognized the problem and corrected it by extending Centaur's own burn enough to make the mission a success.[58]

Titan 34D

By the mid-to-late 1970s, Air Force planners perceived the need for a still more powerful Titan configuration to carry increasingly large payloads such as the Defense Satellite Communications System III satellites. The space shuttle was designed to handle such hefty cargo—the first DSCS III weighed 1,795 pounds, a significant jump from the DSCS II weight of 1,150 pounds— but the shuttle was not ready. And even after the shuttle became fully operational, the Titan 34D, as the new vehicle came to be called, would be useful in a backup role. The Air Force began studies of an advanced launcher concept in 1975. It contracted with Martin Marietta in July 1977 for preliminary design of such a vehicle, with a production contract for five Titan 34D airframes following in January 1978. SAMSO retained program responsibility for the Titan family of vehicles, and it contracted separately with suppliers of the component elements. Martin Marietta would build the two core stages and manage integration of the components into a viable booster.

Aerojet continued as contractor for the two core engines, with a first stage employing the stretched Titan IIIB configuration topped by the standard second stage as on Titans IIIC and IIID. It appears that the long-tank first stage was the driving element in the new vehicle, because a 1978 article in *Aviation Week & Space Technology* stated that Chemical Systems Division's solid rocket motors (SRMs) would each add half a segment "to make them compatible with the long-tank first stage." Thus the SRMs would contain five and a half segments in place of the five used on previous Titans.[59]

None of the components of the basic Titan 34D appears to have posed particular problems in development. The two Aerojet engines and their airframes had already been used on previous Titan configurations. The guidance and control system for the basic Titan 34D was the Western Electric Company radio guidance system. The 5.5–segment solid rocket motor was new, but since CSD had already developed and tested seven-segment motors for the abortive Titan IIIM in 1969 and 1970, the 5.5–segment motor seems to have been fairly easy to develop. On August 25, 1979, the firm tested the 5.5–segment motor at its Coyote Test Facility near San Jose for the first time with complete success.[60]

Except for the extra half segment, which increased the length of the solid rocket motors from 84.65 to 90.33 feet, Titan 34D's stage 0 was not significantly different from Titan IIIC's. The motors retained their 10–foot

Figure 49. Titan 34D launch from Cape Canaveral, October 30, 1982. Official U.S. Air Force photo, courtesy of the 45 Space Wing History Office, Patrick AFB, Fla.

diameter, and the theoretical specific impulse for the ammonium perchlo-rate-PBAN-aluminum propellant remained slightly above 265 lbf-sec/lbm. The grain configuration stayed the same, with the conical internal cavity extended through the extra half segment. The case and nozzle remained basically unchanged as well. With little change in the average thrust, the total impulse of course went up with the added propellant in the extra half segment, from 1.13 to 1.23 million pounds of force. Equipped with these lon-ger solid-rocket motors and a Transtage, the Titan 34D could carry 32,824 pounds to a 115–mile (185–kilometer) orbit, as compared with 28,600 pounds for the IIIC. The 34D could lift 4,081 pounds to geosynchronous orbit, which compared favorably with 3,080 pounds for the IIIC but not with the 7,480 pounds the Titan IIIE–Centaur could carry to the same orbit.[61]

The Transtage version used on Titan 34D was the uprated AJ10–138A, for which SAMSO contracted with Aerojet. Its preliminary design dated from 1975, with a qualification test in 1979 and contractual deliveries for the first three production engines completed in 1980. Little seems to have changed in the uprated version, including the general features and specific impulse, but the "qualified service" was for 700 seconds as against a rated duration of 500 seconds for the AJ10–138. Seven of the fifteen Titan 34D launches employed the Transtage.[62]

Inertial Upper Stage

A quite different but important upper stage had its maiden launch on the first Titan 34D and later launched on several Titan IVs. This was the Inertial Upper Stage (IUS) that sat atop stage 2 on the first Titan 34D launch. Un-like the rest of the booster, this stage was anything but easy to develop. De-signed under SAMSO management primarily for use with the space shuttle to place payloads in geosynchronous orbit, the IUS encountered difficulties for a variety of complicated reasons. Many of these were technical, but the major problems involved management. One factor—though far from the only one—was that the IUS, which initially stood for *Interim* Upper Stage, was conceived as a temporary expedient until a more capable Space Tug could fly with the shuttle. But the Space Tug, while not formally terminated, "just slid out year-by-year under budget pressure," as one Air Force general expressed it in testimony before a congressional committee, so expectations for the Interim Upper Stage shifted, starting about 1978, from a minimal modification of an existing upper stage such as Transtage or Agena to a "first line vehicle in the Space Transportation System." Yet there remained "considerable cost reduction pressure as an outgrowth of the *interim* stage thinking." Moreover, the Air Force was developing the vehicle under "a contract structure which strongly incentivized performance, but only provided limited cost incentives."[63]

The IUS ultimately overcame its birth pangs to become "an integral part of America's access to space for both military and civilian sectors." It had its beginnings in 1969 when presidential direction gave impetus to studies leading to the space shuttle. Since the shuttle would be incapable of reaching geosynchronous and other high orbits, the Space Tug became the intended solution, with the IUS providing interim capability until the tug could be completed. DoD agreed to develop the IUS, proposing in 1975 that it be a solid-propellant stage to hold down costs. In August 1976 the Air Force selected Boeing Aerospace Company as the prime IUS contractor. The contract provided incentives for meeting performance and cost targets, but Boeing was liable for only 10 percent of cost overruns. Moreover, cost projections for the IUS had been based on the assumption, according to Air Force Maj. Gen. William Yost, that "the M-X and Trident missile programs would develop most of the solid rocket motor technology . . . needed by the IUS. Unfortunately, the schedules for those programs slipped far enough that the IUS program became the leader in developing the solid rocket motor technology necessary to meet our performance requirements." Understandably, the contractor "increased spending to insure that he will achieve his performance goals and earn the performance fees." As a condition of revising the contract with Boeing in 1979, the Air Force therefore insisted that the firm's "management deficiencies be resolved." Among other steps to this end, Boeing appointed a new program manager, assigned senior managers to oversee major subcontractors, and instituted formal reviews.[64]

Boeing had begun a planned 18–month preliminary design phase in August 1976 when it won the contract, to be followed by a 28–month development phase. This would have made the IUS available by June 1980. Soon after winning the basic contract, Boeing subcontracted with CSD to design and test the solid motors for the IUS. CSD chose to use a hydroxy-terminated polybutadiene propellant—also being used by Thiokol in the Antares IIIA motor for Scout developed between 1977 and 1979. CSD selected a carbon-carbon material for the nozzle, which would be manufactured using a new process that held costs to a low level. It was making the case out of Kevlar. (Thiokol was using the same or similar materials on the contemporary Antares motor, raising questions about the accuracy of Yost's claim that CSD was taking the lead "in developing the solid rocket motor technology" needed for the IUS. It appears that CSD and Thiokol were competing for that lead from 1977 to 1979, with Thiokol winning the competition.)[65]

At first the motor development seemed to be going well. In 1977 CSD conducted a series of tests on the nozzle and motor, including successful 85–second tests of the nozzle at the Air Force's Rocket Propulsion Laboratory on June 10 and July 15 and a follow-on 145–second test on October 7. Moving to the Arnold Engineering Development Center, CSD subjected a full-scale motor with 21,000 pounds of propellant to a 154–second test, again successful. On May 26, 1978, the nozzle material using carbon-carbon made with the new cost-saving technique cleared a further hurdle in a 140–second test firing back at the Rocket Propulsion Laboratory.

But on October 19, 1978, during a test at Arnold, the motor's Kevlar case burst at only 750 psi of water pressure instead of CSD's predicted 1,050 pounds. The firm decided that the cause was defective manufacturing equipment. After redesigning the equipment and strengthening the structure of the case, it conducted six more tests between January and September 1979. Five were successful, with the cases withstanding higher pressures than specified.

Meanwhile, the IUS had evolved to include two propulsion units, with larger and smaller motors having similar features. A test firing of the large motor was scheduled for October 17, 1978, but was delayed when inspection revealed that some propellant, improperly cured, had softened and blistered. With the propellant recast, the test proceeded on March 16, 1979, in a 145–second firing that generated more than 50,000 pounds of thrust. Engineers vectored the nozzle several times, demonstrating its ability to direct the thrust for course corrections. A follow-on test of the small motor, also successful, occurred on June 25.[66]

Most other tests in 1979 went well in general, but cracks had appeared in the nozzle of the small motor during firing. One special feature of the nozzle for the smaller motor was its extendable exit cone, added in 1978. This was a series of conical sections that, in the final design as of 1983, telescoped out over one another to increase the nozzle expansion ratio from 49.3:1 without the extension to 181.1:1 with the pieces deployed. Although the design, which would be used only on certain missions, increased the weight of the motor, it added about 15 lbf-sec/lbm to the specific impulse. Unfortunately, about half the exit cones for the small motors were defective and had to be rejected. Further, five motors proved to have more "bad" propellant. Boeing and CSD said they could still be tested, but Aerospace Corporation, advising SAMSO, disagreed.

On October 1, 1979, SAMSO formed a tiger team of experts from several organizations including NASA, the Rocket Propulsion Laboratory, and Aerospace Corporation to investigate technical concerns and management. This resulted in the aforementioned management changes at Boeing and also a change in one supplier. CSD had been making the large Kevlar motor cases and Brunswick Corporation the small ones. As the team found the small cases superior, Brunswick became the supplier of both sets.

During 1980 the remaining problems gradually yielded to solution. The cracks in the nozzle of the smaller motor, it turned out, stemmed from unequal expansion of two materials. A silica phenolic insulation material expanded faster than the carbon-carbon next to it; the answer was to wrap the silica phenolic with graphite to limit its expansion. The rejected exit cones came from a supplier, Kaiser, that was still learning about the properties of carbon-carbon; a change in tooling and ply patterns plus improved quality-control procedures ended the deficiencies. The degraded propellant had all come from a single batch and proved to be usable in tests after all. Three altitude-simulation tests of each motor, small and large, during 1980 at the Arnold Engineering Development Center were successful. There were further problems with propellant cracks, delamination of the carbon material in the extendable exit cones, and the mechanism for extending the exit cones, but the development team solved those too.[67]

For the guidance and control system, Boeing had planned to use Teledyne to supply the computer and inertial measurement unit, but SAMSO insisted that the prime contractor solicit competing bids for both pieces of equipment. Delco won the contract to provide the computer because its proposal "offered higher performance and a more proven design." Teledyne's bid to make the inertial measurement unit also lost out, to Hamilton Standard, a sister division of CSD within United Technologies Corporation, because Hamilton Standard's unit "was more accurate and could be developed with less risk." Boeing let both contracts for guidance and control in December 1977, along with one about the same time to Ball Aerospace Division for a star scanner to supplement one of two inertial measurement units (the second one providing redundancy). There were three computers, with the third one serving as a check in case the other two failed to "agree.." The inertial measurement units were strapped down and each consisted of five rate-integrating gyros and five accelerometers. The gyros and accelerometers measured angular movements and rates as well as linear accelerations and sent the data to the computers for processing by resident software. The computers sent guidance commands to control the thrust vector control system in stage 0 plus the nozzle vectoring in the two liquid-propellant stages of the Titan 34D and both motors of the IUS, as well as hydrazine thrusters in the IUS that controlled roll and provided corrections during coasting flight by the IUS. (For use in the shuttle, the IUS had no responsibility for guidance and control until it was launched from lower orbit by the orbiter.)[68]

In the course of these developments, the IUS fell more than two years behind schedule, and overruns had basically doubled the original projected cost. While many of the problems resulted from contractual arrangements and the initial interim character of the upper stage, many were also technical, having to do with fabrication methods and quality control. They showed that despite more than a quarter century of major U.S. rocket development since 1955, rocket engineering still required constant attention to small details and, where new technology was involved, a certain amount of trial and error. (To be sure, Thiokol's success with Antares IIIA showed that the process of innovation could go smoothly sometimes—but not always, as Thiokol's later problems with the shuttle solid rocket boosters would demonstrate.) Because the IUS was designed principally for use on the space shuttle, NASA's delays with that program made the stretchout of the IUS schedule less worrysome than it would otherwise have been.[69]

In any event, on October 30, 1982, the first Titan 34D and the first IUS together successfully launched two Defense Satellite Communications System payloads—a DSCS II and the first DSCS III—into geosynchronous orbit from Cape Canaveral. As planned, the second-stage burn achieved low Earth orbit, with the first IUS motor carrying the third stage and the satellites into transfer orbit. The second IUS motor placed the payloads in geostationary orbit, with hydrazine thrusters making final adjustments in the placement of each satellite. There was a small glitch in the launch, a telemetry failure attributed to a leak in the seal of a switch. But the guidance and control system, flying "blind," autonomously carried out the provisions of the flight plan.[70]

More than three years before this launch, SAMSO had undergone still another transformation. On October 1, 1979, it split in two once again, with Space Division on Los Angeles Air Force Station keeping the space mission and the ballistic missile functions going back to the Ballistic Missile Office on Norton Air Force Base inland from Los Angeles.[71]

As completely designed, the IUS was, not a single-stage vehicle as its name would suggest, but a two-stage, solid- propellant vehicle for placing spacecraft in high Earth orbit or in an escape trajectory. It was roughly 17 feet long and had a maximum diameter of 9.25 feet. Fully loaded, the large, first-stage motor (SRM-1) carried 21,400 pounds of ammonium perchlo-rate-HTPB-aluminum propellant, but the propellant load could be reduced as required for specific missions—as indeed was done with the first launch. The smaller second-stage motor (SRM-2) could carry up to 6,000 pounds of the same propellant. The propellant-delivered specific impulse of SRM-1 was upwards of 295 lbf-sec/lbm and that for SRM-2 about 290, increasing to more than 300 with an extendable exit cone. Except for the last, these performance figures were slightly below those for Transtage. For Titan 34D launch, the IUS carried about 2,200 pounds less propellant than the fully loaded version, so the Titan 34D–IUS combination could deliver only 4,070 pounds to geosynchronous orbit, a trifle less than the 4,081 pounds of the 34D-Transtage combination (both polar launches from Vandenberg). But fully loaded on a typical shuttle mission, the IUS could place up to 5,100 pounds in geosynchronous orbit.[72]

Including its first (and only) IUS mission, the Titan 34D had a total of fifteen launches from both the Eastern and Western Test Ranges between October 1982 and September 4, 1989. Three of these were failures, the first because the stage-1 engines on the August 28, 1985, launch ceased to function about two-thirds of the way into their burn. Investigators could not determine the cause but there may have been a faulty element in the oxidizer supply system. Thereafter technicians inspected turbopump gear cases and used better clamps on joints in propellant lines. This launch and the next, on April 18, 1986, carried reconnaissance satellites, with the second one failing because of a burn-through of the solid-rocket motor casing due to the breakdown of a bond between the propellant and the case. The third launch failure occurred above Cape Canaveral on September 2, 1988; the Transtage's attitude control system apparently sustained damage at separation of the shroud. The payloads for the twelve successful launches varied considerably but included Defense Support Program payloads and another pair of DSCS II and III satellites.[73]

The Commercial Titan III

After an abortive attempt in 1983 to enter the commercial space launch market, Martin Marietta tried again after President Reagan's post-*Challenger* announcement on August 15, 1986, that NASA and the space shuttle would get out of the business of launching commercial satellites. On November 18, 1988, Martin Marietta Commercial Titan of Denver, Colorado, and NASA announced that they had agreed for the firm to use facilities at Kennedy

Space Center in Florida to launch a commercial version of Titan III. This was actually an improved Titan 34D introduced in 1989, with a stretched second stage. The new version continued to use Aerojet LR87–AJ-11 engines in the already stretched first stage and an LR91–AJ-11 engine in the second stage. CSD's 5.5–segment solid rocket motors continued to constitute stage 0. Guidance and control used a system based on the one in Transtage and located in stage 2. However, the commercial Titan III could fly with a variety of upper stages, including the Inertial Upper Stage, the Centaur G-prime, and the Payload Assist Module, which presumably would have provided their own guidance if they had ever been used.[74]

Another upper stage, actually used with the commercial Titan III for one launch, was the Transfer Orbit Stage, designed by Orbital Sciences Corporation using the larger (SRM-1) IUS motor developed by CSD. The commercial Titan III could carry up to two payloads at once, but since Martin Marietta had trouble finding more than one set of customers willing to launch on the same date, it announced in June 1989 that it would halt its efforts to find two separate customers.[75]

The firm did manage to find two customers for its first launch. On January 1, 1990, the initial commercial Titan III successfully placed two communications satellites—the 3,320–pound Skynet 4A for the British Ministry of Defence and the JCSAT 2 for the Japanese Communications Satellite Company—in geosynchronous orbits using kick motors rather than third stages. The second mission of the commercial Titan III on March 14, 1990, booked to launch the 27,425–pound Intelsat 6 F-3 communications satellite into geosynchronous orbit, fared less well. The second stage did not separate from the spacecraft and its perigee kick motor because of a wiring error in the separation hardware. Controllers from Intelsat were able to command the satellite to separate from the second stage but could only place it in a hopelessly low orbit where it was not operational. Fortunately, in early May 1992 the crew of space shuttle *Endeavour*, after four difficult days of extravehicular activity with three astronauts in space at once, managed to retrieve the satellite, replace the kick motor, and send the Intelsat payload to its proper orbit.[76]

Meanwhile, another commercial Titan III had succeeded in launching Intelsat 6 F-4 into geosynchronous orbit on June 23, 1990. This was followed on September 25, 1992, by the final commercial Titan III launch. Equipped with a Transfer Orbit Stage in its first use, the basic Titan III placed the 5,661–pound Mars Observer spacecraft in a low-Earth transfer orbit, and the third stage sent it on its eleven-month journey to study the geology, geophysics, and climate of Mars. Unfortunately, on August 21, 1993, the spacecraft stopped communicating with Earth three days before it was scheduled to enter Mars orbit, never to be heard from again.[77]

The fate of Mars Observer, although it was unrelated to the satisfactory performance of the commercial Titan III on that particular launch, was perhaps a fitting symbol for Martin Marietta's launch vehicle. It too had essentially failed. At a cost of $130–150 million per launch without an upper stage, it was certainly cheaper than the space shuttle, whose per-launch costs by one reckoning ranged between 1992 and 1999 from $411 million in 1996 to $688 million in 1999, with other years' costs falling between those extremes. But the shuttle and the commercial Titan III were never in competition. The Titan's costs were not low enough to make it an attractive alternative to Europe's Ariane launcher, whose two-tier pricing system essentially subsidized launches for customers outside the European Space Agency.[78]

Titan II as a Space-Launch Vehicle

By the mid-1980s the Air Force had become increasingly uncomfortable with its dependence on the space shuttle for delivery of military satellites to orbit. Eventually this discomfort would lead to the procurement a variety of Titan IV, Delta II, and Atlas II expendable launch vehicles, but the air service also had at its disposal 56 deactivated Titan II missiles in storage at Norton Air Force Base. Space Division contracted with Martin Marietta in January 1986 to refurbish a number (soon to become 14) of the Titan IIs for use as launch vehicles. The conversion process included a variety of steps. Aerojet refurbished but did not uprate the stage-1 and stage-2 engines. Martin Marietta itself had to modify the airframe to accommodate various payloads, using a payload fairing and forward skirt from the Titan III program. And Delco Electronics modified the original Titan II guidance systems to incorporate the essential features of the Universal Space Guidance System used on many of the Titan III configurations.[79]

Designated Space-Launch Vehicle 23G, the Titan II had only two launches during the period covered by this book, on September 5, 1988, and the same day in 1989, both classified missions from Vandenberg Air Force Base. It would go on to carry into orbit a variety of payloads ranging from Defense Meteorological Satellite Program spacecraft to the Clementine space probe and both land- and ocean-observing satellites. Launched on a polar orbit from Vandenberg, the Titan II launch vehicle could carry only about 4,190 pounds of payload into a 115–mile orbit, but this compared favorably with the Atlas E. While the Atlas vehicle could launch about 4,600 pounds into the same orbit, it required two Thiokol Star 37E solid rocket motors in addition to its own thrust to do so.[80]

Titan IV

Titan IV grew out of the same concern about dependence on the space shuttle that had led to the conversion of Titan II missiles to space-launch vehicles. In 1984 the Air Force decided that it needed to ensure access to space in case no space shuttle was available when a critical DoD payload needed to be launched. When Space Division asked for bids on a new vehicle, Martin Marietta proposed a modified Titan 34D and won a development contract on February 28, 1985, for ten of the vehicles that were initially called Titan 34D-7s but became Titan IVs. Following the *Challenger* disaster, the Air Force amended the contract to add thirteen more vehicles.[81]

Based on the Titan 34D and, before that, the Titan IIIM, the initial version of the new booster, later called Titan IVA, had twin seven-segment solid rocket motors produced by CSD as a subcontractor to Martin Marietta. These contained substantially the same PBAN propellant and grain configuration as the 34D, but the additional 1.5 segments brought the length to about 122 feet and the thrust per motor to 1.39 million pounds at peak (vacuum) performance. Each motor still used liquid injection of nitrogen tetroxide for thrust vector control in a fixed, canted nozzle, as had the earlier Titan solid rocket motors. But in the Titan IVA motors the initial expansion ratio increased from 8:1 to 10:1, with the addition of an exit cone extension. The cases continued to be made of steel with field joints and a single O-ring. As originally proposed for Titan IIIM, the first version of the seven-segment motors went through a development period from 1965 to 1970 in which there were four static tests in 1969 and 1970. CSD revived the seven-segment design for Titan IV but added reliability measures adopted for the 34D following the April 1986 failure of a solid rocket motor as well as the January 1986 *Challenger* accident. Successful static tests in December 1987 and February 1988 completed qualification of the seven-segment motors.[82]

To allay Air Force concern that CSD would not be able to produce enough solid rocket motors, Martin Marietta subcontracted in 1987 with a second source, Hercules Aerospace, to provide fifteen sets of motors with higher performance than the CSD motors. Known as the Solid Rocket Motor Upgrade (SRMU), the new motor used an HTPB propellant, only three segments, and a filament-wound graphite-epoxy composite case with a movable nozzle, yielding a lift capacity 25 percent greater than the Titan IVA stage 0's. But the motors on what became the Titan IVB were delayed in their development. An initial test in 1990 at the Edwards Air Force Base rocket site, known at the time as Phillips Laboratory, had to be halted when a crane moving one of the segments collapsed. The segment fell to the ground and ignited, resulting in a fatality. On April 1, 1991, the motor undergoing the delayed test produced more internal pressure than predicted because of a propellant grain deformation, and the case ruptured after just two seconds of firing. Ultimately the first Titan IVB with SRMUs did launch on February 23, 1997, long after the period covered by this book.[83]

Aerojet was Martin Marietta's subcontractor for the engines in stages 1 and 2 of Titan IV, the same basic LR87–AJ-11 and LR91–AJ-11 used on Titan 34D. The stages themselves retained the same configurations as for Titan 34D except that stage 1 was stretched about 7.9 feet to allow for more propellant and thus a longer burn time. Stage 2, similarly, added 1.4 feet of propellant tankage. With greater lengths and the attachment of the longer solid rocket motors, there were also new load and stiffness requirements.[84]

There were five different versions of the Titan IV, designated 401 through 405. For the Titan IVA, version 401 mated the basic Titan IV with a Centaur third stage. Launched only from Cape Canaveral, this configuration could employ fairings 66, 76, or 86 feet in length, depending on the payload, with the fairing surrounding the Centaur as well as the payload. This version was intended to be operational as early as 1991, but launch failures and other factors delayed the first flight until February 7, 1994. With CSD's seven-segment solid rocket motors (SRMs), the Titan IV–Centaur could carry up to 10,000 pounds to geosynchronous orbit. Version 402 used the Inertial Upper Stage on top of stage 2. With the CSD SRMs, Titan IVA–IUS could lift 38,780 pounds to low Earth orbit and roughly 5,300 pounds to geosynchchronous transfer orbit. Versions 403 through 405 were all Titan IVs without a third stage, with 403 and 404 designed solely for missions into polar orbit from Vandenberg and 405 for low Earth orbits launched east from Cape Canaveral. What distinguished version 403 from 404 was that the 403 carried either a 56–foot or a 66–foot payload fairing and could lift 31,100 pounds into polar low-Earth orbit, while the 404 version used a 50–foot payload fairing and could carry 29,600 pounds into low Earth orbit with CSD solid-rocket motors as boosters for the Titan IV. Version 405 could carry payload fairings either 56 or 66 feet long and was capable of lifting 39,100 pounds into low-Earth, low-inclination orbits with the seven-segment SRMs on the Titan IVA.[85]

Guidance and Control

There is comparatively little specific or complete information about the guidance and control systems used with the various versions of the Titan IV. Presumably, versions 401 and 402 used the guidance systems on the Centaur and IUS respectively to guide and control all four stages, with ancillary equipment on each stage complementing that in the uppermost stage. Martin Marietta's fact sheet on Titan IV from about 1988 lists both sets of solid rocket motors, but under "guidance" it specifies only an "inertial with digital computer" produced by Delco. Lockheed Martin's fact sheet from early 2000 includes a drawing of Titan IV with a Centaur upper stage showing Honeywell guidance in the Centaur and Delco guidance in the second stage, but an Air Force fact sheet from 1996 states that the Honeywell digital avionics system would replace the Delco inertial guidance system on the twenty-fourth launch. It seems clear that both systems would not have been used together. Indeed, Isakowitz's "International Reference Guide to Space Launch Systems" (1995 version) specifies that "for Titan III and Titan IV, the avionics is based on earlier Transtage avionics using a four gimbaled carousel gyro," adding that "with the 24th flight Titan IV will use a Centaur-based single-string avionics system using a strapdown platform with ring-laser gyros." It goes on to say that "the basic avionics necessary to fly the lower stages are located in Stage 2."[86]

This seems to mean that, through the end of the cold war (occurring well before the twenty-fourth launch in 1998), Titan IV versions 103 through 105 used the configuration of the Transtage guidance and control system with the Magic 352 computer and the Carousel 5B inertial measurement unit. Before 1992 at the latest, Delco Electronics had developed a new hemispherical resonator gyro, "derived from G. H. Bryan's discovery of the rotation-sensing properties of a ringing wine glass. The resonator flexes at its resonant frequency in a low amplitude standing wave. When the gyro is rotated about its resonator axis, the standing wave precesses relative to the case in exact proportion to the rotation angle. . . . Capacitance pickoffs" then sensed the location of the standing wave and provided data about the vehicle's deviation from the intended rotation angle to the guidance and control computer. Delco stated that this was a "highly reliable rate-integrating gyroscope . . . for use in its new Carousel 400 family of inertial systems." Since it had "no bearings or wear surfaces," each gyroscope was "expected to provide undiminished performance for more than 20 years." It was "dimensionally stable and [had] low thermal and vibration sensitivity and negligible magnetic sensitivity." It also could operate on "extremely low power." Apparently the new gyroscopes and the Carousel 400 inertial systems (or modifications thereof) were never used on Titan IVs before 1991, but they are illustrative of the progression of guidance and control technology in that period.[87]

Honeywell's ring laser gyros (RLGs) also were not used on Titan IVs or on Centaur stages, for which they were also designed, until later, but since they were developed during our period, it is appropriate to discuss them here. Honeywell started developing RLGs in 1965 to take advantage of the fact that laser beams travelled at the speed of light and laser devices involved no moving parts. They also were "virtually insensitive" to the vibrations that were a constant feature of launch vehicles in flight and caused difficulties for previous gyroscopes. Lasers devices also required almost no warmup time, whereas more conventional gyros took time to reach a stable speed. Finally, RLGs did not require periodic calibration, unlike the mechanical gyros. Instead of relying on inertial forces acting upon rotating gyroscopes, as the older technology had done, RLG systems measured "changes in counterrotating beams of laser light that flash around in a tight circle. If the laser gyro itself turns a bit, one beam of light will travel slightly farther around the ring in a given instant of time, the other slightly less far. Differences in the time it takes the laser beams to travel around the ring add up to a precise measurement of the gyro's motion." RLGs were reportedly smaller, cheaper, more reliable, and lighter than previous gyros.[88]

For both the Centaur inertial navigation unit and the Titan IV guidance control unit, Honeywell used a three-axis strapdown system featuring three of its Honeywell GG1342 RLGs together with three QA3000 accelerometers provided by Sunstrand. These were accompanied by an unspecified number of 1750A processors. The system was small—20 by 12 by 9 inches—and weighed only 67 pounds.[89]

Compared with stable platforms using gimbals, strapdown systems like those used by Honeywell were less complex, less expensive, more reliable, and smaller. Fast digital computers had, by the late 1980s, met their needs for increased computations to compensate for the otherwise inherently greater accuracy of the gimballed platforms.[90]

Centaur G-prime

The version of Centaur used with Titan IV was the "modified Centaur G-prime." The G and G-prime configurations were not drastically different from the one used with the Atlas G, but they did incorporate some technical changes. They were designed for use on the space shuttle, but following the *Challenger* accident, NASA cancelled the shuttle-Centaur program in June 1986. Meanwhile, the two G-versions had been developed to fit the Centaur into the shuttle cargo bay. Version G was 20 feet in length with a 40–foot payload capability so it would squeeze into the roughly 60 feet of cargo space in the shuttle bay. This version accommodated such a long payload by reducing propellant capacity to 29,000 pounds. The 30–foot-long G-prime version had more propellant but could handle only 30 feet of payload length in consequence.[91]

Despite the same name, the Centaur G-prime used with Titan IV was not identical to the one designed for the shuttle. Instead, it incorporated technology from the Titan IIIE version of Centaur and the Centaur being designed for use with the commercial Atlas as well as from the G-prime version for the shuttle. It used the RL10A-3–3A engines employed with the Atlas G-version of Centaur but increased the propellant load by 50 percent. The stage remained roughly 30 feet long by a bit more than 14 feet in diameter. The two engines each yielded a thrust of 16,500 pounds with a specific impulse of almost 445 lbf- sec/lbm.[92]

Early Launches

The first launch of a Titan IV took place at Cape Canaveral on June 14, 1989, using a version-402 vehicle with an IUS as the upper stage. The payload was the first of an upgraded series of Defense Support Program satellites. There were no problems with the seven-segment solid rocket motors, but during the operation of the first stage, one engine suffered an explosion in a cooling tube resulting from decomposition of hydrazine. This caused a 4–degree deviation and incapacitated the gimballing of the affected engine. Fortunately, the other engine was able to maintain control, and the Titan IV succeeded in placing the DSP satellite in a geosynchronous orbit.[93]

The second launch, on June 8, 1990, also from Cape Canaveral, used a version 405 Titan IV to launch three second-generation Naval Ocean Surveillance System satellites (for tracking ships through their radio transmissions) plus an unidentified fourth payload into a low orbit. Although all details about these satellites are classified, speculation among interested observers suggests that the NOSS-II satellites, at least, went into a higher orbit using a Titan launch dispenser attached to the satellites. This was a liquid-propellant engine using pressure-fed nitrogen tetroxide and monomethyl hydrazine to produce 900 pounds of thrust.[94]

There followed launches of a 402–version Titan IV on November 12, 1990, from Cape Canaveral and of 403–versions on March 8 and November 7, 1991, from Vandenberg. Although the payloads were not announced for any of these missions, it appears that the first of them placed a Defense Support Program satellite in orbit, the second a Lacrosse radar-imaging satellite, and the third another NOSS-II satellite. After the period covered by this history, Titan IV went on to place many more satellites in orbit through the first years of the twenty-first century. More than a few of these payloads were classified, but they certainly included more DSP satellites and several for the new Military Strategic and Tactical Relay System (Milstar), a worldwide voice-communications system for military command and control that was expected to be jam-resistant and to supplement the Fleet Satellite Communications, Ultra High Frequency Follow-on, and Defense Satellite Communications satellites. On February 7, 1994, the first Titan IV version 401 with a Centaur third stage successfully launched the first Milstar satellite. Subsequent launches apparently included more NOSS-II and Lacrosse satellites. The final Titan IV launch, with a B-model, occurred successfully at Vandenberg on October 19, 2005. It was the 39th Titan IV launched, twelve of them from Vendenberg and twenty-seven from Cape Canaveral. This was also the final Titan launch of any kind, since the last Titan II had placed a DMSP satellite in orbit in October 2003.[95]

Analysis and Conclusions

In a sense, Titan III and Titan IV constituted the first fully DoD vehicles specifically designed for space launch rather than converted from missiles. However, since the basic core of the Titan IIIs and IVs was the Titan II missile, in another sense the later Titans simply followed the pattern of earlier launch vehicles. In fact, it could be argued that Thor-Able and Thor-Agena established the pattern for the Titan family of launch vehicles. What was new with the Titan IIIC was the development of large, segmented solid-rocket motors, which evolved further with Titan 34D and Titan IVA and B. These motors bequeathed a legacy to the shuttle's solid rocket boosters as well.

Like other rocket development efforts before and after them, the Titans experienced their share of problems. For example, a problem with nozzle throats for large segmented motors had already manifested itself in Aerojet's earlier testing of such motors. UTC engineers solved the problem with a tape-wrapped carbon-phenolic throat. Even though this was a cut-and-try sort of approach to design problems the firm's engineers perhaps should have anticipated, it proved to be a major contribution. When scaling up the cases for the huge motors presented other problems, UTC, Westinghouse, and the Ladish Company came up with rolled-ring forging.

Similarly, Aerojet experienced difficulties ranging from malfunctioning bipropellant valves through nozzle-extension weakness to faulty extrapolation of ground test data in the development of Transtage. We have little specific information about the technologies engineers used to solve these problems, but apparently they were further instances of trial-and-error engineering. Later, General Dynamics engineers resolved various issues related to adapting the Centaur for use on Titan instead of Atlas, only to have a fundamental issue of quality control—failure to adjust specifications to match best practice, in this case the use of longer rivets to attach a probe—cause the failure of a mission. Later still, a host of problems with developing the Inertial Upper Stage—such as cracks in nozzles, inferior procedures for making Kevlar cases, and flaws in contract structure and basic management—all called for changes in procedures to deal with issues not anticipated in the initial design process, despite decades of experience with rocket development. As late as 1990 and 1991, at the very end of our period, difficulties with a new solid rocket motor for Titan IVB showed once again that engineers did not have the process of rocket development "down to a science." New technologies, scale-ups from one size to another, and the sheer complexity of rockets with large numbers of parts and a multitude of disciplines that had to interact seamlessly all made the design, development, and production of launch vehicles still as much an art as a science.

Despite such instances, the Titan family of launch vehicles had clearly satisfied a need to launch a large variety of increasingly heavy satellites and spacecraft. Including the 14 Titan II missiles reconfigured into launch vehicles after the missiles themselves were retired, a huge variety of Titan vehicles had launched a plethora of payloads into space. From the first launches of actual payloads by Titan II-Gemini and Titan III in 1965 until the last Titan IVB launch in 2005, Titans had served as launch vehicles for forty years.[96] Meanwhile, if the handwriting was not yet quite on the wall by 1991, it had become clear by 1995 that even in its Titan IVB configuration, the Titan family of launch vehicles was simply too expensive to continue very far into the twenty-first century. Based on studies from the late 1980s and early 1990s, the Air Force had come up with what it called the Evolved Expendable Launch Vehicle (EELV) program to replace the then-existing Delta II, Atlas II, Titan II, and Titan IV programs with a family of boosters that would cost 25 to 50 percent less than their predecessors but could launch 2,500 to 45,000 pounds into low Earth orbit with 98 percent reliability, well above that achieved historically by the Titan family.

In August 1995 the Air Force had granted $30 million contracts to four firms—Alliant Techsystems (which had acquired Hercules Aerospace, the former Hercules Powder Company, in March 1995), Boeing, Lockheed Martin, and McDonnell Douglas—to develop proposals for the EELV family. Only Alliant Techsystems used a Titan component as part of its propulsion package, and in December 1996 the Air Force selected McDonnell Douglas (later acquired by Boeing) and Lockheed Martin, not Alliant Techsystems, to continue with development.[97]

Discussion of actual EELV development is beyond the scope of this book. But the launch of Lockheed Martin's Atlas V on August 21, 2002, followed by Boeing's Delta IV on November 20, 2002, offered perspective on the future of the Titan family. These successes and prospects for the Atlas V and Delta IV families of rockets they represented seemed to suggest that the EELV launchers would capture most of the military market. As of August 2002, the Air Force reportedly believed that "through 2020 the two new EELV families will reduce the cost per pound to orbit to $7,000 compared with $20,000 for the old booster fleet." This would "save the U.S. government $10 billion in launch costs—a 50% savings compared with launching the same military payloads on old Delta, Atlas and Titan boosters." Because of these new launchers with their lower costs, the Titan family of launch vehicles ended its life with a successful launch in October 2005, as we have seen. This final success provided a fitting conclusion to an important program.[98] The Titans had played a major role for a long time, but they were too expensive for the fiscally constrained space program of the early 21st century.

7

The Space Shuttle, 1972–1991

The space shuttle marked a radical departure from the general pattern of previous launch vehicles. Not only was it, unlike its predecessors, a (mostly) reusable launch vehicle; it was also part spacecraft and part airplane. In the Mercury, Gemini, and Apollo launch vehicles, astronauts had occupied the payload of the rocket, but astronauts on the shuttle rode in and even piloted from a crew compartment in the launch vehicle itself. Also, the shuttle commander landed the orbiter portion of the craft and did so horizontally on a runway. The orbiter had wings like an airplane and set down on landing gear as airplanes do. Indeed, the very concept of the space shuttle came from the idea of airliners, which were not discarded after each mission the way expendable launch vehicles had been but were refurbished, refueled, and used over and over again, greatly reducing the cost of operations.

Because of the multifaceted character of the space shuttle, its antecedents are much more diverse than those of the expendable launch vehicles and missiles discussed in *Preludes to U.S. Space launch-vehicle Technology* and in the rest of this book. Many aspects of the orbiters thus fall outside our scope. This chapter focuses on features most comparable to those of earlier launch vehicles—propulsion, guidance and control, and, to a lesser extent, structure.[1]

Studies of a reusable launch vehicle date back a long way and continued through the 1960s. But it was not until the early 1970s that budgetary realism forced planners to accept a compromise version of early schemes. In the post-Apollo era, the administration of President Richard M. Nixon faced numerous challenges included the continuing conflict in Vietnam, vociferous antiwar protests, racial unrest, an economic recession, and a budget crisis. While the highly successful but expensive Apollo effort had demonstrated U.S. technological prowess to the world and in some sense defused the Sputnik crisis, the war in Southeast Asia had stimulated a decade of self-criticism in the United States and the rise of a counterculture that rejected the bourgeois values of the 1950s. Although proponents of a fully reusable launch vehicle compared the use of expendable boosters to throwing away an airplane or a railroad locomotive after each trip, in the environment of the early 1970s they had to bow to competing demands for funding and accept a partly reusable vehicle that was cheaper to develop but far less economical to use than they had hoped.[2]

The NASA budget had already been declining since the mid-1960s, dropping from a high of $5.25 billion in 1965 to $3.75 billion in 1970 and $3.31 billion in 1972 (which, inflation-adjusted to 2006 dollars, would equal $32.98 billion, $19.92 billion, and $15.92 billion). By 1972, both the Air Force and NASA had contracted for a number of studies of space shuttle concepts. In 1968 NASA had introduced a phased project planning approach to research and development in which Phase A comprised advanced studies, Phase B project definition, Phase C actual vehicle design, and Phase D production and operations. The Phase B statement of work in February 1970 called for the shuttle to be a two-stage vehicle that was fully reusable. Then in May 1971 the Office of Management and Budget advised NASA not to expect a budget increase over the next five years. At that time, plans for the fully reusable shuttle showed a development cost of nearly $10 billion with a peak annual cost of about $2 billion. OMB's funding projection—which proved to be roughly accurate in dollar terms, with an actual decline in real terms—meant that NASA could afford to spend only about $1 billion per year for five years to develop the shuttle and still fund other programs. This grim reality led NASA in the course of 1971–72 to change to a stage-and-a-half shuttle concept that was only partly reusable.[3]

For the two-stage shuttle, there had been a fly-back booster and a separate fly-back orbiter. The projected reduction in funding made the fly-back booster too expensive, so gradually NASA and its Phase B contractors, working together, shifted their focus to designs featuring an orbiter with an external propellant tank that would not be recoverable. This cut development costs by permitting the orbiter to be smaller and lighter, but it imposed a penalty in the form of additional costs per launch. Then two of the Phase B contractors, McDonnell Douglas and Grumman, separately urged combining the external tank with strap-on solid rocket boosters to augment the thrust of the orbiter's engines. Despite opposition by Marshall Space Flight Center to the use of solids and despite their higher overall cost, solid rocket boosters with a 156–inch diameter offered lower development costs than other options, for a substantial saving in the near term, which was becom-

Figure 50. Space shuttle *Columbia* lifting off on its second mission into space from NASA's Kennedy Space Center, Florida, November 12, 1981. This photograph shows one of the two giant solid rocket boosters firing to provide the lion's share of the orbiter's lift for the first 24 nautical miles of ascent into space. NASA Historical Reference Collection, Washington, D.C. Courtesy of NASA.

ing critical from the budgetary perspective. By mid-March 1972 the basic configuration emerged for the shuttle that would actually be developed: a delta-winged orbiter attached to an external tank flanked by two solid rocket boosters. Meanwhile, on January 5, President Nixon had announced his support for development of a space shuttle that would give the country "routine access to space by sharply reducing costs in dollars and preparation time."[4]

Space Shuttle Main Engines

While the general configuration of the space shuttle was still in flux, on June 10, 1971, NASA's associate administrator for manned space flight, Dale D. Myers, dispatched a memorandum to the directors of the Manned Space-craft Center (MSC), Marshall Space Flight Center, and Kennedy Space Center in which he communicated the management plan for the space shuttle, approved by NASA administrator James C. Fletcher. This gave lead-center responsibilities to MSC in Houston but retained general direction of the program in Washington at NASA Headquarters. MSC would have responsibility for systems engineering and integration of the components, with personnel from the other two centers on assignment in Houston to support that effort. Marshall would take charge of the main propulsion elements, while Kennedy would manage the "implementation of the shuttle orbiter," meaning, presumably, design of launch and recovery infrastructure and management of launch operations.[5]

Myers had managed the Navaho missile effort for North American and had become vice president of its Space Division, where he had been the general manager for the Apollo spacecraft. He had also overseen North American Rockwell's studies for the space shuttle, and he had experience with aircraft projects as well. Thus he came to his new job with a strong background in all aspects of the shuttle—as launch vehicle, spacecraft, and airplane. Marshall's von Braun had moved on in 1970 to become deputy associate administrator for planning at NASA Headquarters, leaving that job in 1972 for a position as vice president for engineering and development with Fairchild Industries in Germantown, Maryland. His deputy director for scientific and technical matters, Eberhard Rees, had taken over as Marshall center director until Rees himself retired in 1973 and was succeeded by Rocco A. Petrone, a mechanical engineer with a doctorate from MIT. Petrone had come from NASA Headquarters and returned there in 1974, to be succeeded in turn by William R. Lucas, a chemist and metallurgist with a doctorate from Vanderbilt University who had worked since 1952 at Redstone Arsenal and then Marshall, where he became deputy director in 1971. Petrone reorganized Marshall, de-emphasizing in-house capabilities to oversee and test large project components and giving more authority to project officers, less to lab directors—a direction Myers approved. As Rees put it, Myers was "somewhat allergic to 'too much' government interference" with contractors, preferring less stringent oversight than Marshall had provided in the past.[6]

In February 1970, Marshall released a request for proposals for the Phase

B study of the space shuttle main engine (SSME). Contracts went to Rocketdyne, Pratt & Whitney, and Aerojet General. The engine was to burn liquid hydrogen and liquid oxygen in a 6:1 ratio at a combustion-chamber pressure of 3,000 psi, well above that of any previous production engine including the Saturn J-2, which had featured a pressure of about 787 pounds at the injector end of the 230,000–pound-thrust version. The shuttle engine was to produce a thrust of 415,000 pounds of force at sea level or 477,000 pounds at altitude. Although Rocketdyne had built the J-2 and a development version, the J-2S, with a thrust of 265,000 pounds and a chamber pressure of 1,246 psi, Pratt & Whitney had been developing an XLR129 engine for the Air Force Rocket Propulsion Laboratory designed for 250,000 pounds of thrust and a chamber pressure of 2,740 psi. The engine actually delivered 350,000 pounds of thrust and operated at 3,000 psi during 1970.[7] Pratt & Whitney thus seemed to have a distinct advantage in the competition.

At Rocketdyne, seasoned rocket engineer Paul Castenholz, who had helped troubleshoot the F-1 combustion-instability and injector problems and had been project manager for the J-2, headed the SSME effort as its first project manager even though he was a corporate vice president. Recognizing that there was not time to build sophisticated turbopumps, he decided to build a complete combustion chamber fed by high-pressure tanks. The NASA study contract did not provide funds for such an effort, so Castenholz persuaded North American Rockwell's president, Robert Anderson, to approve up to $3 million in company funds. By 1971, testing the engine at Nevada Field Laboratory near Reno, Rocketdyne had a cooled combustion chamber that achieved full thrust for 0.45 seconds. The thrust amounted to 505,700 pounds at a chamber pressure of 3,172 psi, exceeding the performance of Pratt & Whitney's XLR129 by a considerable margin.[8]

Although the long hours Castenholz spent on this engine apparently contributed to a divorce, he had provided his company with an engine that would ultimately win the competition. Funding constraints led to combining Phase C and Phase D contracts, so on March 1, 1971, Marshall released to the three Phase B contractors a request for proposals to design, develop, and deliver thirty-six engines. In July NASA announced the selection of Rocketdyne, but Pratt & Whitney soon protested to the General Accounting Office that the choice was "manifestly illegal, arbitrary and capricious, and based upon unsound, imprudent procurement decisions." On March 31, 1972, the GAO finally decided the case in favor of Rocketdyne, with the contract signed on August 14. Pratt & Whitney's protest delayed development of the engine, although Rocketdyne worked under interim and letter contracts until the final signature.[9]

It was not until May 1972 that Rocketdyne could begin significant work on the space shuttle main engine in something close to its final configuration, and some design parameters would change even after that. By then, however, NASA had settled on a "parallel burn" concept in which the main engines and the solid rocket boosters would both ignite at ground level. The space agency had already decided in 1969 that the engine would employ something called "staged combustion." This contrasted with the system in the Saturn engines, where the exhaust from the gas-generator-operated turbines contributed little to thrust. In the shuttle the turbine exhaust—having burned with scant oxygen and thus still being rich in hydrogen—flowed back into the combustion chamber, where the remaining hydrogen burned under the high pressure and contributed to thrust. This was necessary in the shuttle because the turbines had to burn so much fuel to produce the high chamber pressure critical to the SSME's performance.[10]

"It was the high combustion chamber pressure combined with the amplification effect of the staged combustion cycle that made this engine a quantum jump in rocket engine technology and created a significant challenge to the contractor and government team charged with its design and development," according to Robert E. Biggs, who directed the test program for engines and who became a member of the space shuttle main engine management team for Rocketdyne in 1979. Unlike the H-1 and F-1, the SSME appears not to have been plagued by combustion instability. Castenholz and his engineers had started engine development with an injector based on the J-2, which had shown good stability. It was transpiration cooled by a flow of gaseous hydrogen. According to Biggs, Rocketdyne had added "two big preventors [of instability] on an injector that was basically stable to begin with." By preventors, he evidently meant coaxial baffles, which apparently worked well.[11]

The XLR129 had been a staged-combustion engine, and its success had given NASA and industry the confidence to use the same concept on the shuttle. But timing for such an engine was both intricate and sensitive, as Rocketdyne and Marshall would learn. Rocketdyne's design used two preburners with both low-pressure and high-pressure turbopumps to feed each of the propellants to the combustion chamber and provide the high pressure required. The XLR129 had used only a single preburner, but two of them provided finer control in conjunction with an engine-mounted computer, subcontracted to Honeywell for development. This computer would monitor and regulate the propulsion system during start, automatically shut it down if it sensed a problem, throttle the thrust during operation, and shut down the engine once it had completed its mission.[12]

By the winter and spring of 1974, the Honeywell controller had experienced difficulties with its power supply and interconnect circuits. These problems were severe enough to attract the attention of NASA administrator Fletcher and his deputy, George M. Low. Low commented that Rocketdyne had done a "poor job" of controlling Honeywell, which in turn had done a "lousy job" and was in "major cost, schedule, and weight difficulty." Moreover, Rocketdyne had fallen behind in converting test stands at Santa Susana for testing components of the engine itself, including the turbopumps. Not only had the schedule slipped, but there was a cost overrun of about $4 million that required congressional reprogramming. In a program that was underfunded to begin with, this was intolerable, and under pressure from Fletcher and Low, Rockwell International, as the firm had become in 1973, shifted Castenholz to another position, replacing him first with Norman Reuel and then, when Reuel's heart condition forced a change, with Dominick Sanchini, a tough veteran who had led development of the main engine proposal in 1971. Despite twenty-seven years devoted to the rocket business, Castenholz would no longer contribute directly to launch-vehicle development. Meanwhile, about the same time, Marshall made J. R. Thompson its project manager for the SSME. A 1958 aeronautical engineering graduate of Georgia Institute of Technology, Thompson had worked for Pratt & Whitney before becoming a liquid propulsion engineer at Marshall on the Saturn project in 1963, the year he earned his master's in mechanical engineering from the University of Florida. He became the space engine section chief in 1966, chief of the man/systems integration branch in 1969, and main engine project manager in 1974.[13]

In May 1975 component testing began at Santa Susana and prototype engine testing started at NASA's National Space Technology Laboratories (formerly the Mississippi Test Facility, renamed in 1974 and, in 1988, again rechristened the John C. Stennis Space Center). Typically there was about a month between testing of a component and a related engine test in Mississippi. But test personnel soon learned that the highly complicated test hardware at Santa Susana was not adequate for its mission. As the NASA administrator in 1978, Robert Frosch, told the Senate Subcommittee on Science, Technology, and Space, "We have found that the best and truest test bed for all major components, and especially turbopumps, is the engine itself." Given the insufficient equipment, the program gradually ceased component testing at Santa Susana between November 1976 and September 1977.[14]

There were many problems during testing, especially with turbopumps and timing, but the scope of this history limits detailed coverage to a few. The initial problems involved timing, specifically "how to safely start and

Figure 51. A shuttle main engine being placed on the A-2 test stand at what later became the John C. Stennis Space Center in Mississippi in preparation for a test firing. Note the rows of tubing encircling the combustion chamber to provide regenerative cooling. Courtesy of NASA.

shut down the engine." After five years of analysis, Rocketdyne engineers had "sophisticated computer models that attempted to predict the transient behavior of the propellants and engine hardware during start and shutdown," as Biggs explained. Test personnel expected that the engine would be highly sensitive to minute shifts in propellant amounts, with the opening of valves being time-critical. Proceeding very cautiously, testers took twenty-three weeks and nineteen tests (with replacement of eight turbopumps) to reach 2 seconds into a 5–second start process. It took another twelve weeks, eighteen tests, and eight more turbopump changes to momentarily reach the minimum power level, which at that time was 50 percent of rated thrust. Eventually Biggs's people developed a "safe and repeatable start sequence" by using the engine-mounted computer, also called the main engine controller. "Without the precise timing and positioning" it afforded, probably they could not have developed even a satisfactory start process for the engine, so sensitive was it.

The process began with the purging and desiccation of the propulsion system with dry nitrogen and helium, as the cryogenic propellants could freeze any moisture left in the system. Then after a slow cooldown using the propellants, full opening of the main fuel valve started the fuel flow. This initially occurred from the latent heating and expansion the hardware (still warmer than the liquid hydrogen) imparted to the propellant. However, the flow was pulsating with a pressure oscillation of about 2 hertz (cycles per second) until pressure in the main thrust chamber stabilized after 1.5 seconds. Then oxygen flowed to the fuel and oxidizer preburners and the main combustion chamber in a carefully timed sequence such that liquid oxygen arrived at the fuel preburner 1.4 seconds after the full opening of the main fuel valve, at the main combustion chamber at 1.5 seconds (a tenth of a second later), and at the oxygen preburner at 1.6 seconds. Only test experience revealed that a key time was 1.25 seconds into the priming sequence. If the turbine speed in the high-pressure fuel turbopump at that precise moment was not at least 4,600 rpm, the engine could not start safely. So 1.25 seconds became a safety check point.

Also, if any "combustor prime" coincided with a downward oscillation (dip) in the fuel flow, excessively high temperatures could result. For example, if the liquid oxygen arrived late at the fuel preburner or early at the main combustion chamber, there would be "major burning of the engine hardware." Inaccurate timing could also destroy the high-pressure oxidizer turbopump. And an error of 1 or 2 percent in valve position or a timing error of as little as a tenth of a second could seriously damage the engine. So it is not surprising that the first test to achieve 50 percent of rated thrust did not occur until the end of January 1976, and it took another year to reach the rated power level of 375,000 pounds of thrust at sea level (470,000 pounds in the vacuum of space). Only at the end of 1978 did the engineers achieve a final version of the start sequence that would preclude the problems they had encountered in more than three years of testing. There were also difficulties with the "shutdown sequence," but they were less severe than those with starting the engine safely.[15]

One major turbopump problem occurred on March 12, 1976. Earlier tests of the high-pressure liquid-hydrogen pump, both at Santa Susana and in Mississippi, had revealed significant vibration levels, but until the March 12 test, engineers had not appreciated their seriousness. The prototype engine test on that day was supposed to last 65 seconds to demonstrate a 50 percent power level rising to 65 percent for a single second. The test did demonstrate 65 percent power for the first time, but engineers had to halt the test at 45.2

seconds because the high-pressure fuel turbopump was losing power. After the test, the pump could not be rotated with a tool used to test its torque, as was normally done. Investigation revealed a failure of the turbine-end bearings supporting its shaft. When test data showed a major loss in the efficiency of the turbines plus a large vibration with a frequency about half the speed of the pump's rotation, experts immediately recognized the symptoms of an instability in the dynamics of the rotors called subsynchronous whirl.

While recognizing the problem, they did not know what to do about it in a system whose turbine-blade stresses and tip speeds were at the outskirts of the state of the engineering art, so the program assembled a team that ultimately included the premier rotordynamics experts in government, industry, and academia—not only from the United States but also from Great Britain. The pump was centrifugal, driven by a two-stage turbine 11 inches in diameter that was designed to deliver 75,000 horsepower at a ratio of 100 horsepower per pound, an order-of-magnitude improvement over previous turbopumps. The team studied previous liquid-hydrogen turbopumps that had suffered from subsynchronous whirl, such as the one on the J-2. Following a program involving engine and laboratory tests, as well as tests of components and subsystems, the investigators found twenty-two possible causes, of which the most likely appeared to be hydrodynamic problems involving seals that had a coupling effect with the natural frequency of the rotating turbines. Efforts to decrease the coupling effect included damping the seals and stiffening the shaft. The fixes did not totally end the whirl, but they did delay its inception from 18,000 rpm, which was below the *minimum* power level, to 36,000 rpm, which was above the *rated* power level.

As these design improvements increased feasible operating speeds, investigators learned that a mechanism not related to subsynchronous whirl was overheating the turbine bearings, which had no lubrication but were cooled by liquid hydrogen as they operated. Extensive analysis of the cooling revealed that a free vortex was forming at the bottom of the pump's shaft where the coolant flowed. This vortex reduced the pressure, hence the flow of coolant. In a piece of cut-and-try engineering, designers introduced a quarter-sized baffle that changed the nature of the vortex and allowed more coolant to flow. With this fix and the relegation of the whirl problem to a power level above what was rated, by mid-January 1977 the high-pressure fuel turbopump could support long-duration tests of the engine for the first time.[16]

This problem with the fuel pump had delayed the program, but it was not as devilish as explosions in the high-pressure liquid-oxygen turbopump. If a fire started in the presence of liquid oxygen under high pressure, it quickly burned up the metal parts—including whatever part may have caused the problem—thus removing all evidence that could lead to a solution. Following resolution of the fuel-pump whirl problem, there were four fires in the high-pressure oxygen turbopump between March 1977 and the end of July 1980. The high-pressure oxygen turbopump was on the same shaft as the low-pressure oxygen turbopump that supplied liquid oxygen to the preburners. The common shaft rotated at nearly 30,000 rpm. The high-pressure pump was centrifugal and provided as much as 7,500 gallons of liquid oxygen at a pressure higher than 4,500 psi. The turbine supplied 28,000 horsepower with an efficiency of nearly 80 percent, just slightly less than that of its hydrogen counterpart. An essential feature of the pump's design was to keep the pumped liquid oxygen fully separated from the hydrogen-rich gas that drove its turbines. To ensure that separation, engineers and technicians had used a variety of seals, drains, and purges.

Despite such precautions, on March 24, 1977, an engine caught on fire so severely it removed most physical evidence. Investigators determined from instrumentation that the fire had started near a complex liquid-oxygen seal. Since it was not clear what a redesign should involve, testing on other engines resumed, indicating that one of the purges did not prevent the mixing of liquid oxygen and fluids draining from hot gas. On July 25, the team tried out a new seal intended as an interim fix while redesign went on, but the interim seal worked so well that it became the permanent solution, along with increasing the flow rate of the helium purge and other measures.[17]

On September 8, 1977, there was another disastrous fire originating in the high-pressure oxygen turbopump. Test data pointed to a gradual breakdown of bearings on each end of the turbopump's shaft, but there was no clear indication of why they failed. Fixes included enhanced coolant flow, better balance in the rotors, heavier-duty bearings, and new bearing supports. The other two fires did not involve design flaws but did produce delays.

In 1972 the shuttle program had expected to launch a flight to orbit by the beginning of March 1978. The engine and turbopump problems discussed above, and many others involving the propulsion system, were not the sole causes for missing that deadline, but engines certainly would have kept the shuttle from flying anywhere near that early if all else had gone as planned. By March 1978 the expected first-flight date had slipped to March 1979, and then an engine fire at the end of 1978 and other problems caused even a September 1979 launch to be postponed. While turbopumps were demonstrating longer trouble-free periods by early 1979, there continued to be failures in July and November 1980. Thus it was not until early 1981 that the space shuttle main engine fully qualified for flight. Problems had included turbine-blade failures in the high-pressure fuel turbopump, a fire involving the main oxidizer valve, failures of nozzle feed lines, a burn-through of the fuel preburner, and a rupture in the housing of the main fuel valve. But finally on April 12, 1981, the first space shuttle lifted off and the main engines performed with only a minor anomaly, a small change in mixture ratio caused by radiant heating in the vacuum of space. Some insulation and a radiation shield prevented the problem on subsequent flights.[18]

As a report evaluating the shuttle program concluded in 1981, "In assessing the technical difficulties that have been causing delays in the development and flight certification of the SSME at full power, it is important to understand that the engine is the most advanced liquid rocket motor ever attempted. . . . Chamber pressures of more than 3,000 psi, pump pressures of 7,000–8,000 psi, and an operating life of 7.5 hours have not been approached in previous designs of large liquid rocket motors."[19]

Each shuttle had three main engines, which could be gimballed 10.5 degrees in each direction in pitch but only 8.5 degrees in yaw. The engines could be throttled over a range from 65 to 109 percent of their rated power level, although there had been so many problems demonstrating the 109 percent level in testing that it was not available on a routine basis until 2001. Moreover, the 65 percent minimum power level (changed from an original 50 percent) was unavailable at sea level because of flow separation. During launch, the three main engines ignited before the solid rocket boosters. When computers and sensors verified that the engines were providing the proper thrust level, the SRBs ignited. To reduce vehicle loads during the period of maximum dynamic pressure (reached at about 33,600 feet some 60 seconds after liftoff) and to keep vehicle acceleration at a maximum of three Gs, the flight control system throttled back the engines during this phase of the flight. Throttling also made it feasible to abort the mission either with all engines functioning or with one of them out.[20]

At 100 percent of the rated power level, each main engine provided 375,000 pounds of thrust at sea level and 470,000 pounds at altitude. The minimum specific impulse was above 360 lbf-sec/lbm at sea level and 450 at altitude. This was substantially higher than the J-2 second- and third-stage Saturn engine, which had a specific impulse of somewhat more than 290 lbf-sec/lbm at sea level and 420 at altitude. The J-2's thrust levels were also substantially lower—161,400 pounds at sea level (where they weren't used) and 230,000 at altitude. Of course, the space shuttle main engines were considerably less powerful than the Saturn V's F-1s with their 1.522 million pounds of thrust per engine, but the SSMEs were much more sophisticated. At a length of 13.9 feet and a diameter of 8.75 feet, the shuttle engines were also substantially smaller than the F-1s, which had a length of 19.67 feet and a diameter of 12.25 feet. Still, they were impressively large, standing twice as tall as most centers in the National Basketball Association.[21]

Solid Rocket Boosters

Because they ignited before launch, the space shuttle main engines did perform some of the same functions for the shuttle as the F-1s did for the Saturn V, but in most respects it was the twin solid rocket boosters on the orbiters that served as the principal sources of thrust. They provided 71.4 percent of the shuttle's power at liftoff and during the initial stage of ascent until about 75 seconds into the mission, when they separated from the orbiter for later recovery and reuse.[22]

The decision in March 1972 to use solid rocket boosters on the shuttle placed Marshall Space Flight Center in a difficult position. Although von Braun's engineers had been involved in developing the solid-propellant Pershing missile in the two years before they transferred to NASA and to some extent thereafter, their experience was primarily with liquid propellants, which they preferred. However, facing threats from the Office of Management and Budget to close the center, Marshall could hardly avoid cheerfully accepting the development of the solid boosters to bolster its value to the nation. Probably to that end, NASA administrator Fletcher gave Marshall responsibility for integrating the components of the boosters rather than contracting that function out as some NASA officials proposed.[23]

Even before the decision to use solid rocket boosters, Marshall had provided contracts of $150,000 each to the Lockheed Propulsion Company, Thiokol, United Technology Center, and Aerojet General to study configurations of such motors. Using information from these studies, NASA issued a request for proposals (RFP) on July 16, 1973, to which all four companies responded with initial technical and cost proposals in late August 1973, followed by final versions on October 15. Because the booster cases would be recoverable, unlike previous strap-on motors like those for Titan III, and because they had to be rated to carry astronauts, they needed to be sturdier than their predecessors. Lockheed, UTC, and Thiokol all proposed segmented cases without welding. But although Aerojet had been an early developer of such cases, it ignored a requirement in the RFP and proposed a welded case without segmentation. It argued that such a case would be lighter, less costly, and safer, with transportation to launch sites being provided from Aerojet's production site by barge. (At that time, shuttle launches were expected from both Kennedy Space Center and Vandenberg, although Vandenberg was in fact never used.) Had Aerojet's proposal won the competition, the *Challenger* disaster of 1986 might never have occurred, because its cause involved the SRBs' segmented joints. However, the source evaluation board with representatives from five NASA centers and the three military services ranked Aerojet last, with a score of 655 for mission suitability, while Lockheed, Thiokol, and UTC got scores of 714, 710, and 710. With such close technical ratings, the board selected Thiokol as winner of the competition, based on its quoting the lowest cost of the three, and also on its managerial strengths. NASA announced the selection on November 20, 1973.[24]

Since Thiokol had plants in Utah and NASA administrator Fletcher was from that state, the decision was controversial. Lockheed protested the award, but the General Accounting Office decided on June 24, 1974, that "no reasonable basis" existed to question the validity of NASA's decision. Thiokol, meanwhile, had proceeded with design and development on the basis of interim contracts, with the final one for design awarded on June 26, 1974, followed by one for development, testing, and production on May 15, 1975.[25]

Part of the reason why it would be cheaper to develop large solid-propellant boosters rather than liquid-propellant boosters for the shuttle was the prior existence of the solid rocket motors for Titan III. But an earlier contributing factor was the Air Force's Large Segmented Solid Rocket Motor Program, of which Aerojet's testing of 100–inch-diameter solids in the early 1960s had been a forerunner. In late 1962 what soon became the Air Force Rocket Propulsion Laboratory at Edwards Air Force Base inaugurated Air Force Program 623A to develop large solid motors that DoD and NASA could use as space-launch vehicles. The initial funding for 120– and 156–inch-diameter segmented motors and for continuing work on systems for thrust vector control came from the Air Force. NASA then paid for part of the 156–inch and all of a 260–inch program. In the course of the testing of thrust vector control systems, Lockheed had developed a Lockseal mounting structure that allowed the nozzle to gimbal, and Thiokol later scaled it up to the size required for large motors, redesignating it Flexseal.

Lockheed tested both 120– and 156–inch motors in the program, and Thiokol tested 156–inch motors with both gimballed (Flexseal) and fixed nozzles. These tests concluded in 1967, as did those for 260–inch-diameter motors by Aerojet and Thiokol. There were no direct applications of the 260–inch technologies, but Thiokol's participation in the 120– and 156– inch portions of the Large Segmented Solid Rocket Motor Program gave it experience and access to designs, materials, fabrication methods, and test results that contributed to its development of the solid rocket boosters for the space shuttle. It also drew upon its experience with Minuteman.[26]

The design for the solid rocket booster was intentionally conservative, using a steel case of the same type (D6AC) used on Minuteman and Titan IIIC. The Ladish Company of Cudahy, Wisconsin, made the cases for each segment without welding, using the process called rolled-ring forging that it had helped to develop for Titan IIIC. In this process, technicians punched a hole in a hot piece of metal and then rolled it to the correct diameter. For the shuttle, the booster's diameter turned out to be 12.17 feet (146 inches) and the overall length 149 feet. Each booster had four segments plus fore and aft sections. The propellant consisted of the same three principal ingredients used in the first stage of the Minuteman missile, ammonium perchlorate for the oxidizer, aluminum for the fuel, and PBAN polymer as the binder. Its grain configuration was an eleven-point star in the forward end merging into a large, smooth, tapered cylindrical shape. This yielded a theoretical specific impulse of more than 260 lbf-sec/lbm.[27]

Thiokol was not responsible for the entire solid rocket booster. Fletcher had decided that the "SRB is to be designed in-house with the exception of the SRM," meaning in practice that Marshall "did design a number of the systems in-house, such as the structural components," according to George Hardy, project manager for the SRB at Marshall from 1974 to 1982. But Thiokol was responsible for "development, fabrication, and qualification of the motor," according to W. P. Horton, Marshall's chief engineer on the project; "we did not do that in-house, but we integrated it into the rest of the system like the structures, the thrust vector control, the electronic black boxes, and the parachutes" used for recovery of the cases.[28]

Further complicating the mix of responsibilities for the booster, on December 21, 1973, NASA selected United Space Boosters—a subsidiary of UTC's parent firm but formed with people from UTC and operating under its aegis—to assemble, check out, launch, and refurbish the solid rocket boosters. For the first six shuttle flights, Marshall retained a significant oversight role as, in effect, its own prime contractor, but thereafter United Space Boosters assumed responsibility for "engineering and flight evaluation."[29]

Within this complex arrangement involving Thiokol and United Space

Boosters, Marshall strove essentially "to avoid inventing anything new," in Hardy's words. The best example of this approach was the PBAN propellant. Many subsequent innovations had provided higher performance, but now, with cost and human-rating being prime considerations, Thiokol employed a tried-and-true propellant used on the first stage of Minuteman and also in the Navy's Poseidon missile. As John Thirkill, Thiokol deputy director for the booster, said in 1973, "Over the last fifteen years, we've loaded more than 2,500 first stage Minuteman motors and around 500 Poseidon motors with this propellant."[30]

The configuration of the propellant grain provided for the thrust to vary so as to provide the boost required for the planned trajectory but to limit the acceleration to 3 Gs for the astronauts. For the first six shuttle missions, the initial thrust was 3.15 million pounds per booster. The eleven-point star in the forward section of the SRB had long, narrow points, providing an extensive burning surface. This declined progressively as the points burned, reducing the thrust as the point of maximum dynamic pressure approached at about 60 seconds into the launch. At 52 seconds after liftoff, flames had consumed the star points to form a cylindrical perforation in the forward as well as the rear segments of the booster. As this burned, expanding its diameter, the thrust increased slightly from the 52nd to about the 80th second. Thereafter it tapered off to zero as the burning consumed the propellant at about the 120th second. Then the SRBs separated from the rest of the shuttle and fell oceanward, slowed by parachutes.[31]

A major drawback of the PBAN propellant was that about 20 percent of the weight of its exhaust consisted of toxic and corrosive hydrogen chloride, which could damage the ozone layer that protected Earth from most ultraviolet radiation. NASA studies showed, however, that ozone depletion would be slight, so a shift to a less powerful propellant that produced less hydrogen chloride was deemed unnecessary.[32]

Once the Ladish Company had forged the motor cases in Wisconsin, the segments traveled by rail to a firm named Cal Doran near Los Angeles. There, heat treatment imparted greater strength and toughness to the D6AC steel. Then the segments went further south to Rohr Industries in Chula Vista, close to San Diego. That firm added tang-and-clevis joints to the ends of the segments. For these critical elements, shuttle designers had departed from the Marshall avoid-inventing-anything-new mantra. Although superficially the shuttle "field joints"—so called because they were assembled in the field, at the launch site—resembled those for Titan IIIC, in many respects they were different.

One key difference lay in the orientation of the joint. For the Titan solid rocket motor, the single tang pointed upward from a lower segment of the case and fit into the downward-projecting two-pronged clevis of the segment above, which encased it. This arrangement protected the joint from rain or dew dripping down the case and entering the joint. For reasons that are not clear, the shuttle reversed this direction.

A second major difference was that, where the Titan joint had used only one O-ring in an indentation on the inside of one segment of the clevis, encompassing the tang and sealing against escaping gases, the shuttle employed two O-rings. (See figure 55, p. 309.) Insulation on the inside of the Titan motor case protected the case, and with it the O-ring, from excessive heating. To keep this insulation from shrinking in cold temperatures and allowing a gas blow-by when the motor was firing, there were heating strips on the Titan motor. Both the Titan and the shuttle also used putty to improve the seal provided by the O-ring(s), but when the shuttle added the second O-ring for supposed further insurance, it did not include heating strips. A further difference in the joints was in the number of pins fitting into holes in the tang and clevis and linking them. The Titan motor used 240 such pins; the shuttle, despite its larger diameter, had only 177.[33] There is no certainty in counterfactual history, but perhaps if the shuttle designers had simply accepted the basic design of the Titan tang-and-clevis joints, the *Challenger* accident, in which hot gases leaking through a field joint ignited the external tank, would not have occurred.

Unlike the field joint, the nozzle for the solid rocket boosters did follow the precedents of the Titan solid rocket motors and the Large Segmented Solid Rocket Motor Program. Like the Titan motors, the shuttle employed carbon phenolic throats to ablate from the extremely hot flow of combustion products. In the shuttle motor, the propellants burned at 5,700°F, so ablation was needed for the carbon phenolic to become vaporized and thereby carry off heat so as to prevent thermal-stress cracking followed by ejection of portions of the nozzle. As of June 1979, the expansion ratio of the nozzle was 7.16:1, a ratio used for the first seven missions. Starting with the eighth mission, in August 1983, nozzle modifications improved the initial thrust of each motor from 3.15 to 3.3 million pounds. The exit cone was extended by 10 inches, while a 4-inch decrease in the diameter of the nozzle throat raised the expansion ratio to 7.72:1 and thus raised the velocity of the gases from the motor passing through the nozzle.[34]

The nozzle was partially submerged, and for gimballing it used the Thiokol Flexseal design. It was capable of 8 degrees of deflection, necessitated among other reasons by the shuttle's now-familiar roll soon after liftoff to achieve its proper trajectory. Having less thrust than the solid rocket boosters, the space shuttle main engines were incapable of achieving the necessary amount of roll. Nor would the liquid-injection thrust vector control used on the Titan solid rocket motors have met the more demanding requirements of the shuttle. Hence the importance of the Lockseal-Flexseal development during the Large Segmented Solid Rocket Motor Program supported by both NASA and the Air Force.[35]

Although only four segments of the solid rocket boosters were linked by field joints, there were actually eleven sections tied together by tang-and-clevis joints. Once they had been through machining and fitting processes, they were assembled at the factory into four segments into which Thiokol loaded the propellant. The joints assembled at the factory were called factory joints, as distinguished from the field joints that technicians assembled at Kennedy Space Center. Thiokol poured and cast the propellant at its factory in Brigham City, Utah, usually doing so in matched pairs from the same batches of propellant in order to reduce thrust imbalances. At different times the solid rocket motors used four different types of D6AC steel cases, which varied slightly in thickness.[36]

In part because of its comparative simplicity, the solid rocket booster required far less testing than the liquid-propellant main engine. Whereas the SSME had needed 726 hot-fire tests and 110,000 seconds of operation for certification, the solid rocket booster required only four development and three qualification tests, with operation of less than 1,000 seconds total. However, there were other tests. One was a hydroburst test on September 30, 1977, at Thiokol's Wasatch Division, also in Utah, which demonstrated that an empty case could withstand the pressures it would encounter during launch without cracking. A second hydroburst test on September 19, 1980, with only the aft dome, two segments, and the forward dome, was also successful. There were other tests of the tang-and-clevis joints that put them under pressure until they burst. Required to survive stresses up to 1.4 times those expected during launch, they actually withstood pressures of between 1.72 and 2.27 times the maximum expected in flight.[37]

The first developmental static test, DM-1 at Wasatch on July 18, 1977, was successful, but the maximum thrust the motor delivered was only 2.9 million of an expected 3.1 million pounds. There were other anomalies, including excessive erosion in parts of the nozzle, leading to such modifications as more ammonium perchlorate in the propellant and changes in nozzle coatings. On January 18, 1978, DM-2 was another success but again led to adjustments in the design. It turned out that the rubber insulation and polymer liner protecting the case were thicker than necessary. Thinning them lowered their weight from 23,900 to 19,000 pounds. There were also modifications in the igniter, grain design, and nozzle coating to reduce the flame intensity of the igniter, the rate of thrust increase for the motor, and erosion of portions of the nozzle. As the motor for DM-3 was being assembled, study of the DM-2 casing revealed that an area with propellant had been burning between segments. This required disassembling the DM-3 motor and increasing the thickness of a noncombustible inhibitor on the end of each segment. Designers also extended the rubber insulation to protect the case at the joints. This delayed the DM-3 test from July to October 19, 1978.

Again the test was satisfactory, but while the thermal protection on the nozzle was effective, the igniter still caused the thrust to rise too quickly. Designers could see no evident solution to the rapid rate of thrust increase, an apparent tacit admission that engineers did not fully understand the complex combustion process. It did seem evident, though, that the rate had to rise quickly to preclude thrust imbalances between the two motors, so they went back to an igniter design closer to that used in the DM-1 test and simply accepted the rapid thrust rise, at least for the moment. On Febru-

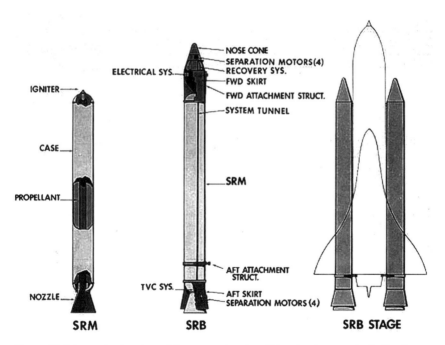

Figure 52. Schematic drawing of the space shuttle solid rocket booster including its components. Courtesy of NASA.

ary 17, 1979, DM-4 ended the four developmental tests with a successful firing. The qualification tests, QM-1 through QM-3 from June 13, 1979, to February 13, 1980, were all successful. These seven tests furnished the data needed to qualify the solid rocket motor for launch—excluding the electronics, hydraulics, and other components not the responsibility of Thiokol. Other tests on booster recovery mechanisms, complete booster assemblies, loads on the launchpad and in flight, and internal pressure took place at Marshall and at the National Parachute Test Range, El Centro, California. These other tests were all completed by late May 1980, well before the first shuttle flight in 1981.[38] Of course, this was well after the first planned flight originally scheduled for March 1978. So if main-engine development had not delayed the flights, presumably the booster development could have done so on its own.

External Tank

The third part of the main shuttle propulsion system was the external tank (ET), the only major nonreusable part of the launch vehicle. It was the largest and, when loaded, the heaviest part of the space shuttle at about 154 feet in length and 27.5 feet in diameter. To find a contractor to design and build it, NASA issued a request for proposals to Chrysler, McDonnell Douglas, Boeing, and Martin Marietta on April 2, 1973. Rockwell had already won the contract for the orbiter, so NASA excluded it from the competition, but Rockwell nevertheless teamed with Chrysler. All four bidders submitted their proposals on May 17. The source selection board gave the highest technical ratings to Martin Marietta and McDonnell Douglas. Martin argued that it alone among the bidders had relevant experience, since Titan III, with a core vehicle between two large solid rocket motors, had a design similar to the shuttle's. Martin's bid was by far the lowest of the four, although the board recognized that the firm was bidding below true expected costs—"buying in," as it was called. But as NASA deputy administrator George Low said, "We nevertheless strongly felt that in the end Martin Marietta costs would, indeed, be lower than those of any of the other contenders." Consequently, on August 16, 1973, NASA selected Martin Marietta (Denver Division) to negotiate a contract for the design, development, and testing of the external tank, a selection that the other competitors did not protest. NASA required assembly of the structure to occur at the Michoud facility near New Orleans.[39]

The external tank seemed—in some people's views—to pose few technological demands. James Kingsbury, head of Marshall's Science and Engineering Directorate, said, "There was nothing really challenging technologically in the Tank. . . . The challenge was to drive down the cost." Similarly, Larry Mulloy, who was Marshall's project manager for the solid rocket booster but also worked on the tank, said, "There was no technological challenge in the building of the External Tank. The only challenge was building it to sustain the very large loads that it has to carry, and the thermal environment that it is exposed to during ascent" within a weight limit of about 75,000 pounds. As it turned out, however, this was in fact a major hurdle, which only came to be fully appreciated after the loss of space shuttle *Columbia* on February 1, 2003, to a "breach in the Thermal Protection System on the leading edge of the left wing" resulting from its being struck by "a piece of insulating foam" from an area of the external tank known as the bipod ramp (according to the diagnosis of the Columbia Accident Investigation Board). During reentry into the atmosphere, this breach allowed aerodynamic superheating of the wing's aluminum structure, its melting, and the subsequent breakup of the orbiter under increasing aerodynamic forces.[40]

As Mulloy's statement suggested, the external tank had to carry the cryogenic liquid-hydrogen and liquid-oxygen propellants for the three shuttle main engines. It also served as the "structural backbone" for the shuttle stack and had to withstand substantial heating as the shuttle accelerated to supersonic speeds through the lower atmosphere where the dynamic pressures were high. This heating was much more complex than on a launch vehicle like the Saturn V. At the top, the tank needed only to withstand the effects of high-speed airflow. But farther down the stack, the tank's insulation had to encounter complex shock waves as it passed through the transonic speed range (roughly Mach 0.8 to 1.2). As the airflow became supersonic, these shock waves came from the boosters, the nose of the orbiter, and the structural attachments connecting the tank, boosters, and orbiter. As the waves impinged on the sides of the external tank, they created heating rates up to 40 BTUs per square foot per second. This was a much smaller heating load than that facing a nose cone reentering the atmosphere, but it was substantial for the thin aluminum sheeting of which the external tank was formed in order to reduce the weight.[41]

As designers examined the requirements for the external tank, they found that not even the arrangement of the hydrogen and oxygen tanks involved the simple application of lessons from the Centaur and the Saturn upper stages. In both, the liquid-hydrogen tank was above the liquid-oxygen tank. Since liquid oxygen was six times as heavy as liquid hydrogen, this arrangement made it unnecessary for the hydrogen tank to be strengthened to support the heavier oxygen during the strains of liftoff. For the shuttle, however, the engines were not directly under the tanks, as they were for the Saturn and Centaur, but were off to one side. With the heavy oxygen tank on the bottom of the external tank, its weight would have created an inertial force difficult to overcome by gimballing the SSME and the SRB nozzles. Especially after separation of the solid boosters, the weight of the oxygen tank would have tended to cause the orbiter to spin around the tank's center of gravity. Placing the oxygen tank on top moved the shuttle stack's center of gravity well forward, making steering much more feasible. But it also forced designers to make the liquid-hydrogen tank—and also an intertank structure between it and the oxygen tank—much sturdier than had been necessary on the Saturn upper stages.[42]

This, in turn, compounded a problem with the external tank's weight. The initial empty weight allowance for the ET had been 78,000 pounds, but in 1974, Johnson Space Center in Houston (renamed from the Manned Spacecraft Center in 1973) lowered the goal to 75,000 pounds. Moreover, NASA asked Martin Marietta not only to reduce the weight but to do so at no additional cost. In fact, the space agency suggested that it would be helpful actually to reduce the cost. Even though Marshall lowered the safety factor for the ET, the initial standard-weight tank used on shuttle flights 1 through 5 and 7 weighed some 77,100 pounds. But through concerted efforts, Martin Marietta was able to achieve a 10,300–pound weight reduction for the lightweight tanks first used on flight 6. It did this through a variety of design changes that included eliminating some portions of stringers (longitudinal structural stiffeners) in the hydrogen tank, using fewer circumferential stiffeners, milling some portions of the tank to a lower thickness, using a stronger type of aluminum that allowed thinner sections, and redesigning antislosh baffling.[43]

The resultant external tank included a liquid-hydrogen tank that comprised 96.66 feet of the ET's roughly 154–foot length. It had semi-monocoque design with fusion-welded barrel sections, forward and aft domes, and five ring frames. It operated at a pressure range of 32–34 psi and contained an antivortex baffle but no elaborate antislosh baffles because the lightness of the liquid hydrogen made its sloshing less significant than that of liquid oxygen. The feed line from the tank allowed a maximum flow rate of 48,724 gallons per minute from its 385,265–gallon capacity. The intertank structure was much shorter at 22.5 feet. Made of both steel and aluminum, it too was semi-monocoque with a thrust beam, skin, stringers, and panels. It contained instrumentation and a device called an umbilical plate for supply of purge gas, detection of hazardous gas escaping from the tanks, and boil-off of hydrogen gas while on the ground. The intertank also had a purge system that removed the highly combustible propellants if they escaped from their tanks or plumbing fixtures.

Above the intertank was the liquid-oxygen tank. Its 49.33 feet of length, combined with those of the intertank and the liquid-hydrogen tank, exceeded the total length of the ET because it and its liquid-hydrogen counterpart extended into the intertank. The liquid-oxygen tank was an aluminum monocoque structure operating with a pressure range of 20–22 psi. It allowed a maximum of 19,017 gallons of liquid oxygen to flow to the main engines when they were operating at 104 percent of their rated thrust. Containing both antislosh and antivortex mechanisms, the tank had a capacity of 143,351 gallons of oxidizer.[44]

The thermal protection system for the external tank not only had to with-

Figure 53. The first external tank for the space shuttle rolling off the assembly line at the Michoud Assembly Facility in New Orleans on September 9, 1977. The external tank contains separate compartments for liquid hydrogen and liquid hydrogen, the propellants for the space shuttle main engine. Courtesy of NASA.

stand the complex aerodynamic heating generated by the shuttle structure as the vehicle passed through the atmosphere and the speed range of maximum dynamic pressure; it also had to keep the cryogenic propellants from boiling. The tank was coated with a "superlight" ablator, on top of which was an inch of foam similar to that used on the Saturn S-II. Unlike the S-II insulation, however, which only had to protect against boil-off, the ET's insulation also had to prevent the formation of ice on the foam from the −423°F liquid hydrogen and the −297°F liquid oxygen, because if ice came off the tank during launch, it could easily damage the critical thermal protection system on the orbiter. Thus the ET needed thicker insulation than the S-II. It was in fact so effective that despite the extremely low temperatures inside the tanks, the surface of the insulation felt "only slightly cool to the touch." For the first two shuttle flights, there was a white fire-retardant latex coating on top of the foam; thereafter, following testing to determine that the foam alone provided enough protection during ascent, the shuttle team dispensed with this coating, saving 595 pounds and leaving the orange foam to add its distinctive color to the white of the orbiter and solid rocket boosters at launch.[45]

Like the main engines and the boosters, the external tank underwent extensive testing before the first shuttle launch. The entire propulsion system was designed under Marshall oversight, with center director Lucas continuing von Braun's practice of using weekly notes to provide overall communication and thus aid systems engineering. In view of this, the Columbia Accident Investigation Board was perhaps unfairly critical in 2003 when it wrote:

> In the 1970s, engineers often developed particular facets of a design (structural, thermal, and so on) one after another and in relative isolation from other engineers working on different facets. Today, engineers usually work together on all aspects of a design as an integrated team. The bipod fitting [in the area where foam separated on *Columbia*'s last flight] was designed from a structural standpoint, and the application process for foam (to prevent ice formation) and Super Lightweight Ablator (to protect from high heating) were developed separately.

However, the board went on to note in all fairness:

> It was—and still is—impossible to conduct a ground-based, simultaneous, full-scale simulation of the combination of loads, airflows, temperatures, pressures, vibration, and acoustics the External Tank experiences during launch and ascent. Therefore, the qualification testing

did not truly reflect the combination of factors the bipod would experience during flight. Engineers and designers used the best methods available at the time: test the bipod and foam under as many severe combinations as could be simulated and then interpolate the results. Various analyses determined stresses, thermal gradients, air loads, and other conditions that could not be obtained through testing.[46]

Design requirements specified that the space shuttle system not shed any debris, but on the initial shuttle flight, the external tank produced a shower of it, causing engineers to state that they would have been hard pressed to clear *Columbia* for its first flight if they had known that would happen. When the shuttle lost foam from the bipod ramp on flight 7, wind-tunnel testing showed that the ramp area was designed with an angle that was aerodynamically too steep, and designers changed the ramp angle from 45 degrees to a shallower 22 to 30 degrees. However, this and a later "slight modification to the ramp impingement profile" failed to prevent the destruction of space shuttle *Columbia* on February 1, 2003. It is beyond the scope of this history to discuss the *Columbia* accident further, but its investigation revealed that despite advances in analytical capabilities down to 2003, the board was unable to pinpoint the "precise reasons why the left bipod foam ramp was lost" on that flight.[47]

This inability is the more striking in that the board included a staff of more than 120 people aided by about 400 NASA engineers who did an intensive investigation lasting months. One reason for the lingering uncertainty was the fact that foam did not "have the same properties in all directions" or the "same composition at every point." It was "extremely difficult to model analytically or to characterize physically . . . in even relatively static conditions, much less during the launch and ascent of the Shuttle." Factors that may have caused the foam to separate and damage the thermal protection of the wing included "aerodynamic loads, thermal and vacuum effects, vibrations, stress in the External Tank structure, and myriad other conditions" such as "wind shear, associated Solid Rocket Booster and Space Shuttle Main Engine responses, and liquid oxygen sloshing in the External Tank." Even in 2003, "Non-destructive evaluation techniques for determining External Tank foam strength have not been perfected or qualified."[48]

With statements like "In our view, the NASA organizational culture had as much to do with this accident as the foam," the board clearly indicted more than technology in the *Columbia* disaster. But it seems certain that a major cause was NASA and contractor engineers' failure to understand the reasons for and full implications of foam shedding from the external tank. As well-known space commentator John Pike commented, "The more they study the foam, the less they understand it." And as a newspaper article in the *Los Angeles Times* accurately stated, "Getting every ounce of the foam to stick to the external tank has bedeviled NASA engineers for 22 years. . . . Why foam falls off any area of the tank remains a scientific mystery." In the more sober language of the investigation report, "Although engineers have made numerous changes in foam design and application in the 25 years the External Tank has been in production, the problem of foam-shedding has not been solved."[49]

Whatever the larger causes of the accident, from the perspective of this book, the *Columbia* accident was but one more instance in which engineers did not have the design, development, and operation of rockets "down to a science." Despite countless billions of dollars spent on researching, developing, and operating a large number of missiles and rockets, and despite a great deal of effort on NASA's and contractors' parts to understand and correct this particular problem, there were aspects of rocketry that eluded the understanding of engineers and even of scientists such as investigation board member Douglas D. Osheroff, a Nobel Prize–winning physicist from Stanford University. Osheroff had conducted some simple experiments with foam that helped to illuminate the "basic physical properties of the foam itself" but also demonstrated "the difficulty of understanding why foam falls off the external tank." As he said, "Attempts to understand [the] complex behavior and failure modes" of the components of the shuttle stack were "hampered by their strong interactions with other systems in the stack."[50]

The external tank connected to the orbiter at one forward and two aft attachment points. The solid rocket boosters attached to the ET at the boosters' forward and aft ends. At the rear end of the shuttle stack, there were also umbilicals that conveyed fluids, gases, electrical power, and electrical signals between the ET and the orbiter. Electrical signals among the orbiter and the two boosters also passed through these umbilicals. The external tank provided the space shuttle main engines with propellants until about eight minutes after launch, when the engines usually shut down just before reaching orbital velocity. The orbiter's guidance and control system then commanded the jettison of the tank, which continued on a ballistic trajectory until it disintegrated in the atmosphere. At main-engine cutoff, two orbit maneuvering systems on the rear end of the shuttle completed the insertion of the vehicle into Earth orbit and circularized the orbital path.[51]

The Orbiter

The orbiter itself included many features related primarily to its roles as a spacecraft and an airplane that landed on a runway. While these features are certainly relevant to its character as a reusable launch vehicle, the scope of this book dictates that they be addressed only briefly here. This section will focus on those aspects of the orbiter related to its actual launching of satellites and other spacecraft.[52]

These features were primarily a structure that could withstand the aerodynamic pressures, vibration, and heating associated with launch; a payload bay and associated equipment used for releasing satellites or spacecraft from the shuttle's near-Earth orbit (about 130–350 miles in altitude); the engines; and a guidance and control system. The wings, tail structure, landing gear, flight deck equipment, and thermal protection system were mainly for reentry and landing. The many elaborate life-support systems that enabled the crew to live and work in space related principally to the orbiter's role as a spacecraft.[53]

As part of the overall space shuttle effort, NASA issued the request for proposals (RFP) to design and build the orbiter (Phase C/D) on March 17, 1972. DoD, especially the Air Force, had a large say in the requirements for the vehicle: NASA had needed military support to get the shuttle program approved, because without a commitment from DoD to use the shuttle for all of its launch needs, the vehicle would not be economically viable. To satisfy DoD requirements, the shuttle had to be able to handle payloads 60 feet long with weights of 40,000 pounds for polar orbits or 65,000 pounds for orbits at the latitude of Kennedy Space Center where Earth's rotation gave additional impetus to the launch. Although the RFP did not specify the distance a reentering orbiter had to be able to deviate from its reentry trajectory (its so-called cross-range), in fact the Air Force's demand for a 1,265–mile cross-range ultimately dictated the wing shape. The RFP went to Grumman/Boeing, Lockheed, McDonnell Douglas (teamed with Martin Marietta), and North American Rockwell. On July 26, 1972, NASA announced that the Space Transportation Systems Division of North American Rockwell had won the contract. Rockwell had prevailed largely on the strength of its management capabilities and lower cost.[54]

Having won, Rockwell did not keep all of the work at its own plants. It did retain the orbiter's nose, crew compartment, and forward and aft fuselage. But in November the California firm sent out RFPs on the midfuselage, wings, and vertical stabilizer. On March 29, 1973, it subcontracted with General Dynamics' Convair Division for the midfuselage, Grumman for the wings, and Fairchild Industries' Republic Division for the stabilizer. Another subcontract on the same date went to McDonnell Douglas for the orbital maneuvering system. For the wings, Air Force requirements for cross-range dictated a delta planform. The particular design had to offer adequate control from reentry to landing, to keep aerodynamic heating within acceptable limits, to furnish a high enough lift-to-drag ratio at speeds above Mach 5 to satisfy the cross-range criteria, and to generate enough lift at lower speeds to enable safe landings. To meet these criteria, NASA decided upon a design derived from a Lockheed proposal and called double-delta. The term referred to a wing in which the forward portion was swept more heavily than the rear part. In December 1972 NASA decided that the forward section should have a 79–degree sweep that tapered rather abruptly to 45 degrees before shifting to a slightly greater sweep at the trailing edges. Wind tunnel testing showed that this and other features of the shuttle design were not optimal, so in May 1973 the initial sweep changed to 81 degrees. This became the basic configuration for the wing.[55]

Fine-tuning continued. In June 1974 there were changes to the wingtips and aerodynamic control surfaces. Throughout the development of the shuttle, wind tunnel testing at a variety of facilities, including those at NASA Langley and NASA Ames Research Centers plus the Air Force's Arnold Engineering Development Center, provided a great deal of data. Before the first shuttle flight in 1981, there were 46,000 hours of testing in various wind tunnels, of which 24,900 involved models of the orbiter itself, 17,200 the overall launch configuration, and 3,900 the orbiter and the Boeing 747 aircraft that would ferry it back to Kennedy Space Center from other landing sites (in practice, almost exclusively Edwards AFB). The art of wind tunnel testing had evolved significantly since the birth of supersonic airplane flight in the mid-1940s, but it could not accurately predict all aspects of flight, especially at the shuttle's Mach 27–28 reentry speeds. The largest shuttle model to be placed in a wind tunnel was Rockwell's 0.36–scale version built in 1975 and tested in Ames's huge 40–by–80–foot subsonic tunnel. But this tunnel could not come close to shuttle reentry speeds, and Langley's helium tunnels, which were capable of high speeds, could accommodate models of only a few inches in length. Moreover, their flows were cold and could not simulate heating conditions on ascent and, more especially, on reentry. Shock tunnels, which provided temperatures for speeds up to Mach 17 and could test slightly larger models, operated for only milliseconds. Problems of scaling up and other limitations of wind tunnel testing including partial flow separation restricted the accuracy of the data tunnels could provide, but their findings were nevertheless critical and stimulated design changes.[56]

Using available data and existing theory, engineers could evaluate designs computationally, but computers were as yet not capable of predicting complex nonlinear airflows, heating, and atmospheric chemistry. They supplemented wind tunnel data but were no substitute for measurements from actual flight testing. Computational fluid dynamics (CFD) could use computer software and theory to predict flow behavior. But the shuttle came too early to take full advantage of this advance in the tools of the aerodynamic and structural dynamic disciplines, CFD being in its infancy in the 1970s. Moreover, even later computer predictions were no substitute for actual data from instrumentation on real flights. Except for the subsonic and low suborbital approach and landing tests of the partially equipped orbital test vehicle *Enterprise* in 1977, the only orbital test flights would occur with astronauts aboard. Thus designers had to take data where they could get them before the first shuttle flights.[57]

At NASA's Dryden Flight Research Center (from 1959 until 1976, known simply as the Flight Research Center), researchers Lawrence Taylor and Kenneth Iliff had developed an automated technique using differential equations that converted measurements of dynamic aircraft behavior in flight into numerical values that could be compared with wind tunnel predictions. Their computer programs, developed in the mid-1960s and later improved by Richard Maine, allowed wind tunnel predictions to be corrected according to actual data collected on the early shuttle flights (including the approach and landing tests) using instrumentation on the orbiters. Known as parameter identification, this technique helped to inform future flight planning and design modifications, although it was dependent on the quality and completeness of the recorded data, which were often lower than researchers desired. By 1981 engineers at Rockwell, in the Air Force, and at Johnson Space Center, which managed development of the orbiters, as well as at Dryden were all using parameter identification to analyze data from actual shuttle flights and then to correct information from wind tunnels and to revise CFD codes so they could better predict shuttle behavior. At Johnson, engineers incorporated changes in the database resulting from parameter identification into shuttle simulators and ultimately into the shuttle's flight control system. Such changes primarily affected reentry and landing rather than launch, but they also corrected a misprediction of the shuttle's center of pressure that was important for calculating the center of gravity for certification of shuttle payloads.[58]

Meanwhile, design and development proceeded with available data from wind tunnels, computer predictions, and numerous structural tests on models and on *Enterprise* itself after it had completed the approach and landing tests. The forward fuselage of the orbiters consisted of skin-stringer panels, bulkheads, and frames made of aluminum alloy. There were separate upper and lower sections so that the pressurized crew module could fit between them in the course of assembly. The crew compartment had two hatches, one in the side and one into the payload bay through an airlock.

The midfuselage, manufactured by Convair, contained a load-carrying structure between the forward and aft fuselage sections, the "wing carry-through structure" connecting the two external wing sections, and the payload bay and payload bay doors. Made mostly of aluminum and aluminum honeycomb, the structure also included titanium end fittings and boron-aluminum tubes, with the payload bay doors consisting of a graphite epoxy Nomex composite some 23 percent lighter than aluminum honeycomb. The external hinges of the door were of Inconel construction. The payload bay was 60 feet long by 15 feet wide, the 15–foot dimension being the bay's only design parameter that came primarily from NASA rather than the Air Force. NASA had projected that the future space station would have modules of that width. A unique feature of the shuttle compared with previous U.S. launch vehicles was the remote manipulator system, with a mechanical arm that could deploy or retrieve payloads of as much as 65,000 pounds. Canada's Spar Aerospace designed and constructed the arm under an agreement between NASA and the National Research Council of Canada, providing an international component to the shuttle.[59]

The blended double-delta wings, built by Grumman on Long Island and shipped to California through the Panama Canal, were made mostly of conventional aluminum. The aft fuselage supported the three space shuttle main engines, the orbital maneuvering system pods, and the vertical stabilizer. The internal thrust structure supporting the main engines was made of titanium, reinforced in places by boron-epoxy tubular struts for greater stiffness and lightness. An upper thrust structure, supporting the removable pods for the orbital maneuvering system and the reaction control system, was made of aluminum except for a support frame for the vertical stabilizer, which had a titanium structure. The pods themselves were mainly of graphite-epoxy and aluminum. McDonnell Douglas Aeronautics made these along with the propellant tanks for the orbital maneuvering system. An aft heat shield was of aluminum construction, with a main engine-mounted heat shield of Inconel honeycomb. The vertical stabilizer with a sweep along its leading edge of 45 degrees included a rudder that separated into two halves to function as a speedbrake. Fairchild Republic on Long Island made the vertical tails for the first four orbiters, with Grumman Aerospace producing the one for *Endeavour* after Fairchild Republic went out of business in 1987.[60]

Small Engines

Located in the pods on either side of the orbiter's vertical stabilizer, the two orbital maneuvering systems used engines designed and developed by Aerojet under a February 1974 subcontract to Rockwell. Since these engines used storable, hypergolic propellants—monomethyl hydrazine as fuel and nitrogen tetroxide as oxidizer—Aerojet, with its extensive experience using storables, was a logical choice to design them. The propellants had to be storable because the engines would be needed throughout the mission, including in the deorbit burn that slowed the orbiter for reentry, so cryogenic propellants with their continual boil-off were not suitable. To keep the engines simple and reliable, Aerojet designed them with pressurized tanks rather than turbopumps to feed the propellants. Since the engines were small, with only 6,000 pounds of rated thrust, and operated at a low chamber pressure of 125 psi, the use of comparatively heavy pressurized tanks added little weight. Aerojet engineers opted not to use baffles on the injectors, as they were difficult to cool, lowered engine performance, and had an inclination to crack. It took about six months, but the designers succeeded in using acoustic cavities to suppress oscillations and came up with a regeneratively cooled, gimballed engine that yielded a specific impulse of almost 315 lbf-sec/lbm. Designed by October 1975 and qualified on October 16, 1980, the engines proved to be highly reliable. They performed mission after mission, requiring no major overhauls, unlike the more sophisticated space shuttle main engines.[61]

The orbiter also used a reaction control system consisting of thrusters for attitude control in space and for minor translation maneuvers. Mounted in both the nose and aft sections of the orbiter were 38 primary thrusters rated at 870 pounds of thrust and 6 vernier (fine adjustment) thrusters with 25 pounds of thrust apiece. Rockwell subcontracted for these engines with the Marquardt Company, which specialized in small rocket engines for spacecraft and satellite attitude control. Both large and small thrusters were pressure fed and burned nitrogen tetroxide and monomethyl hydrazine. Their thrust chambers and nozzles were film cooled, and the engines presented few problems in development.[62]

Thermal Protection System

Like the external tanks, the orbiters required a thermal protection system, with the major difference that the spacecraft required the protection more for the extreme heat of reentry than for launch. For that reason, this account will not dwell on the many complexities of the orbiters' tiles. During the early studies leading up to the space shuttle, the choices for thermal protection were basically a hot structure using rare and costly materials like René 41, columbium, and zirconia—all of which retained structural strength at the high temperatures the shuttle would encounter—plus ceramic reusable surface insulation. With the shuttle on a tight budgetary leash, there was a reluctance to risk the delays and costs that would ensue if metalworkers had trouble working with the rare materials. The alternative did not prove to be an optimal solution, but since the early 1960s Lockheed Missiles and Space Company had been working on ceramic fibers that could be used for protection against heat.[63]

Already in December 1960, Lockheed had applied for a patent on a ceramic insulation material that was reusable. Conventional ceramics like porcelain were brittle, but Lockheed's material, while still inelastic, was made of fibers matted together rather than a glasslike substance. Lockheed made a radome for Apollo of silica fibers in 1962, although it was not actually used. Three years later Lockheed had a material known as LI-1500 made of silica fibers. Highly porous and weighing only 15 pounds per cubic foot, it underwent testing on an Air Force reentry test vehicle that subjected it to 2,300°F without its cracking, melting, or shrinking. Lockheed continued experimenting at its Palo Alto Research Center, increasing the porosity and reducing the weight of its reusable surface insulation, as it was called, to 9 pounds per cubic foot in a material it called LI-900. The matted silica fiber was white, which meant that it tended to reflect but not to radiate heat away from it, so it had to be thicker than a substance that would radiate more freely. A black coating would improve the ability to emit heat and provide protection in areas of high heating. For the coating of the black material, Lockheed chose a mixture of tetrasilicide and borosilicate glass. The resultant high-temperature reusable surface insulation, except for higher emissivity, retained the construction and characteristics of the white low-temperature reusable surface insulation.[64]

Lockheed's LI-900 and LI-1500 held up well in both thermal and acoustic testing during the latter part of 1972. Finally, in what Lockheed called a "sudden death shootout," that firm's materials competed with others from General Electric, Martin Marietta, and McDonnell Douglas in a series of tests at Johnson Space Center in December 1972. Only Lockheed's tiles survived the testing. When Lockheed won the thermal-protection subcontract in 1973, the firm established a production facility in Sunnyvale, California. To bond each of a multiplicity of sizes, thicknesses, and types of tile to the aluminum skin of the orbiter, designers used a nylon felt made by Du Pont, which stretched with the aluminum without imparting too much strain to the brittle tiles. Workers bonded the felt to the skin and the tile to the felt.[65]

While the tiles were undergoing further development and refinement, there was a significant amount of testing in wind tunnels, acoustic facilities, and actual flight. In flight tests in 1980, both an F-104 and an F-15 research aircraft at NASA's Dryden Flight Research Center subjected tiles to a variety of tests in some sixty flights simulating maneuvers of the shuttle and its flight loads at speeds up to Mach 1.4 and dynamic pressures of 1,140 pounds per square foot. The tests showed that methods of adhesion were in need of improvement. Also tested at Dryden were new substances called felt reusable surface insulation and advanced flexible reusable surface insulation, which held up well.[66]

All of the testing resulted in changes, including a strengthening of the bond between the tiles and the felt pads. The tiles were fragile as well as brittle. There were more than 30,000 of them, and each had to be precisely installed by hand. Space shuttle *Columbia* was the first orbiter to be outfitted with real tiles as well as the first to be launched. It arrived during 1978 at Air Force Plant 42 in Palmdale, California, near Edwards AFB, and installation of tiles began along with checkout of other systems. As plans for a 1979 launch of *Columbia* fell victim to delays in testing and qualification of the main engines, among other issues, Rockwell had a surfeit of workers at Kennedy Space Center whom it did not want to lay off. So it decided to fly *Columbia* to KSC aboard the 747 shuttle carrier aircraft and turn some of the workers there into tile installers. It flew with about 6,000 temporary tiles glued in place in March 1979, in addition to some 24,000 regular tiles.

Starting with a workforce of 260 at KSC, Rockwell had been obliged to increase to 940 by early 1980. Installation went more slowly than anticipated, and problems with bonding and other issues required the troublesome removal and replacement of individual tiles. Proof-testing of tiles showed that 13 percent of them could not withstand 125 percent of maximum flight loads. For some twenty months, installers worked three shifts a day, six days a week, installing tiles, testing them, and removing those that failed the proof test. For a time, more tiles got removed than installed. This was cut-and-try engineering with a vengeance! With multiple differences in size, type, and thickness of tiles to fit on the varied contour of the orbiter, the process was a nightmare. But eventually it got done. One estimate placed the delay in launching *Columbia* due to tile problems at "close to a year."[67]

Among the different types of reusable surface insulation, the high-temperature black tiles covered most of the underside of the vehicle, where temperatures rose to between 1,200°F and 2,300°F during reentry, while white low-temperature tiles were used in selected areas where temperatures were below 1,200°F. There was felt reusable surface insulation in blankets applied to areas where temperatures would be less than 700°F and aerodynamic loads would be minimal. After delivery of *Columbia*, engineers developed the advanced flexible reusable surface insulation. Applied in blankets, AFRSI was both lighter than tiles and easier and cheaper to install. For *Challenger* and later shuttles, it protected areas where temperatures were below 1,200°F and aerodynamic loads were minimal. Data from *Columbia*'s first flights helped determine where the newer blankets could be applied. AFRSI blankets replaced most of the low-temperature tiles on space shuttles *Discovery* and *Atlantis* and were added to parts of *Columbia* after 1986. They were also used on *Challenger*.[68]

Other parts of the orbiters' thermal protection system included thermal windowpanes; thermal seals; and reinforced carbon-carbon on the wings' leading edges, the nose cap, and other areas needing protection above 2,300°F. Limitations on space preclude discussion of all of these types of protection, but the reinforced carbon-carbon had an important history going back to 1958. Supported by the Dyna-Soar and Apollo programs, the Vought Missiles and Space Company began developing a composite made up of graphite fibers in a matrix of carbon. The process started with laying pieces of a carbon-based cloth such as rayon, impregnated with a phenolic resin, in stacks inside a mold and curing them in an autoclave. Heating then converted the resin to a more basic form of carbon, resulting in a very porous all-carbon product. Addition of furfuryl alcohol under pressure plus further heating, done three times, yielded the desired strength and density. The result was light, resistant to high temperatures, and (when coated with alumina, silicon, and silicon carbide) resistant to oxidation. Between 1964 and 1968, Vought produced a carbon-carbon nose cap, a heat shield for the Apollo spacecraft that fit behind its principal ablative shield, and a wind tunnel model of a Mars probe designed to enter that planet's atmosphere. For the shuttle, Vought used a Union Carbide carbon cloth and a Hexcel phenolic resin, but the basic process remained unchanged.[69]

The thermal protection system for the shuttle continued to be fragile. More than 300 tiles had to be replaced after the first flight of *Columbia*. Over the course of the space shuttle's history down to 2003, the tiles suffered an average of 143 "divots" (areas of damage), 31 of them more than an inch in one direction. From 1985 to 1988 an F-104 at Dryden Flight Research Center and a WP-3D aircraft off Florida's southeast coast tested shuttle tiles and advanced flexible reusable surface insulation for rain damage. The flight tests showed that rain did damage the insulating materials, although launch or landing of the shuttle in light rain *"may* be permissible without extensive tile damage." In the *Columbia* accident, however, it was not the fragile tiles but the reinforced carbon-carbon on the left wing's leading edge that was damaged by foam from the external tank, according to the accident investigation board. "During re-entry this breach in the Thermal Protection System allowed superheated air to penetrate through the leading edge insulation and progressively melt the aluminum structure of the left wing" until aerodynamic forces caused it to fail and the orbiter to break up, killing the crew in the process.[70]

Guidance, Navigation, and Control System

The space shuttle had a much more complex guidance, navigation, and control (GN&C) system than previous launch vehicles. The system not only had to control the vehicle during ascent to orbit but also was responsible for operating the orbiter in space and then during reentry and landing on a runway. Moreover, it had to do these things both automatically and in response to manual commands from the flight crew. Since the orbiter was a fly-by-wire vehicle, even manual commands had to pass through the four general-purpose computers that provided the main flight control for the vehicle. (A fifth computer served as backup.) There were no mechanical linkages between control devices moved by the crew and flight-control devices such as actuators for engine gimballing, reaction-control jets, or aerodynamic control surfaces like elevons and the rudder.[71]

At the heart of the GN&C system were the five general-purpose computers. Although computers had been around a long time by the 1970s, the early versions were massive. Minicomputers had come on the market in the mid-1960s and continued to flourish into the 1980s, but microprocessors and personal computers were scarcely on the horizon in the early 1970s when NASA had to select a computer for use in the space shuttles. (Since the system would be needed for the approach and landing tests in 1977, development had to begin quickly and could not wait for microprocessors to go through their growing pains.) One candidate for the shuttle computer was the Singer-Kearfott SKC-2000, a 32–bit computer with 16,000 words of memory that could be expanded to 24,000. (The 32–bit word size allowed more accurate calculations than smaller sizes.) But the SKC-2000 required 430 watts of electricity, weighed 90 pounds, and cost $185,000 per unit. So it was fortunate that, in November 1972, shuttle program personnel learned about IBM's AP-101. NASA had used IBM 4Pi computers in the Skylab program, and the AP-101 was a variant of those machines. It too was a 32–bit machine, but it had 32,000 words of memory, consumed only 370 watts of electricity, weighed less than 50 pounds, and cost only $87,000. Like the 4Pi, it was an off-the-shelf product, and Skylab had convinced the agency that using such products was advantageous. So on September 25, 1973, NASA issued a separate contract to IBM for "design and maintenance of the primary avionics system software (PASS)," related development of associated software-production tools, and the setting up of an Avionics Development Laboratory and Shuttle Avionics Integration Laboratory at Johnson Space Center. Since PASS was the software used in all four primary AP-101s, development of this system was critical to guidance, navigation, and control.[72]

NASA kept its hand on software development, but since contractual arrangements called for Rockwell to provide the actual computers, the orbiter's prime contractor subcontracted with IBM for AP-101B computers plus input-output processors to go with them. While NASA set up a Software Development Laboratory at Johnson Space Center for "debugging, integrating, and verifying" computer codes, IBM's work in developing software was eased significantly by a high-level computer language developed by Intermetrics, a firm located in Cambridge, Massachusetts, and founded by five MIT programmers who had worked on the software for the Apollo spacecraft computer. Known as HAL/S, the language may have been named in honor of J. Halcombe "Hal" Laning Jr., a professor at the university and part of the Draper Laboratory who had done work on computer languages that prepared the way for HAL/S. (Laning had also worked on the Atlas and Thor guidance systems, including the Q-guidance that was central to the Thor and Polaris guidance systems.) Selecting this high-level language for most of the computer coding was critical because it was much easier to modify programs in such a language. Between 1975 and 1980 there were at least two thousand changes in requirements affecting the software. On June 16, 1983, more than two years after the first shuttle flight, Richard Parten, Johnson Space Center's first chief for software development, observed, "We'd still be trying to get the thing off the ground if we'd used assembly language," harder to modify but a powerful tool whose programs were expensive to develop and required extra care to verify, but which in fact was used for the shuttle's operating system software because that was rarely changed.[73]

NASA decided for at least a couple of reasons that there would be four primary computers. Three computers were needed to determine which computer was wrong if two of them disagreed in their outputs. With the computers synchronized, the third computer would "vote" which of the two was correct. Warning lights would flash in the cockpit, and the crew could turn off the errant computer. They could later restore it to functioning if possible, but that left them at least for a time with only two computers and no way of determining which was right in a disagreement. With four computers, if one went out, three were left to vote on disagreements. Relatedly, in projections for fly-by-wire aircraft, analysts anticipated that in a system of three computers, failures would cause the loss of an airplane "three times in a million flights, whereas quadruple computer system failures would cause loss of aircraft only four times in a *thousand million* flights." To protect the astronauts, the odds were clearly much better with four computers. Initially, the shuttle program planned to use the fifth computer as a backup system only for the approach and landing tests in 1977, extended to include the first four shuttle flights into orbit, which were considered orbital test flights. But after considering the number of changes in software, the program kept the backup computer for added safety.[74]

Confidence in many of the decisions about the AP-101 computers and their use in the shuttle came from a digital fly-by-wire (DFBW) flight research project conducted at NASA's Dryden Flight Research Center from 1972 to 1985 in an F-8C aircraft. In its phase II flights starting in 1976, the F-8 flew with three AP-101 digital computers and an analog backup, and one key contribution of the project to the space shuttle lay in testing and correcting problems in the computer itself. The DFBW project received the first nine AP-101s and found that they failed much more frequently than predicted, causing IBM to troubleshoot. At least equally critical for the shuttle, since it would use fly-by-wire technology, was Dryden's demonstration that the concept could work safely without a mechanical (nonautomated) backup system. The F-8 also demonstrated that multiple computers could be synchronized and flown safely. Already on May 4, 1972, John "Jake" Garman, a future manager of shuttle software development, came to Dryden for a briefing on "failure analysis, software control, [and] flight software readiness." Later, on eight flights in 1977, the F-8 carried a test package of the shuttle's backup-system software. It is a measure of the DFBW project's contributions that the cash-strapped shuttle program contributed about a million dollars in funding for the F-8 effort.[75]

The proper memory size for the shuttle computers proved to be a thorny question. On previous programs, memory requirements had grown by several hundred percent, but NASA continued to underestimate with the orbiters, not learning from history. In the period from 1969 to 1971, most estimates of memory needs had been in the neighborhood of 28,000 words. Rockwell had set the figure at 32,000 words in its bid for the orbiter contract, and in its subcontract with IBM, each computer plus its input-output processor (IOP) had 64,000 words of memory between them, with the entire memory shared. However, the ascent software alone occupied 140,000 words of memory on its first iteration, and IBM never got it below 98,840 words—by itself exceeding the 64,000 words by a large margin. By replacing the original memory with double-density modules, the contractor increased the memory in each computer/IOP combination to 106,496 32–bit words by the time of the shuttle's first flight in 1981, but that was not nearly enough. The solution came in the form of mass memory units built by Odetics, two of which together provided a total of 16 million 16–bit words of memory on magnetic tape.[76] Clearly, the NASA/industry shuttle engineers did not have the process of estimating memory down to a science, but they were able to adapt.

There were other problems with the general-purpose computers. For example, to guard against a flaw in the software common to the four primary systems, the backup flight control system needed to have a completely different software system. Rockwell won the backup software contract. That firm had never liked the fact that IBM's PASS system for the primary computers was a priority-interrupt, or asynchronous, system that did computations as required in a predefined order of priority. For the backup computer, Rockwell instead used a time-sliced, or synchronous, system in which the program allocated periods of time for each task and then moved on to the next one, regardless of whether the previous task was finished. The advantage of the priority-interrupt system was that if the computer got overloaded, it degraded smoothly. Although there were problems making the two diametrically opposed systems work together, a bigger problem was that Rockwell divided the task of software development. It provided the operating system and did the integration. But it contracted with Intermetrics for, among other things, systems-management and sequencing software, and it had the Draper Laboratory furnish the programming for guidance, navigation, and flight control. Coordination among the three was poor. Edward Chevers at Johnson Space Center described the system at the time of early simulations as "a virtual basket case." Nevertheless, although not perfected, it was considered ready if needed by the first shuttle flight. Fortunately, the shuttle never needed to use the backup system during the period of this history.[77]

The AP-101Bs remained in service in the shuttles until 1991, when much smaller and more capable AP-101Ss began to replace them. As late as 1989 a NASA review maintained, no doubt with some justification, that the electronics system in the shuttle was "the most sophisticated, most advanced, most integrated system in operational use." Through the Honeywell HDC-601 dual-redundant engine controllers on each of the main engines, it issued the commands for the engines to fire and later controlled their gimballing through actuators, then commanded main engine cutoff at the appropriate time just before orbit insertion. The AP-101s also controlled the amount of thrust produced by the main engines through the HDC-601s. Via electronic assemblies in each solid rocket booster, the general-purpose computers directed the ignition of the solid propellant motors and later the gimballing of the nozzles and the jettisoning of the boosters. Following cutoff of the main engines, the IBM computers controlled the firing of the orbital maneuvering system's engines and those in the reaction control system as needed. Even during ascent, the crew could manually command thrust vector control and engine throttling, but these override commands still went through the computers.[78]

Of course, there were other mechanisms in the GN&C system that provided data to the AP-101s and enabled them to perform their functions. These included the inertial measurement units (IMUs), the rate gyro assemblies for sensing changes in angular velocities like roll rate, the accelerometers, and, at speeds below Mach 5 on the descent, the air data system probes. Singer-Kearfott of Little Falls, New Jersey—known as Singer Electronics System Division by 1988—provided the initial IMU, which remained in service until 1991. At 10.28 by 11.5 by 22 inches and weighing 58 pounds, this was a four-gimbal platform stabilized by only two gyros. One of them could sense changes in pitch and roll and kept the platform aligned in those two axes—not with respect to the orbiter but to the platform's position in space. The other provided stabilization in the yaw axis and also served as the platform rate detector. Two accelerometers in each of three IMUs (all located in the crew compartment) provided the AP-101s with data on linear accelerations in three axes, while the platform itself furnished information about attitude to provide guidance for steering commands. Since the accuracy of the IMUs declined over time, software took into account preflight calibrations to help correct for this. Once the shuttle was in orbit, star trackers and optical alignments from stars by the crew allowed adjustments to remove the remaining inaccuracies.[79]

There were three rate gyro assemblies near the top (forward skirt) of each solid rocket booster, each with one pitch and one yaw gyro that provided the AP-101s with data about angular rates around those two axes for computing steering commands while the SRBs were burning. The computers calculated roll from a combination of rate and attitude information and furnished a common signal to both the SRB nozzles and the main-engine gimbals for steering. Just before the shuttle jettisoned the SRBs, the GN&C system switched over to use of rate gyros on the orbiter. There were four orbiter rate gyro assemblies, located under the payload bay at midfuselage. Each assembly contained three rate gyros for sensing rates of motion about the pitch, yaw, and roll axes. Besides the accelerometers in the IMU, the orbiters had four body-mounted accelerometer assemblies in the crew compartment. These used a combination of pendulous light beams and photo diodes to provide data about acceleration during ascent and—especially crucial—during approach and touchdown, when the shuttle was coming in without power, unlike conventional airplanes, so that "energy management," or control of the speed and altitude, was vital to safe landing.[80]

The Shuttles and Their Flights

The launch vehicle that emerged from the involved and cost-constrained development of its many components was, as the Columbia Accident Investigation Board noted, "one of the most complex machines ever devised." It included "2.5 million parts, 230 miles of wire, 1,060 valves, and 1,440 circuit breakers." Although it weighed 4.5 million pounds at launch, its solid rocket boosters and main engines accelerated it to 17,500 mph (Mach 25) in slightly more than 8 minutes. The three main engines burned propellants fast enough to drain an average swimming pool in some 20 seconds. At its tallest point on the launchpad (the tip of the external tank), a space shuttle stood about 154 feet high. Laid on its belly on a football field, it would have stretched from one goal line to past midfield. An orbiter itself was no small vehicle—122.17 feet long with a wingspan of 78 feet and a height at the top of its vertical stabilizer of 56.58 feet. In its payload bay, it could accommodate a spacecraft roughly the size of a school bus, weighing anywhere from 38,000 to 56,300 pounds, depending on the orbit selected. Although the thermal protection system was one of the shuttle's Achilles' heels and a cause of endless trouble, even it was remarkable. With a blowtorch placed on one side of a shuttle tile, a person could touch the other side with impunity. Despite its flaws, the space shuttle was a magnificent engineering achievement.[81]

From its first orbital test flight on April 12, 1981, to the end of 1991, there were forty-four shuttle launches with only one failure, an almost 98 percent success rate (however tragic the loss of the *Challenger* crew). On these missions, the shuttles had launched many communications satellites, several tracking and data relay satellites to exchange information with spacecraft flying in low Earth orbits, a number of DoD payloads (including a pair of Defense Satellite Communications System satellites, two signals intelligence satellites, another military communications satellite, and a Defense Support Program satellite), many scientific and technological experiments, and several important NASA spacecraft, among others. One notable payload was Magellan, a deep space probe to Venus equipped with a synthetic aperture radar designed to map 70 percent of the planet's surface. Released from *Atlantis*'s payload bay on May 5, 1989, Magellan used a two-stage IUS to travel from Earth orbit to a rendezvous with Venus. There it began mapping on September 15, 1990, and ultimately achieved coverage of a highly gratifying 96 percent. The 1,200 gigabits of data it sent back to Earth augmented the information scientists had about our sister planet by a significant amount.[82]

The ambitious deep-space probe Galileo was the payload on another *Atlantis* mission launched on October 18, 1989. Released from the payload bay 6.5 hours after launch, this spacecraft also used a two-stage IUS to fly past Venus, where it began employing the gravity of two planets (Venus once, Earth twice) to assist it in its trajectory. On the way, it surveyed Venus and observed the Moon as well as the asteroids Gaspra and Ida before it headed to Jupiter. There it explored the planet and its moons, vastly increasing our knowledge of planetary science. Another *Atlantis* flight, on October 6, 1990, sent Ulysses on an interplanetary course boosted by a single-stage IUS coupled with a Payload Assist Module-S (the S standing for shuttle). It circled and studied Jupiter while getting a gravity assist from that planet and headed back to the Sun, about which it gathered extensive data. Built by the European Space Agency, Ulysses found, among its other discoveries, that the solar wind blew faster in the region of the Sun's poles than at the Sun's equator.[83]

Arguably the single most important of the shuttle's first forty-four missions was that of *Discovery* launched on April 24, 1990. The day after the launch, it successfully deployed the Hubble Space Telescope. Named for distinguished American astronomer Edwin Hubble, the telescope could view the universe from above Earth's atmosphere, which hindered the clarity of telescopes mounted in strategic locations on the planet. It also had the power to "observe objects twenty-five times fainter and ten times more distant than were visible with the largest telescopes on Earth." Unfortunately, as we have seen, the primary mirror on the telescope proved to have a "spherical aberration" that, while "only 1/25th the width of a human hair," significantly interfered with the telescope's focus. Fortunately, scientists found a way with computer enhancement to provide a temporary "fix" for the defect that enabled the telescope to yield important data despite the flaw. Then, in December 1993, a subsequent shuttle crew repaired the telescope, affording countless stellar images that enhanced astronomers' understanding of the universe to an appreciable degree. For example, the telescope returned data from distances billions of light years from Earth, providing support for the widely held view that the universe began with a Big Bang. Hubble's images from galaxies more distant than any seen before provided data about the way the structures of galaxies evolved over a majority of the history of the universe.[84]

The *Challenger* Accident

Before the launching of Magellan, Ulysses, or the Hubble Space Telescope, however, NASA had endured the loss of space shuttle *Challenger* and its seven-person crew to an explosion. Since this book is not an operational history, it is not the place for a detailed analysis of the tragedy, but because the accident cast a spotlight on the technology of the solid rocket boosters and resulted in a partial redesign, it requires some discussion. On the twenty-fifth shuttle launch, *Challenger* lifted off at 11:38 a.m. on January 28, 1986. Even that late in the day, the temperature had risen only to 36°F, 15 degrees below the temperature for any previous shuttle launch. Engineers at Morton Thiokol (as the builder of the SRBs had become in 1982 after the Morton Salt Company took over Thiokol Corporation) had voiced reservations about launching in cold temperatures. But under pressure to launch in a year scheduled for fifteen flights, six more than ever before, NASA and Morton Thiokol agreed to proceed. Almost immediately after launch, smoke began escaping from the bottommost field joint of one solid booster in the direction of the external tank, though this was not noticed until postflight analysis. By 64 seconds into the launch, flames from the joint began to encounter leaking hydrogen from the ET, and some 73 seconds after launch, the vehicle exploded and broke apart.[85]

On February 3, 1986, President Ronald Reagan appointed a commission to investigate the accident, headed by former secretary of state William P. Rogers and including two astronauts, a world-famous test pilot, and a Nobel

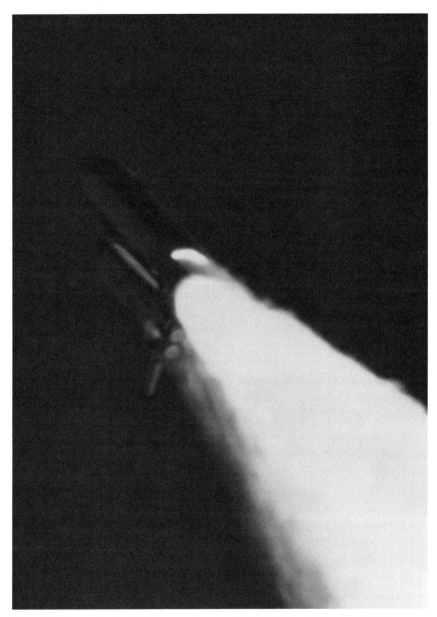

Figure 54. Space shuttle *Challenger* on January 28, 1986, almost 59 seconds into launch, with a plume of flame escaping from a joint in the solid rocket booster just above the exhaust nozzle. *Challenger* exploded less than 15 seconds later. Courtesy of NASA.

Prize–winning physicist, Richard P. Feynman. The commission determined that the cause of the accident was "the destruction of the seals [O-rings] that are intended to prevent hot gases from leaking through the joint during the propellant burn of the rocket motor." NASA had detected and recorded erosion of either the primary or secondary O-rings on twelve flights, starting with flight 2 and including four of the five flights immediately preceding the fateful *Challenger* launch. On nine of these flights, there had been actual blow-by of hot gases past one of the O-rings. Known at the flight readiness review for the *Challenger* launch, these data did not cause NASA to delay the launch even though the worst examples of O-ring problems had occurred in cold weather. On the other hand, historian Stephen Waring argued from information in the presidential commission's report that when Morton Thiokol engineers had expressed reservations about launching in cold weather because of stiffness in the O-rings, Marshall engineers had doubted a correlation between temperature and erosion since blow-by had taken place at 75°F. Both Marshall and Morton Thiokol engineers had come to believe that the O-rings were safe because they had worked before. This was similar to the fallacious belief before the later *Columbia* accident that because pieces of insulating foam had frequently struck the orbiter, they were no longer a significant danger.[86]

It is possible to maintain that the cause of the *Challenger* accident was faulty assembly of the particular field joint that failed rather than a problem with the design of the joint. The Rogers Commission may have erred in judging that "neither NASA nor Thiokol fully understood the mechanism by which the joint sealing action took place."[87] But it does seem clear that neither NASA nor most Morton Thiokol engineers believed that the launch could lead to disaster. If they had, surely they would have postponed it. The fact that they went ahead shows, one more time, that rocket engineers still did not have the launching of such complex vehicles completely "down to a science." Some engineers had concerns, but they were not convinced enough of their validity to insist that the launch be postponed.[88]

In any event, following the accident there was an extensive redesign of many aspects of the shuttle, notably the field joints. The tang, instead of being a single cylindrical piece that fit down into the slot in the clevis, now added a tang capture feature, creating in effect a slot in the tang with the capture feature enveloping one side of the clevis. A third (or capture-feature) O-ring improved the sealing capability of the new design, with an additional leak check port that gave extra insurance that the primary O-ring

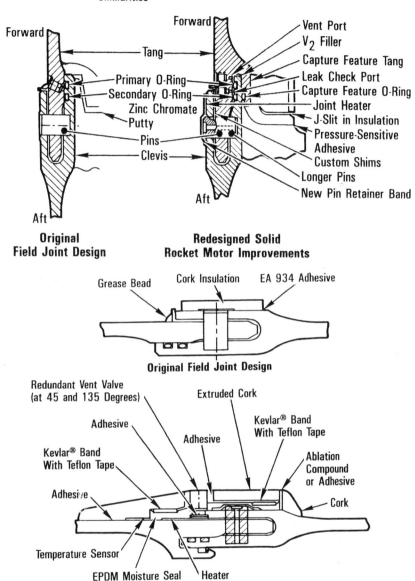

Similarities

Forward — Tang

Forward

Vent Port
V₂ Filler
Capture Feature Tang
Leak Check Port
Capture Feature O-Ring
Joint Heater
J-Slit in Insulation
Pressure-Sensitive Adhesive
Custom Shims
Longer Pins
New Pin Retainer Band

Primary O-Ring
Secondary O-Ring
Zinc Chromate Putty
Pins
Clevis

Aft

Aft

Original Field Joint Design

Redesigned Solid Rocket Motor Improvements

Grease Bead   Cork Insulation   EA 934 Adhesive

Original Field Joint Design

Redundant Vent Valve (at 45 and 135 Degrees)

Extruded Cork

Adhesive

Adhesive

Kevlar® Band With Teflon Tape

Kevlar® Band With Teflon Tape

Ablation Compound or Adhesive

Adhesive

Cork

Temperature Sensor

EPDM Moisture Seal   Heater

Redesigned Solid Rocket Motor Field Joint

Field Joint Protection System

Figure 55. Two sets of cross sections of the original and redesigned field joint for the space shuttle solid rocket boosters showing details of both. Taken from NASA, *National Space Transportation System Reference*, vol. 1, *Systems and Facilities*, 33a, 33b. Courtesy of NASA.

was properly aligned at ignition. The tang capture feature limited "the deflection between the tang and clevis O-ring sealing surfaces caused by motor pressure and structural loads." Custom shims "between the outer surface of the tang and inner surface of the outer clevis leg" compressed the O-rings. This new design allegedly ensured that the seals would not leak under twice the anticipated structural deflection. Also added to the field joint were external heaters to keep the joint and O-ring no cooler than 75°F plus weather seals to help maintain that temperature and prevent water from intruding into the joint.[89] See figure 55 for these and other features of the redesign.

Following *Challenger*, both U.S. law and policy changed, essentially forbidding the shuttle to carry commercial satellites and largely restricting the vehicle to missions using the shuttle's unique capabilities *and* requiring people to be present. A concomitant result was the rejuvenation of the Air Force's expendable launch vehicle program. While the Delta II was the only launcher directly attributable to the 32–month hiatus in shuttle launches following the accident, the Air Force also ordered more Titan IVs and eventually produced other expendable launch vehicles. The shuttle became a very expensive launch option, since its economic viability had been premised on rapid turn-around and large numbers of launches every year, whereas in 1989 it flew only five missions, and six each in 1990 and 1991. It was able to fly seven missions in 1992, but at a cost of $553 million per mission by one calculation. This was slightly more expensive than a Titan IV expendable launch vehicle. Admittedly, the shuttle offered more flexibility and slightly greater capability, but its initial promise of "sharply reducing costs" was not borne out in reality.[90]

Analysis and Conclusions

The space shuttle had the most demanding requirements of any launch vehicle yet developed. Taking off vertically, carrying humans not as passengers—as they were, in essence, for Mercury, Gemini, and Apollo—but as an integral part of the vehicle's operation, reentering and landing horizontally on a runway, and being reusable all called for sophisticated hardware and software. Yet the shuttle was developed under conditions of unusual fiscal constraint and operated with unprecedented expectations for safety, reliability, and cost efficiency. Under the circumstances, it is not surprising that it failed to live up to those expectations. Does that mean that the shuttle was a failed experiment?

Different criteria will produce different answers to that question. The shuttle admittedly did not sharply reduce costs, so by that measure it failed. This failure largely stemmed from the shuttle's nature as an outgrowth of heterogeneous engineering, involving negotiations of NASA managers with the Air Force, the Office of Management and Budget, and the White House, among other entities. Funding restrictions during its development and other compromises led directly to higher operational costs. Indeed, compromises on reusability (the external tank) and the employment of solid rocket motors, plus unrealistic projections of how many flights per year the shuttles could achieve, virtually ensured failure in this area from the beginning.[91] The deaths of two crews and the loss of two of five orbiters from the fleet earn another negative vote. In these and other ways the space shuttle underlined that rocketry is not a science, if that means it can accurately predict performance, safety, and costs early in the development process—or even later, in the case of safety. As the Columbia Accident Investigation Board pointed out, "Launching rockets is still a very dangerous business, and will continue to be so for the foreseeable future as we gain experience at it. It is unlikely that launching a space vehicle will ever be as routine an undertaking as commercial air travel."[92]

And yet, for all its flaws, the shuttle represents a notable engineering achievement that can perform significant feats expendable launch vehicles cannot perform. Shuttle crews have rescued satellites from unsatisfactory orbits, repaired the Hubble Space Telescope, and helped build the International Space Station. These are remarkable accomplishments that yield a resounding vote for the success of the shuttle.

Along the way, at numerous points in the development process, questions arose for which engineers had no ready answers. When the known phenomenon of subsynchronous whirl afflicted the high-pressure fuel turbopump, there were no solutions at hand in the rocket engineering database or theoretical literature. It took considerable effort by a team of international rotordynamics experts in government, industry, and academia to arrive at a "fix" that, while it did not end the whirl, did relocate it to operating levels above the rated power level of the main engines. Other problems with turbopumps and the advanced propulsion system also were beyond the existing state of the art and required much effort to solve. Moreover, although in 1986 the Rogers Commission could conclude that the engines had "performed extremely well," it also noted that they continued "to be highly complex and critical components of the Shuttle that involve an element of risk principally because important components of the engines degrade more rapidly with flight than anticipated."[93] In other words, even after the engines were developed, the engineers could not predict how long they would hold up.

Development of the solid rocket boosters benefited from previous re-search funded by the Air Force and NASA in the Large Segmented Solid Rocket Motor program from the 1960s and from the solid rocket motors' greater simplicity compared with those using liquid propellants. But though the solid rockets had fewer problems and many fewer tests, there were un-knowns here as well, including an incomplete understanding of the complex combustion process. Moreover, when Thiokol departed from the tang-and-clevis design used successfully on the Titan III solid boosters, it may have contributed to the *Challenger* disaster, which ironically might have been prevented in advance by using Aerojet's proposal for unsegmented boosters. If these assumptions are correct, again the state of engineering knowledge at the time of the solid-rocket selection process was not fully adequate to predict future performance.

Predictive techniques were also less than perfect in the design of the or-biters. Wind tunnels and an infant computational fluid dynamics capability could not fully foretell how the vehicles would operate at extremely high speeds in space, during reentry, and in the coast through the atmosphere to landing. One instance was a misprediction of the shuttle's center of pres-sure, which could only be corrected after the shuttle's early orbital flights provided actual flight data that allowed researchers using parameter identi-fication to ascertain the error and provide corrections for both the operation of the vehicle and for future use in wind-tunnel and CFD testing.

Similarly, in developing the shuttle's highly sophisticated guidance, navi-gation, and control system, an underestimation of the computers' memory needs and a large number of software changes again showed limitations in the state of engineering knowledge. In all of these examples, engineers were ultimately able to solve or work around the problems, so their difficult na-ture actually provides a powerful testimony to the ability of the engineers involved to make the system work.

Less impressive in some respects was Lockheed's discovery and develop-ment of insulation materials for the orbiter. Although those materials had marvelous insulation properties and furnished a workable solution to a re-calcitrant problem of reentry heating, they were always troublesome and altogether too fragile for repeated use without expensive replacement of many of them after every mission. This was not so much a reflection on Lockheed engineers as on the immaturity, despite many years of effort, of the state of engineering knowledge in the area of high-temperature thermal protection—and perhaps aerothermodynamics itself. Lockheed, inciden-tally, made many more contributions to shuttle technology than its share of the production workload would suggest. These contributions ranged from the initial technology for a gimballing nozzle on the solid rocket boosters to the suggestion of the double-delta planform used for the orbiters' wings.

One final point that might be made about the design and development of the space shuttle concerns the large number of firms and institutions that contributed to the process. Wind tunnel testing took place not only at NASA's Ames and Langley Research Centers but at the Air Force's Arnold Engineering Development Center. Air Force as well as NASA funding had contributed technology for the solid rocket boosters, and production occurred at the Air Force's Plant 42 as well as at NASA's Michoud facility and elsewhere. Besides the NASA centers just mentioned, Johnson and Marshall had key management and development responsibilities, Kennedy Space Center was where the vehicles launched, and both the National Space Technology Laboratories and Dryden Flight Research Center played key roles in various aspects of shuttle support and development. Besides Rockwell, Martin Marietta, Grumman, Lockheed, and McDonnell Douglas, many other industrial firms made major contributions, as did academics and other research centers such as the Air Force's Rocket Propulsion Laboratory. The shuttle was a major NASA undertaking, but almost countless other agencies, institutions, and individuals contributed to it, including some outside the United States. Both its successes and its shortcomings were the result of a highly collective and cooperative effort.

8

The Art of Rocket Engineering

Through the satellites and spacecraft that they have lifted into the heavens, since 1958 launch vehicles have brought remarkable changes to life on Earth. From what we watch on television to the way we wage war, Americans and people throughout the globe have come to depend on satellites.

How did launch vehicle technology and access to space evolve so quickly? One key factor was the cold war, whose ending from 1989 to 1991 provides the terminus for this book. It is scarcely credible that the huge investments necessary for developing American missiles, launch vehicles, and satellites would have been forthcoming without the Sputnik launch in 1957 and the fears it aroused about the Soviet Union's ability to send missiles into the U.S. heartland.[1]

The total amount of money expended on missiles and launch vehicles during the cold war has probably never been credibly estimated. Clearly it was enormous and constituted a major factor in the rapid development of the requisite rocket technology. One early estimate put total costs for ballistic missiles up to about 1965 at $17 billion (or, converted to 2006 dollars, some $112 billion.) This cost included missile sites, which had no relevance for space-launch vehicles. But it also encompassed factories for producing propellants, engines, airframes, and guidance and control systems; test facilities; ranges with their testing, tracking, and control equipment; laboratories, and much else.[2] To give but one other indicator of the enormous expenditures for missiles and launch vehicles, the total cost of the Saturn launch vehicles from 1959 to 1973 amounted to $9.3 billion. Adjusted for inflation, that came to $58.6 billion in 2006 dollars.[3]

But the cold war and the spending it stimulated were not enough, by themselves, to bring about the rapid development of missile and launch-vehicle technology. Fears of Soviet missile attacks provided the context in which the technology could develop. But Congress, a succession of presidents, and the American public would not have invested the many billions of dollars necessary for such rapid development without the prodding of heterogeneous engineers such as Trevor Gardner, John von Neumann, Bernard Schriever, Theodore von Kármán, Wernher von Braun, and William F. "Red" Raborn. This was especially true in the light of the many failures of missiles and rockets in the early years from the late 1950s to the mid-1960s.

Given the complexity and increasing size and thrust of the missiles and launch vehicles developed even in the first decade after Sputnik, it is not surprising that many of them failed. The sophistication of these vehicles engendered the popular phrase "rocket science" to characterize the arcane knowledge that developers and operators had to possess.

But from the beginning to the present, developers have not always been able to foresee the problems that could arise when rockets operated in the harsh environments of launch and flight through the atmosphere into space. The body of literature, mathematical formulas, ground-testing facilities and equipment, computer tools, and other infrastructure supporting rocket development continued to grow. Yet as recently as the *Columbia* disaster in February 2003, we learned once more that aspects of rocket behavior in flight defied understanding and prediction. This was far from an isolated instance in the history of rocketry, for lack of predictability had been a part of rocket development and operation from Robert Goddard's unavailing attempts to reach high altitudes until the present. Could it be that there had never really been such a thing as rocket science, at least if that term is defined to mean a body of knowledge complete enough to permit such predictions?

Of course, uncertainty was by no means foreign to science. Like rocketry, twentieth-century science had flourished amid uncertainties. These ranged from the Heisenberg uncertainty principle (that it was impossible to determine both the precise position and the energy of an electron) to the questions that still swirl around the Big Bang theory about the origin of the universe. And the continued success of rocket developers in overcoming unexpected problems and making their rockets work compares favorably with scientists' abilities to accommodate unexpected data and adjust their theories accordingly.[4] If scientists sought basically to understand the universe or its components, rocket developers primarily tried to make their vehicles work as designed. They employed science and any other resources that would contribute to this endeavor, and they certainly wanted to understand how their creations worked. But often they had to use trial and error methods to "fix" a problem without necessarily understanding its causes or why one solution worked when another didn't.

This book has argued that the basic process of developing rocket technology involves engineering more than rocket science. In making this point, it follows the general theses of Edwin Layton, Walter Vincenti, and Eugene S. Ferguson about the differences between engineering and science—in particular their discussions of engineers' emphasis on *doing* as distinguished from scientists' *knowing*, on the centrality of design to engineering, and on the role of art in what engineers do.

Vincenti, in particular, called for further historical analysis of the complicated process by which engineers must sometimes make decisions based on knowledge that is not complete or certain.[5] I did not begin my research on launch vehicle technology with the thought of applying the three scholars' ideas to the field of rocket technology. But as I studied the process of developing rockets, it became increasingly evident that rocket history illustrated the ideas of Layton, Vincenti, and Ferguson in innumerable ways. Their concepts apply most cogently to issues involving injection, ignition, and combustion of propellants. German engineers encountered problems in these areas while designing the V-2, and related difficulties recurred on the United States's H-1, F-1, and, in different ways, space shuttle main engine, among many others. In 1962 NASA's assistant director for propulsion, A. O. Tischler, called injector design, for example, "more a black art than a science."[6] The process became less "black" as time went on, but it remained very much an art, suggesting its nature as engineering rather than science.

The uncertainties about rocket design weren't restricted purely to propulsion issues. As engineers scaled up earlier designs to larger sizes, used new materials, or in other ways sought to improve performance and reliability, unexpected problems continued to arise down to the end of the period covered in this book and beyond. NASA engineers' lack of understanding of the forces operating on the foam protecting the space shuttle's external tank even after the *Columbia* disaster was one dramatic example.[7] But the problems with developing the solid rocket motor upgrade for Titan IV that persisted until after the end of the cold war also illustrate the uncertainties engineers contended with.

A great many other examples of failure to predict problems, resulting in trial-and-error engineering, appear throughout this book. But perhaps the best case for the uncertainties comes from a speech titled "Accepting Risk" given by N. Wayne Hale Jr. on October 25, 2004. At the time, Hale was deputy manager of NASA's space shuttle program. (He became manager in

2005.) But informing his comments were the fifteen years he spent as the shuttle's entry flight director. In that job, he had to make the go/no-go decision for each shuttle reentry into the atmosphere from orbit for an unpowered landing.

"I have given the Go 28 times. Every time was the toughest thing I have ever done," Hale wrote. "And I have never ever been 100% certain, it has always been gray, never a sure thing." More directly relevant to rocket engineering, Hale commented, "When our predecessors invented the shuttle, based on their aircraft test experience and previous space programs, they set up a standard that everything should work properly in . . . 99.7% of the cases." This left three chances in a thousand for things to go awry. Why didn't the engineers design for 100 percent reliability? "Because," Hale stated, "to try to cover everything . . . would require a vehicle design that probably is too heavy to get off the ground, and would require a set of proof testing that would take a lifetime to accomplish. . . ."

"It has always been part of the engineer's job to determine when enough has been done," Hale went on. "Knowing when we have done enough is the art of engineering . . . knowing when the testing is adequate and it is time to . . . move on." But, he conceded, "Humankind collectively does not know enough to scientifically drive the risk of space flight to zero. A hundred years would not provide enough time for all of us working together to positively eliminate any risk."[8]

Hale's observations mesh nicely with the arguments above about the art of rocket engineering and the uncertainties inherent in design and operation. In view of such problems, it is especially remarkable that rocket technology developed as rapidly as it did. Rocket engineers deserve a lot of credit for the many ways they fixed problems and the numerous innovations they developed to advance the technology rapidly and, for the most part, successfully.

Unfortunately, the available sources don't shed as much light as scholars would like on the processes and individuals involved in producing innovations. Through the account of Charles B. Henderson, we have a perspective on how he and Keith Rumbel went against contemporary theory to discover that large percentages of aluminum added significantly to the performance of some solid propellants.[9] This led to the propellants used in Polaris, Minuteman, the Titan solid rocket motors, the shuttle solid rocket boosters, and other solid stages. Similarly, we know a good deal about Karel Bossart's "steel balloon" structures for the Atlas and Centaur, but we still don't fully understand how he conceived the idea. For many other innovations, we know even less. Large teams of engineers worked on designing and developing various parts of missiles and rockets. And for some innovations—such as carboxy-terminated polybutadiene and hydroxy-terminated polybutadiene binders for solid propellants—not just a lot of people but many firms seem to have been involved in perfecting the technology.[10] No known sources spell out exactly how they did this, although the general outlines of the process emerge from a number of different accounts.

Other key contributions that we know quite a bit about, and have discussed in this volume and its predecessor, *Preludes to U.S. Space launch-vehicle Technology*, include those by Walter Thiel, John Parsons, Charles Bartley, Levering Smith, Karl Klager, and Robert Corley. But many innovators who remain anonymous clearly participated in technological solutions. Their talent, technical knowledge, and abilities at problem solving must have been a prerequisite to the rapid development of launch vehicle technology even though, in many cases, it can only be inferred from documentation uncovered thus far.

According to the Air Force's Otto Glasser, the significant innovation of all-up testing—flight testing all stages at once instead of each major component in succession—used on Minuteman and other vehicles, including the Saturn launch vehicles, resulted from an error by Secretary of the Air Force James H. Douglas that required Glasser to speed up Minuteman development by a year. All-up testing was the only way he could think of to do that, even though it violated "all normal, sensible standards." But it worked and became an accepted and valuable technique.

Glasser also provided an interesting analogy for the difficulty of sorting out who in large organizations came up with a particular innovation: "If you were to back into a buzz saw, could you tell me which tooth it is that cut you?"[11] One example of innovations that did not occur during initial design but arose as responses to problems during testing and that fit Glasser's "which tooth?" model was UTC's tape-wrapped carbon-phenolic nozzle throat for the Titan solid rocket motors. Another was the process of rolled-ring forging developed by UTC, Westinghouse, and the Ladish Company. We know which companies developed the technologies involved, but not particular individuals or groups who made these discoveries, let alone how they did it. Similarly, we know little about how Aerojet engineers solved problems with Transtage. These problems ranged from malfunctioning bi-propellant valves to a weakness in a nozzle extension. Once again, though, the resultant technologies appear to provide further examples of trial-and-error engineering.

To the extent that we do know much about innovations in rocketry, they seem not to follow any single pattern. Instead, in line with Hugh Aitken's view about radio technology, they appear to have entailed "a process extending over time in which information from several sources came to be combined in new ways."[12] For missiles and launch vehicles, a great many organizations, institutions, and firms provided that necessary information. The list includes the Air Force's Western Development Division and its successors, the Army's organizations at Redstone Arsenal, the Navy's Special Projects Office and its successors, NASA Headquarters and various centers (notably JPL, Langley, Lewis, and Marshall for rocketry, but also the Dryden Flight Research Center and Wallops Flight Facility), the Naval Ordnance Test Station and its successors, the Air Force Rocket Propulsion Laboratory and its predecessors as well as successors, the Arnold Engineering Development Center, the Armour Institute (later Illinois Institute of Technology), the Chemical Propulsion Information Agency (CPIA) and its predecessors, firms such as Aerojet, Rocketdyne, Pratt & Whitney, Lockheed, Douglas, the Martin Company, UTC/CSD, and Thiokol. But this tally doesn't begin to exhaust the number of contributors.

Representatives from many of these entities came together during development planning to exchange information and ideas. Particularly when problems arose, groups consisting of participating engineers and experts from elsewhere engaged in troubleshooting to discover workable "fixes." For the endemic problem of combustion instability, many university researchers also got involved over a long period of time.[13] The process of rocket development was complex and usually not recorded except in technical reports that typically identify only the author(s), if we are to judge by a wide selection of sources. People who were involved sometimes remembered (accurately or not) how they or someone else arrived at a solution, but more often they could not recall the precise process.

Among the factors that contributed to the rapid development of missile technology, the literature emphasizes interservice competition, usually as a negative factor but one that could spur innovation.[14] Much less noticed but perhaps more important in promoting rocket development were interservice and interagency cooperation. The CPIA constituted one important focus of such cooperation, but there were many others. For example, the Air Force regarded the Navy and its Polaris missile as a competitor for funding, roles, and missions. This helped to promote development of the air arm's own solid-propellant missile, Minuteman. At the same time, Polaris would not have been possible without some technology developed by and for the Air Force, and Minuteman borrowed Polaris's use of a large percentage of aluminum as a fuel in its own propellants. Similarly, the Air Force at best accepted NASA with great reluctance as another developer of rocket technology, and in 1958–60 the Army did not like losing the von Braun group and JPL to the new civilian agency. Yet both services did cooperate with the space agency (and vice versa), even to the extent of loaning valuable talents such as the Air Force's Samuel Phillips to help NASA's programs. And many astronauts came to NASA from the Navy, Marine Corps, and Air Force.[15]

Almost as important and elusive as the history of innovations themselves was technology transfer. Here, again, clear evidence of the process is hard to find. We know that federal contracting agencies often prevented contractors from withholding information about innovations developed under government contract. Thus Lockheed could not protect the technology in its Lockseal bearing. As a result, Thiokol scaled it up and used it to swivel and direct the thrust of the shuttle solid rocket boosters under the name Flexseal.

We also know that people who learned rocket technology in one organization frequently moved to another organization. They took their knowledge with them, thus transferring technology and furthering rocket development. For example, Charles Bartley, who had developed early rubberized, composite solid propellants at JPL, later founded Grand Central Rocket Company, an early contributor to solid rockets before it became part of Lockheed Propulsion in 1960–61.[16] Even earlier, Bartley shared his knowledge with Thiokol when it went into the rocket business. Similarly, Barnet R. Adelman had experimented on both liquid and solid propellants at JPL before becoming technical director of the Rocket Fuel Division at Phillips Petroleum and then director of vehicle engineering for Ramo-Wooldridge. In the latter position he joined Col. Edward N. Hall as a principal proponent of the Minuteman missile. From there he went on to become one of the founders and second president of United Research Corporation (later UTC/CSD). The knowledge he gained in his previous positions, especially with regard to Minuteman, clearly helped his firm develop the solid rocket motors for Titan III and IV.[17] Adelman and other founders' knowledge of the people involved in solid propellant rocketry also enabled them to attract experienced engineers who contributed to the successes of the new firm.

Numerous other examples of technology transfer through the movement of personnel from one organization to another appear in these two books. One such case is Ray C. Stiff Jr., who as an ensign working in Lt. Robert C. Truax's group at Annapolis about 1942, had discovered that aniline ignited hypergolically with nitric acid, providing the basis for later storable propellants. After transferring to Aerojet, he became manager of its Liquid Rocket Operations near Sacramento and played a major role in developing the self-igniting propellants for the Titan II engine.[18]

The von Braun group, including Krafft Ehricke and many others, brought German technology to this country, and important contributions came from such unconnected immigrants as Karel Bossart and Karl Klager. But their innovations and transferred technology also led to independent and important contributions by many Americans. For example, Rocketdyne engineers learned much from the V-2 and its German creators, yet the American engineers also made critical innovations of their own in developing subsequent rocket engines, from the ones used on the Navaho and Redstone missiles to the space shuttle main engine. The process was complex and did not proceed in linear fashion from the V-2 to, for instance, the F-1 engine for the Saturn V, although there were linkages between the two projects and others.

Another key factor in the comparatively rapid and successful development of missiles and rockets consisted of management systems and the abilities of individual managers to provide leadership and use systems engineering to ensure that the various components of rockets and their payloads worked together as designed. This was especially necessary as the number of contributors from government (including the military), industry, and universities grew and became interdependent.

Management systems came quickly. In 1955 the Atlas missile had only 56 major contractors. In five years, the number had grown to about 2,000. To cope with these numbers and the need to keep the various components of the missile on schedule and ensure that they worked together, Western Development Division's Bernard Schriever resorted to the unique solution of a systems engineering–technical direction contractor, Ramo-Wooldridge. His organization also put in place a management control system to monitor schedules and to deal with problems as they arose. Charts, graphs, and computer tracking all allowed WDD and its successors to keep their various missiles on schedule and to limit costly launch failures.

Similarly for the Navy, Red Raborn developed the Program Evaluation and Review Technique, which resembled Schriever's system in many respects, to keep the various components of the Polaris program on track for rapid development. Both Raborn's and Schriever's systems had their critics, but without something like them, it is hard to conceive how early missile development could have been so rapid and successful. The two systems also contributed to Secretary of Defense McNamara's planning, programming, and budgeting (PPB) system. While PPB too had its detractors, it did contribute to cost and configuration control over DoD programs.

Transferred to NASA through George Mueller and Sam Phillips, these sorts of reforms in turn enabled the civilian space agency to land humans on the Moon within the decade of the 1960s as President Kennedy had proposed. Mueller and Phillips brought concepts for staying on schedule, keeping within budget, and controlling configurations to a NASA that sorely needed them to succeed in the ambitious Apollo program. These and other management systems were absolutely critical to the success of missile and launch-vehicle development.

With the end of the cold war about 1990, the fundamental nature of rocket engineering did not change, but its context became drastically different. The urgency for new technologies declined significantly, and costs became much more a central concern. In response to the new environment, based on studies from the late 1980s and early 1990s, the Air Force's Evolved Expendable Launch Vehicle (EELV) program replaced the Delta II, Atlas II, Titan II, and Titan IV programs with some boosters projected to cost 25 to 50 percent less. The new vehicles could launch 2,500 to 45,000 pounds into low Earth orbit with a 98 percent reliability rate—significantly above that achieved historically by most launch vehicles.

In December 1996, the Air Force selected McDonnell Douglas (later acquired by Boeing) and Lockheed Martin to develop the new EELV family of launch vehicles.[19] The launch of Lockheed Martin's Atlas V on August 21, 2002, followed by Boeing's Delta IV on November 20, 2002, suggested that the EELV launchers would predominate in the military market, although other projects later came along. With the Atlas V featuring a Russian RD-180 engine, it was obvious that a radical change had occurred since the end of the cold war. As of August 2002, the Air Force reportedly believed that the new EELVs would significantly reduce costs of launching satellites. Soon, however, costs for the program had significantly increased because of low commercial demand, making the 2002 estimates invalid. But the Delta IV and Atlas V continued to operate despite the higher costs, which as of fiscal year 2006 apparently still were much lower than those for older launch vehicles.[20]

Similar concerns about cost prevailed in NASA's efforts to develop new launch vehicles.[21] NASA also had to contend with the issue of safety in the wake of the *Columbia* disaster. In 2005 under new administrator Michael Griffin, the space agency developed a concept for a safer pair of launch vehicles that would build on the technologies used in the space shuttle without the liabilities of a landable orbiter with wings that could be struck by debris from the external tank during launch. A crew launch vehicle named Ares I would consist of a single solid rocket booster derived from the shuttle as the first stage, a second stage powered by a Rocketdyne J-2X engine derived from the J-2 used on the Saturn V (plus a follow-on J-2S version), and a capsule on top for the crew, equipped with an escape rocket that could quickly separate the crew from the launch vehicle in the event of problems. The crew would face little danger from debris such as that separating from the shuttle with tragic consequences during the launch of *Columbia* in 2003 and again, though without significant damage, during the launch of space shuttle *Discovery* in the summer of 2005 as well as later missions.

For future space exploration, NASA also envisioned a heavy-lift launch vehicle names Ares V consisting of two solid rocket boosters, a central booster tank derived from the shuttle's external tank, five Rocketdyne RS-68 engines modified from those used on the Delta IV, and an Earth departure stage powered by a J-2X engine. Both launch vehicles would be configured in stages like expendable launch vehicles such as the Saturn V and Titan III or IV, with Ares V resembling a Titan III or IV since the two solid rocket boosters would flank the central booster tank.[22] It should be noted, however, that the vehicle configurations were by no means fixed but in a state of continuing evolution as of this writing.

Despite these new Air Force and NASA launch vehicles, according to one recent observer if a rocket engineer from the mid-1970s were "transported into a present-day rocket-design office," he or she "would have no difficulty in coping with the technology of present-day [2007] rocket motors as progress in this field has been slow in recent decades." There were some new developments in the years before 2006, but a DoD review in that year stated that usable results from one program—called Integrated High-Payoff Rocket Propulsion Technology—were "somewhat unrealistic for application in the medium term." Thus, for the time being it appeared that older technologies would continue to be used.

How the struggle for cheaper access to space will play out in the new environment ushered in by the fall of the Soviet Union remains to be seen. But this history of the difficulties and uncertainties of developing launch vehicles in a different environment should help current and future rocket engineers understand the kinds of problems they can expect to encounter. Over the roughly fifty years from the serious beginnings of U.S. missile and rocket development to the present, the reliability of rockets improved vastly. In the recent past, launch vehicles have failed only 2 to 5 percent of the time, compared with, for example, more than 30 percent failures for the first 227 Atlas launches. But even with reduced failure rates, as the Columbia Accident Investigation Board stated in 2003, "Building and launching rockets is still a very dangerous business and will continue to be so for the foreseeable future while we gain experience at it. It is unlikely that launching a space vehicle will ever be as routine an undertaking as commercial air travel."[23]

Knowing this and gaining familiarity with the problems encountered in the past should help the present generation of rocket engineers design improved launch vehicles under the cost constraints that they currently face. Perhaps an understanding of the history of U.S. rocketry will also help Congress and the American public understand how difficult it will be to develop rockets that are both better and cheaper.

Appendix

Notable Technological Achievements

1. During the 1930s in New Mexico, Robert H. Goddard pioneered the use of gyroscopic control of vanes in the engine exhaust, film cooling of the combustion chamber, parachutes for recovery of the rocket and its instruments, streamlined casing, pendular stabilization, clustered engines, lightweight propellant tanks, baffles to reduce sloshing, various instruments for measuring aspects of the rocket's performance, thermal insulation for the extremely cold liquid-oxygen tanks, a gimballed tail section for stabilization, a mechanical catapult to give the rocket greater initial velocity, lightweight centrifugal pumps to force propellants into the combustion chamber, a gas generator, igniters, injection heads, and launch controls. However, because he did not publish details of his innovations during his lifetime, others had to reinvent them.
2. In 1942 John W. Parsons of the Guggenheim Aeronautical Laboratory at the California Institute of Technology (GALCIT) developed the first castable composite solid propellant, consisting of asphalt as a binder/fuel and potassium perchlorate as an oxidizer. Castable composite technology, when further developed, found many uses in a huge number of solid-propellant motors.
3. In the early 1940s the German V-2 demonstrated a number of important technologies including film cooling, an eighteen-pot (mixing-compartment) propellant injector system that prevented combustion instability, a steam (hydrogen-peroxide) turbine and a turbopump to feed propellants to the combustion chamber, a fuel cutoff system using the Doppler effect, a control system using an analog computer, a stabilized platform for guidance and control, double-integrating accelerometers, air-bearing gyroscopes, and a pendulous integrating gyro accelerometer.

4. Following the discovery, sometime before February 1942, by Ens. Ray C. Stiff Jr. at the Naval Engineering Experiment Station in Annapolis, that aniline ignited hypergolically (spontaneously on contact) with nitric acid, GALCIT and Aerojet developed storable-propellant technology that was later used on the Titan and other engines.

5. To prevent combustion instability, Caltech researcher Edward W. Price used a star-perforated internal-burning grain with a double-base propellant on the "White Whizzer" in 1946 at the Naval Ordnance Test Station, Inyokern, California.

6. By 1947 Charles Bartley, John I. Shafer, and H. Lawrence Thackwell Jr. at the Jet Propulsion Laboratory had developed a castable composite propellant more sophisticated than Parsons's, using a polysulfide polymer, LP-2, developed by the Thiokol Chemical Corporation. The three engineers used it with an internal-burning star-shaped grain design on the small Thunderbird rocket.

7. By the late 1940s, General Electric had developed a new type of doublet injector for the A-1 missile that was modified for use on North American Aviation's engines, such as those for the Navaho and Redstone.

8. Karel J. Bossart and the Consolidated Vultee Aircraft Corporation used aluminum skin and integral monocoque propellant tanks pressurized with nitrogen on the MX-774B missile—never fully developed but flight-tested starting on July 13, 1948—to lighten the structure significantly. Aluminum later found significant use in rocket structures, and the integral propellant tanks later evolved into the extremely light "steel balloon" structure for the Atlas missile and launch vehicle as well as the Centaur upper stage, although both used stainless steel rather than aluminum.

9. On February 24, 1949, a Bumper WAC reached a reported altitude of 244 miles and a maximum speed of 7,553 feet per second, both records at that time. More important, the two-stage vehicle demonstrated conclusively that a rocket's velocity could be increased with a second stage and that a second stage could be ignited at a high altitude (98,813 feet, or 18.7 miles).

10. The Navy's Viking rocket, which began launching on May 3, 1949, pioneered the use of a gimballed engine for steering. (Robert H. Goddard had flight-tested a different sort of gimbal for the entire tail end of a smaller rocket in 1937, but the Viking team and later the Vanguard perfected the technique used today.)

11. By 1950, when he applied for a patent finally granted in 1965, Edward

A. Neu Jr. of Reaction Motors had invented the "spaghetti" construction for combustion chambers, which involved preforming cooling tubes so that, when joined together, they became the shell of the combustion chamber. This technology created a strong yet light chamber, later adopted by other firms.

12. Starting with German V-2 technology, North American Aviation improved it substantially by 1951 on a 75,000–pound-thrust engine for the Navaho cruise missile. Lighter and more powerful than the V-2's, the Navaho power plant featured improved injectors and turbopumps, plus a more elegant solution to combustion instability than the eighteen pots of the V-2, providing a basis for the engine of the Redstone missile.

13. In 1951 Willy Fiedler developed a jet thrust deflector at Point Mugu Naval Air Missile Test Center to correct thrust misalignment in JATOs. Working for Lockheed, he later adapted the concept to create the jetavators used to steer the Polaris missile, which had its first functional flight on July 20, 1960, from USS *George Washington*.

14. In the early 1950s, JPL began to use vibration test tables before launching the Corporal missile to ensure that components could withstand the vibrations that accompanied launch. Other missile developers soon followed suit.

15. By early 1953, Thiokol had succeeded in scaling up solid-propellant rocket technology to produce a motor 31 inches in diameter and 14 feet 4 inches long for the RV-A-10 missile that was successfully flight tested. Although never produced, the missile prepared the way for even larger motors.

16. Between 1952 and 1954, Thiokol chemists developed polybutadiene-acrylic acid (PBAA), a new solid-propellant binder that permitted higher concentrations of solid ingredients and greater fuel content. First used in Minuteman, PBAA proved to have a lower tear strength than Thiokol's earlier polysulfide propellant, so Thiokol added 10 percent acrylonitrile to the PBAA, creating polybutadiene-acrylic acid-acrylonitrile (PBAN), later used in the space shuttle's solid rocket boosters.

17. Refuting contemporary theory, Atlantic Research Corporation chemical engineers Keith Rumbel and Charles B. Henderson discovered in 1954–55 that significant additions of aluminum to a solid propellant added substantially to the specific impulse. Aluminum as a fuel became a critical ingredient in the propellants used in Polaris, Minuteman, Titan solid rocket motors, and space shuttle solid rocket boosters, among other applications.

18. With a gimballing system further developed from the Viking rocket's, the Vanguard launch vehicle, first flown on October 23, 1957, was able to operate successfully without use of fins for stabilization—a precedent later followed by some other launch vehicles including many Deltas and Titans.

19. Departing from the German V-2's use of a stabilized platform for guidance and control, Vanguard provided an early postwar example of a strapped-down guidance and control system that found use in other rockets, especially after small but powerful digital computers became available in the late 1970s to work in conjunction with the strapdown equipment.

20. After World War II, the Hercules Powder Company's operation at the Allegany Ballistics Laboratory (ABL) in Maryland built upon the wartime research of Drs. John F. Kincaid and Henry M. Shuey at the National Defense Research Committee's Explosives Research Laboratory at Bruceton, Pennsylvania, to develop cast double-base propellants first used in jet-assisted takeoff units in 1947 and applied in one of the third-stage motors on Vanguard, used in the launch of a 52.25–pound Vanguard satellite on September 18, 1959. This technology was later used on upper stages for the Delta and Scout launch vehicles and also Minuteman I.

21. In 1956, Young Development Laboratories in New Jersey developed a method of wrapping threads of fiberglass soaked in epoxy resin around a liner made of phenolic asbestos. With curing, this process produced a strong, rigid shell of Spiralloy that ABL used for the case and nozzle of its third-stage motor for the Vanguard launch vehicle. This type of strong but light composite-structure technology came to be used and further developed for other upper stages.

22. By December 18, 1956, engineers at Redstone Arsenal had greatly improved on the guidance and control technology of the V-2 to produce a system featuring an ST-80 stabilized platform, pendulous integrating gyro accelerometers, and air-bearing gyroscopes for the Redstone missile.

23. At Air Force insistence, in 1953 North American Aviation initiated the Rocket Engine Advancement Program, which developed the RP-1 kerosene rocket fuel, much better suited for rocket engines than the kerosene used as jet fuel. The specifications for RP-1 were not available until January 1957, but it found extensive use in Atlas, Thor, and Jupiter engines as well as other kerosene-burning propulsion systems.

24. By March–April 1957, Aerojet had produced a 30–foot-by-40–inch

motor called Jupiter Senior that used a propellant of polyurethane, ammonium perchlorate, and aluminum. This prepared the way for the Aerojet motors used in Polaris and Minuteman and became the Algol I, first-stage motor for the Scout launch vehicle.

25. In the late 1950s, engineers working on the Polaris missile developed a thrust termination system that opened six ports in the front of the second stage by blowing out plugs with pyrotechnics, permitting expanding gases to escape and halt the acceleration so that the warhead would proceed on a predetermined ballistic path to the target area.

26. In the same period, after static tests of Polaris motors showed that early carbon-throat and exit-cone liners for the nozzles were inadequate, molybdenum throats and liners were employed in the production version of the missile, with an exit-cone liner made of molded silica phenolic, preparing the way for liners used on later missiles and rockets.

27. In the spring and summer of 1957 the Arma Division of American Bosch tested its inertial guidance system, used on the Atlas missile, on the supersonic sled track at the Naval Ordnance Test Station, Inyokern, California, to see if it could withstand the heavy forces and vibrations of the launch process—a new departure in evaluating components for ballistic missiles and rockets, with the Air Force already beginning to establish a Central Inertial Guidance Test Facility (CIGTF) at the service's Missile Development Center at Holloman Air Force Base, New Mexico, with its own sled.

28. On May 31, 1957, a launch of the Jupiter missile showed, among other things, the successful functioning of its inertial guidance system, featuring smaller air-bearing gyros than those on the Redstone and an improved stabilized platform. This guidance and control system weighed only 250–300 pounds, less than half Thor's 650–700 pounds, yet it provided a circular error probable of only 0.8 nautical miles as against Thor's roughly 2 nautical miles, although the Thor system (see entry 31) was also innovative in a different way.

29. By August 8, 1957, the Army's Jupiter C had successfully tested an ablative coating for warheads that soon became the standard method of protecting reentry vehicles from aerodynamic heating.

30. In the early 1950s, Richard H. Battin and his boss, Hal Laning, at the MIT Instrumentation Laboratory developed the delta inertial guidance system, used on the Atlas E and F missiles, in which the position and velocity of the missile at engine cut-off were planned in advance but the guidance computer had flexibility in the ways it achieved the objective.

31. On December 9, 1957, a launch of the Thor missile proved the effectiveness of its innovative inertial guidance-and-control system, developed by the MIT Instrumentation Laboratory and AC Spark Plug using
liquid-floated gyros and the Instrumentation Lab's Q-guidance system,
which was more sophisticated than the delta guidance system used on
the Atlas in that it permitted much of the computation for guidance (the
Q-matrix) to be performed long before the missile was fired, leaving
only a small amount of calculation to be done by the analog computer
on the missile.

32. In 1957, by slicing a Regulus II missile of 20–inch diameter into three
portions, reattaching them, and firing them, Aerojet proved that seg-
menting a rocket motor did not reduce performance. Aerojet had done
the reassembling with bolted flange joints and had fired the resultant
motor using the original Regulus propellant and nozzle. Further tests on
larger motors by Aerojet demonstrated the feasibility of a large segmented
design, permitting ground transportation of huge rocket motors.

33. After 1957 an experimental engine group at Rocketdyne developed a
solid-propellant initiator for the propulsion system that greatly simplified the X-1 and later engines. About the same time, Bell engineers
developed a similar solid-propellant starter cartridge for the Hustler
engine, which evolved into the model 8001 engine for Agena A.

34. Following a failed launch of a Jupiter missile in November 1957, Army
Ballistic Missile Agency engineers installed additional instrumentation
for a December launch and diagnosed a problem with the turbopump
gearcase, which Rocketdyne engineers solved with a bearing retainer.
Rocketdyne subsequently redesigned the turbopumps on the Atlas and
Thor to resolve similar problems.

35. By 1958, North American Rocketdyne engineers had improved and
simplified the Air Force's 75,000–pound-thrust Navaho engine to create a slightly more powerful power plant for the Redstone missile with
78,000 pounds of thrust.

36. Also by 1958, when the Navaho cruise missile was cancelled, Rocketdyne engineers had incorporated a number of innovations in a 120,000–
pound-thrust Navaho engine that contributed to the next generation
of Rocketdyne engines for other vehicles. These innovations included:
(1) conversion of the combustion chamber from welded sheet metal to
lighter-weight brazed tubes for regenerative cooling along the lines first
developed by Ed Neu at Reaction Motors; (2) increased chamber pressure and nozzle area ratio; (3) the powering of the turbopumps with a

gas generator burning the same propellants as the combustion chamber instead of with a separate hydrogen-peroxide system; (4) and smaller, more efficient turbopumps.

37. On January 31, 1958, a Juno I launch vehicle featuring a liquid-propellant first stage developed from the Redstone missile and solid-propellant upper stages using scaled-down Sergeant motors launched the first U.S. satellite, Explorer I, developed like the Sergeant motors by a team headed by JPL engineers.

38. By 1958, researchers at the Allegany Ballistics Laboratory had added ammonium perchlorate to the cast double-base propellant used in the ABL third stage for Vanguard, creating a composite-modified double base (CMDB) propellant used in the second stage of the Polaris A2 missile, which had its first successful launch from a submerged submarine on October 23, 1961. A CMDB process had developed separately at Atlantic Research Corporation, and further modifications occurred at ABL and Hercules' plant in Kenvil, New Jersey, leading to other uses on missiles and upper stages of launch vehicles.

39. Beginning in 1958, the Vought Missiles and Space Company developed a composite named carbon-carbon consisting of a carbon-based cloth impregnated with phenolic resin, cured and heated under pressure. This and other composites found many uses in rockets and spacecraft, providing thermal protection for nozzle throats and even aerodynamic surfaces, such as the leading edges of the shuttle orbiter's wings.

40. The Bell model 8048 engine for the Agena A, initially flown in early 1959, became the first to be tested in a high-altitude test chamber at the Air Force's Arnold Engineering Development Center, which simulated altitudes of 80,000 to 100,000 feet.

41. On January 7, 1960, The Polaris flew its first guided flight with a guidance and control system featuring the Q-guidance developed for the Air Force, an inertial rate-integrating gyroscope with a pendulous integrating gyro accelerometer using the same type of gyro, and the first fully transistorized, digital differential analyzer used in a missile as well as the first use of a digital computer for an inertial guidance system on a missile. The entire "black box" with its inertial components and electronics, developed by the MIT Instrumentation Laboratory and GE, weighed only 225 pounds—even less than the Jupiter's 250–300 pounds.

42. A number of technologies from the Air Force's Thor missile and the Navy's Vanguard launch vehicle allowed NASA to develop the long-lived Delta launch vehicle, first launched in May 1960.

43. Between 1959 and 1960, Aerojet engineers developed an optional nozzle extension for the Able-Star upper stage that increased the expansion ratio from 20:1 to 40:1. Nozzle extensions found extensive use on other vehicles.

44. On October 4, 1960, the Scout team at Wallops Island launched the first successful Scout vehicle, also the first completely solid-propellant launch vehicle in the U.S. inventory.

45. Beginning in 1960, United Technology Corporation (UTC)—in cooperation with sister Pratt & Whitney Aircraft Division of United Aircraft—developed a clevis joint for the large segmented solid-rocket motors for Titan IIIC that featured "male" and "female" segments projecting from the ends of each section of the motor, with the "male" clevis portion facing upward (when the vehicle was on the launchpad) and fitting into the "female" portion that formed a slot facing downward, to create a "tongue-and-groove" arrangement. Some 237 hand-placed pins and 3 fixed pins fit through the two joints and held them together. An O-ring in a slot around the inside of the "female" part of the joint circled the "male" portion and kept hot gases from escaping. This joint performed successfully on several models of Titan III and also on Titan IVA.

46. Before 1961 the Minuteman program had developed vectorable nozzles for Minuteman I's stage 1. These were rotated by a battery-powered hydraulic control unit.

47. By February 1, 1961, the Autonetics Division of North American Aviation had developed the guidance system for Minuteman I with a stable platform supported by two gas-bearing, two-degree-of-freedom gyroscopes and use of gas-bearing accelerometers, considered to be "the first mass production application of semi-conductors to high reliability military electronics."

48. On February 9, 1961, UTC tested a fixed nozzle using nitrogen tetroxide to provide thrust vector control, demonstrating the basic type of thrust vector control used on Titan IIIC for steering.

49. On May 5, 1961, a Mercury-Redstone launch vehicle boosted astronaut Alan Shepard into a successful suborbital flight, proving the "man-rating" of the vehicle by Joachim P. Kuettner and his team from the Development Operations Division of the Army Ballistic Missile Division, which became NASA's George C. Marshall Space Flight Center on July 1, 1960. Rating the Mercury-Redstone for human flight involved some eight hundred changes including an automatic in-flight abort system and three solid-propellant rockets in an "escape tower" with parachutes to carry the astronaut safely to the land or water below.

50. On September 29, 1961, the Polaris A3 team successfully tested a method of thrust vector control (steering) on stage 2, using freon injected into the exhaust from the nozzles to create a shock pattern that deflected the stream and achieved the same results as movable nozzles at a much smaller weight penalty. As with the system used on Titan IIIC, similar systems were later used on other missiles and rockets.

51. In 1961–62, flight tests demonstrated the effectiveness of a new guidance system for Polaris A3, featuring better transistors, wires that were welded instead of soldered, smaller gimbals made of aluminum instead of beryllium, and smaller accelerometers, resulting in a system about half as large and a third as heavy as the earlier Polaris system, yet accurate enough to carry the warhead 2,500 nautical miles with considerably greater precision.

52. Between 1958 and 1963, Rocketdyne engineers solved the combustion instability problems on the Atlas sustainer engine by using rectangular pieces of metal called baffles, attached to a circular ring near the center of the injector face and extending from the ring to the combustion chamber walls. Subsequent engines often used this solution, although in each individual case this and related solutions had to be redesigned.

53. A number of firms contributed by the early 1960s to the development of the carboxy-terminated polybutadiene solid-propellant binder used, for example, on Minuteman II's stage 2 and on the Castor II stage for the Scout and Delta launch vehicles.

54. In the early 1960s, Lockheed developed a Lockseal bearing that allowed the nozzle on a large solid-propellant motor to gimbal. Thiokol subsequently scaled it up to a 156–inch-diameter motor, calling it Flexseal, and later applied it to the space shuttle solid rocket boosters.

55. Aerojet's Transtage engine, developed between 1963 and April 1964, featured an ablatively cooled thrust chamber and a radiatively cooled nozzle assembly.

56. On November 1, 1963, launch of Titan missile N-25 showed that a team from Aerojet, Martin, STL, and Aerospace Corporation had solved the pogo (longitudinal oscillation) problem in the Titan-Gemini first-stage launch vehicle through a standpipe in the oxidizer feed lines, increase in pressure, and a piston accumulator in the fuel feed lines.

57. In September 1964, UTC tested a technology featuring a graphite cloth bonded with a phenolic resin to protect the throat section of a nozzle for the 120–inch-diameter segmented solid rocket motor for Titan IIIC, successfully solving a nozzle-failure problem. This tape-wrapped carbon-phenolic throat marked a major advance in large solid-rocket

technology, working flawlessly on the Titans and paving the road for the giant solid-rocket boosters on the space shuttle, which used the same basic technology for its nozzle throats.

58. By January 1965, engineers from Marshall Space Flight Center, other NASA centers, Rocketdyne, universities, and the Air Force had used cut-and-try engineering and their best data to come up with a combination of baffles, enlarged fuel-injection orifices, and changed impingement angles that solved the combustion instability in the huge F-1 engines used in a cluster for first-stage propulsion in the Saturn V launch vehicle.

59. By early 1965, Boeing and Marshall Space Flight Center engineers had solved welding problems for the propellant tanks and other parts of the S-IC stage for Saturn V, with Marshall devising an electromagnetic hammer that delicately smoothed out bulges produced by welding the huge tanks. The welders used a jig to hold the aluminum pieces in place while a welding tool moved along the seam of the weld, instead of having the pieces move through a stationary welding device as was normal.

60. Between 1962 and 1966, Autonetics Division of North American Aviation upgraded the guidance and control system from Minuteman I to Minuteman II, including semiconductor integrated circuits and significant miniaturization, plus gas-bearing gyros filled with hydrogen instead of helium, to yield three times the accuracy of Minuteman I in a system roughly one-quarter the size.

61. In 1965 Honeywell began developing ring laser gyros (RLGs), which had the advantage over earlier types of gyroscopes that they were not sensitive to the vibrations in launch vehicles that degraded more conventional gyroscopes. Lasers required almost no warm-up time and no periodic calibration. Honeywell used three of its GG1342 RLGs in a three-axis strapdown guidance-and-control system for the Centaur Inertial Navigation Unit that apparently came into use only after the end of the period covered by this history.

62. By 1966, together, Convair/General Dynamics and Pratt & Whitney tamed liquid hydrogen for use as a fuel in the Centaur upper stage, preparing the way for its use in the upper stages of the Saturn launch vehicles and in the space shuttle main engine.

63. By the late 1960s propellant chemists at the Air Force Rocket Propulsion Laboratory, Thiokol, the Army's Redstone Arsenal, Atlantic Research, Hercules, and the Navy had developed hydroxy-terminated polybutadiene as a binder for solid propellants. Besides tactical rockets, it was

used on the Payload Assist Module (employed as a third stage for the Delta launch vehicle and as an upper stage for the space shuttle) and the Antares IIIA rocket motor for the Scout, among other places.

64. On March 3, 1969, the Saturn team succeeded in launching the first Saturn V, which stood some 363 feet tall with its 80 feet of payload included, contained about five million parts (some 5,000 transistors and diodes, five miles of tubing, an acre of adhesive bonds, and 2.5 miles of welding), and had enough power to launch a roughly 95,000–pound payload with three astronauts on board into Earth orbit and, on later missions, into a trajectory toward the Moon.

65. Engineers at NASA's Lewis Research Center and at General Dynamics developed the Automatic Determination and Dissemination of Just Updated Steering Terms (ADDJUST) system for the guidance-and-control computer used on Centaur D-1, available by April 5, 1973, to predict winds during launch from very recent weather data, greatly reducing postponements of launches because of high upper-level winds. The space shuttles adopted a similar system after it proved its worth on Centaur.

66. Between 1972 and 1981, the space shuttle team developed the fly-by-wire guidance, navigation, and control (GN&C) system that still in 1989 was called "the most sophisticated, most advanced, most integrated system in operational use." The system not only had to control the vehicle during ascent to orbit but also was responsible for operating the orbiter in space and then during reentry and landing on a runway.

67. Between 1972 and 1981, Rocketdyne developed the space shuttle main engine, featuring extremely high combustion chamber pressure and a staged combustion cycle "that made this engine a quantum jump in rocket engine technology."

68. Sometime before 1992 at the latest, Delco Electronics had developed a new, highly reliable hemispherical resonator gyro with "no bearings or wear surfaces" that could operate on "extremely low power."

Notes

Notes are in shortened format, with full information provided in the list of sources. Archival and private sources are identified by an abbreviation at the end of the citation denoting the repository, such as NHRC for the NASA Historical Reference Collection or NASM for the Smithsonian National Air and Space Museum; these abbreviations are given at the head of the Archival and Private Sources section of the list of sources. Oral history interviews, denoted OHI, are also listed in this section. All other cited works, without such a repository abbreviation, are listed under Published Sources.

Preface

1. The two books are being published simultaneously by the University Press of Florida.

Introduction

1. Tatarewicz, "Telescope Servicing Mission" (quotations, 392, 365); Bilstein, *Testing Aircraft, Exploring Space*, 156–57; Crouch, *Aiming for the Stars*, 278–79; Launius, *NASA*, 126.

2. See Layton, "Mirror-Image Twins," 562–63, 565, 575–76, 578, 580; Layton, "Technology as Knowledge," 40; Layton, "Presidential Address," 602, 605; Vincenti, *What Engineers Know*, 4, 6–7, 161; Ferguson, *Mind's Eye*, xi, 1, 3, 9, 12, 194. Of course, there are many ways in which science and engineering overlap, as emphasized in Latour, *Science in Action*, 107, 130–31, 174, and by Layton himself as quoted in Bijker, Hughes, and Pinch, *Social Construction of Technological Systems*, 20.

3. Simmons, résumé, n.d., and e-mail messages, July 15, 2002 (quotations), UP. Actually, black powder has probably been known since before 1100.

4. Even people who know better often refer to liquid- or solid-fuel rockets. But the liquids or solids in question include not just fuel but an oxidizer. This is what distinguishes rockets from jet engines; jets use oxygen from the atmosphere to combine with their fuel and permit combustion, while rockets carry their own oxidizer. Hence, the proper terminology is liquid- or solid-*propellant* rockets.

5. Incidentally, the proper technical designation of liquid-propellant combustion systems is engines, while their solid-propellant counterparts are called motors.

6. This section is based on far too many sources to cite here. One source that covers much the same material in language comprehensible to nonexperts is NASA Education Division, *Rockets*, 12–18. Guidance involves selection of a maneuvering sequence to move a rocket from a particular location and direction along its trajectory to the place and attitude needed to carry it to its destination; control executes the maneuvers dictated by the guidance function. See Haeussermann, "Guidance and Control," 225.

Chapter 1. Viking and Vanguard, 1945–1959

1. See Hall, "Vanguard and Explorer," 101–4; Hall, "Earth Satellites," 111; and sources cited in this chapter.

2. Biographical information on Rosen is from his official NASA biography, February 6, 1962, in "Rosen, Milton W.," NHRC.

3. Ibid.; Rosen, OHI; Rosen, *Viking*, 18–20. The Pendray book Rosen mentioned was *The Coming Age of Rocket Power* (1945), which Rosen described as "a proposal that rockets be used for exploring the upper atmosphere" (18). On NRL, the first quotation is from a pamphlet with the NRL seal on its cover and "1923–1973" inscribed inside, in "Naval Research Laboratory," folder 012164, NHRC; the others are from Hevly, "Tools of Science," 221–22.

4. Rosen, OHI, 29–30 (quotation, 30); Rosen, *Viking*, 20–21.

5. Rosen, OHI, 31; Rosen, *Viking*, 22–23, 66 (quotation). Here and below in this chapter I have made corrections based on comments by Rosen written and telephoned in May 2002.

6. Rosen, *Viking*, 26; Rosen, OHI, 31.

7. Rosen, *Viking*, 26–27; Martin, "Design Summary," 5, NASM.

8. See Hunley, *Preludes*, 153.

9. Rosen, OHI, 38–40; Rosen, *Viking*, 27–28; Rosen, comments, UP. On Anderson, see Goodstein, *Millikan's School*, 105–7, 252, 271.

10. Rosen, OHI, 44, 52–53; Rosen, *Viking*, 28, 64; Martin, "Design Summary," 36–37, NASM; Harwood, *Raise Heaven and Earth*, 256; Rosen, comments, UP.

11. Harwood, *Raise Heaven and Earth*, 253; Martin, "Design Summary," 5, 6, 99, NASM. For V-2's alcohol percentage, see Hunley, *Preludes*, chap. 2; for hydrogen peroxide concentration, Kennedy, *Vengeance Weapon 2*, 77; Rosen, comments, UP.

12. Rosen, *Viking*, 58–62, 236–37; Winter and Ordway, "Pioneering Commercial Rocketry," 162–63; Martin, "Design Summary," 104–10, NASM; Scala, "Viking Rocket," 34, which contains a handy launch chronology; Rosen, comments, UP (quotation).

13. For Rosen's claim, see his OHI, 44. On Bossart's relationship to swiveling, J. Neufeld, *Ballistic Missiles*, 47, says the Germans (meaning the von Braun group) first conceived the idea and discarded it but then Bossart came up with the design independently. Swenson, Grimwood, and Alexander, *This New Ocean*, 22, says, "Bossart and associates proposed a technique basically new to American rocketry (although patented by Goddard and tried on some German V-2s)—controlling the rocket by swiveling the engines, using hydraulic actuators responding to commands from the autopilot and gy-

roscope. This technique was the precursor of the gimballed engine method employed to control Atlas and other later rockets." Chapman earlier wrote, "Bossart's 'swinging' powerplant, forerunner of the gimbal control system (movable in *any* direction) prevalent in today's liquid-propellant rockets, was the first of its type anywhere" (*Atlas*, 34), then noted, "He was to learn a few years later that the Germans had experimented with the idea on the V-2" but had discarded it because of the "plumbing problem" with the 18–topf V-2 engine, which would have been difficult to rotate.

14. J. Neufeld, *Ballistic Missiles*, 26, 45–49; Winter and Ordway, "Pioneering Commercial Rocketry," 161–62; Convair, "MX-774 Flight Test Report," NASM.

15. See quotations in note 13; cf. Convair, "MX-774 Flight Test Report," NASM, with diagrams of swiveling cylinders; Rosen, *Viking*, 63, with diagram of Viking gimballed motor. G. Sutton, *Liquid Propellant Rocket Engines*, 220, agrees that the Viking had the first gimballed rocket engine to fly.

16. Hagen, "Viking and the Vanguard," 124.

17. Martin, "Design Summary," 5, 60–68, 210, NASM; Harwood, *Raise Heaven and Earth*, 254.

18. According to Harold F. Klock, "Feedback Circuit," in *McGraw-Hill Encyclopedia of Science and Technology*, 7: 36, "In automatic control systems, feedback is used to compare the actual output of a system with a desired output, the difference being used as the input signal to a controller." And, "When the feedback signal is of opposite phase to that of the input signal, the feedback is negative . . ."

19. Rosen, comments, UP (quotation, May 8). He called the methodology for adjusting the system the Nyquist Diagram.

20. Rosen, *Viking*, 172–73 and passim; Rosen, Bridger, and Jones, "Viking 8 Firing," 2–5, NASM.

21. Rosen, *Viking*, 232, 236; Rosen, Bridger, and Snodgrass, "Viking 9 Firings," 5–6, NASM; Harwood, *Raise Heaven and Earth*, 25.

22. Rosen, "Viking and Vanguard," 7, NHRC.

23. Rosen wrote (comments, UP): "This was our idea and I proposed it to Gen. Schriever, but we could not possibly meet [the Air Force's] time schedule, as he patiently explained to me." For background on IGY, see Hunley, *Preludes*, chap. 6.

24. See Hunley, *Preludes*.

25. Green and Lomask, *Vanguard*, 43–47, 49–51; "Project Vanguard Report . . . 1 June 1957," 2–6, NHRC; [U.S. Army Ordnance/GE], "Hermes," 68, NASM, and note 34 below for thrust of original A3–B propulsion system. See Hunley, *Preludes*, chap. 6, on the Stewart Committee.

26. Green and Lomask, *Vanguard*, 54, 262; M. Neufeld, "Orbiter," 249. Rosen's memo was to the NRL director, but Green and Lomask make it clear the director forwarded it to the Stewart Committee.

27. Letter, Milton W. Rosen, Executive Secretary, Committee on Underground Coal Mine Safety, National Research Council, to Monte D. Wright, Director, NASA History Office, May 7, 1981, in "Rosen, Milton W.," NHRC. Despite Rosen's comments, the original statement remained unchanged in the published volume, Ezell, *NASA Historical Data Book* 2: 86.

28. Rosen, comments, UP.

29. "Project Vanguard Report No. 1," 1 (first quotation), NHRC; U.S. Congress, "Project Vanguard," 60 (second quotation), whose authors say they could not document the quoted intent but were assured by "a representative of the Secretary of Defense" that this was one objective of the project, as indeed it was.

30. U.S. Congress, "Project Vanguard," 61; NASA bio, "John P. Hagen," in "Vanguard II (Feb. 17, 1959)," NHRC.

31. U.S. Congress, "Project Vanguard," 61; "Project Vanguard Report . . . 1 June 1957," 2–6 to 2–7, NHRC; Stehling, *Project Vanguard*, 301. Mass fraction is defined in "Rocket Performance: Mass." Cf. the discussion of these issues and others in Green and Lomask, *Vanguard*, 57–90. Stehling was a propulsion engineer whom Rosen lured from Bell Aircraft to be chief of propulsion at NRL. His is thus an insider's history, and he is a clear example of an engineer who contributed to technology transfer by moving from a commercial firm to a government project. His book (and Rosen's on Viking) also abetted technology transfer.

32. Hagen, "Viking and the Vanguard," 127–28 (quotation, 128); Hall, "Vanguard and Explorer," 109; Harwood, *Raise Heaven and Earth*, 286–87; Green and Lomask, *Vanguard*, 66–67; Stumpf, *Titan II*, 15, for dates of Titan I letter and final contracts; Stehling, *Project Vanguard*, 64–65.

33. Hall, "Vanguard and Explorer," 109; Green and Lomask, *Vanguard*, 62–68; Stehling, *Project Vanguard*, 64–66.

34. Early measurements of specific impulse were not standardized. [U.S. Army Ordnance/GE], "Hermes," 63, NASM, says the A3B (as it is rendered there) had a sea-level specific impulse of 242 lbf-sec/lbm, but Vanguard sources list it at 225. See, e.g., "Project Vanguard Report . . . 1 June 1957," 2–25, NHRC.

35. Heppenheimer, *Countdown*, 72–73; Gibson, *Navaho Missile*, 40; "Project Vanguard Report . . . 1 June 1957," 2–34, NHRC; J. Clark, *Ignition!* 32–33, 104–5. C. William Schnare, who worked with Hall at Wright-Patterson AFB, told the author by telephone on March 14, 2002, that he was involved in insisting that NAA convert to kerosene fuel because it was more readily available than alcohol and more energetic. Clark, a propellant chemist working for the Naval Air Rocket Test Station, Lake Denmark, N.J. (taken over by the Army in 1960 as the Liquid Rocket Propulsion Laboratory of Picatinny Arsenal), until his retirement in 1970, wrote:

> A turbojet has a remarkably undiscriminating appetite, and will run, or can be made to run, on just about anything that will burn and can be made to flow, from coal dust to hydrogen. But the services decided, in setting up the specifications for . . . jet fuel . . . , that the most important consideration would be availability and ease of handling. So since petroleum was the most readily available source of thermal energy in the country, and since they had been handling petroleum products for years, and knew all about it, the services decided that jet fuel should be a petroleum derivative—a kerosene. . . . But the permitted fractions of aromatics and olefins [were] 25 and 5 percent respectively [in JP-4]. (32–33)

Aromatics and olefins were what caused the problems in rockets, and as Clark points out (105), the RP-1 specifications limited olefins to 1 percent and aromatics to 5 percent, solving the problems with kerosene use in rocket engines.

36. Martin, "Vanguard," 4, suggests this interpretation is possible by saying of the "state of the art" in 1955, "Engines were under development, using hydrocarbon fuels, which would increase specific impulse to the 240– to 250–second level." On the other hand, the report lists the fuel for the GE engine as jet fuel, suggesting that it probably was not RP-1.

37. "Project Vanguard Report . . . 1 June 1957," 2–25 to 2–32, NHRC; Stehling, *Project Vanguard*, 129–30; Martin, "Vanguard," 49–50, NHRC; Stehling, "Aspects of Vanguard Propulsion," 45–46. In his comments of May 8, 2002, Milton Rosen wrote, with regard to the burn-through problems, that "GE wanted to redesign the entire cooling system, which would have caused a six month delay in deliveries. Kurt [Stehling] recommended running a ¼ inch d[iameter] copper wire through the existing cooling passages, and that solved the problem." Stehling does not discuss his own contribution.

38. "Project Vanguard Report . . . 1 June 1957," 2–37, NHRC; "Vanguard Vehicle Characteristics," n.d. [after December 16, 1959], in "Vanguard Project, History," NASM.

39. Red fuming nitric acid (RFNA) with 6.5 percent nitrogen dioxide plus aniline with the addition of 20 percent furfuryl alcohol.

40. With 35 percent furfuryl alcohol instead of 20 percent.

41. For WAC Corporal's propellants, see Hunley, *Preludes*, chap. 3; for Aerobee's, Newell, *Sounding Rockets*, 63; for rest of last three paragraphs, J. Clark, *Ignition!* 21, 26, 41–45, 47–65.

42. "Project Vanguard Report . . . 1 June 1957," 2–37 to 2–38 (quotation, 2–37), NHRC; "Project Vanguard Report No. 9," 5, NHRC; Martin, "Vanguard," 51–55, NHRC; Green and Lomask, *Vanguard*, 89, 204; Stehling, *Project Vanguard*, 132–33; George E. Pelletier, Director of Public Relations, Aerojet General, "Details of the Aerojet-General Second-Stage Propulsion System for Vanguard Launching Vehicle," February 11, 1959, in "Vanguard II (Feb. 17, 1959)," NHRC. The coolant for the aluminum tubes was IWFNA.

43. "Project Vanguard Report . . . 1 June 1957," 2–41 to 2–42, NHRC; Green and Lomask, *Vanguard*, 89; Martin Company, "Trip to Aerojet-General, Azusa, California, December 18–20, 1956," in NASA, Vanguard Division records, box 1, NA.

44. Green and Lomask, *Vanguard*, 204 (also first quotation); Stehling, *Project Vanguard*, 134–35 (second quotation, 135); and three items in NASA, Vanguard Division records, box 1, NA: memo, Director, U.S. Naval Research Laboratory, to Martin Company, July 25, 1957, signed J. P. Hagen; memo, Martin Company to Director, Naval Research Laboratory, July 29, 1957, signed N. E. Felt Jr.; memo, Director, U.S. Naval Research Laboratory to Bureau of Aeronautics Representative, Azusa, California, November 27, 1957, signed M. W. Rosen. Stehling refers to "burnouts" rather than erosion, but U.S. Congress, "Project Vanguard," 62, specifically states that the problem was "gradual erosion (not burn-through)" and that the "problem was not solved until October 1957." In his comments of May 8, 2002, Rosen pointed out that Stehling was the one who suggested "coating the inner side of the liner with a ceramic and he found the ceramic we used," but as with the first-stage burnout problem (see note 37), Stehling is silent about his own contribution in this regard.

45. Stehling, *Project Vanguard*, 134 (first quotation); Rosen, "Rocket Development

for Project Vanguard" (undated, annotated "Dod approval 23 Jan 57"), 5 (second quotation), in "Rosen, Milton W.," NHRC.

46. Bartley and Bramscher, "Grand Central Rocket Company," 267–68, 271, 273–76; Bartley, letters to *American Heritage of Invention and Technology* and to Hunley, UP. See also Bartley, OHI, passim. In the article and the paper from which it was derived, Bartley spelled Settlemire as Settlemine, but in the letter to Hunley and in the OHI, it is spelled with an *r*.

47. "Project Vanguard Report . . . 1 June 1957," 2–45 to 2–49, NHRC; Bartley and Bramscher, "Grand Central Rocket Company," 273–74; Green and Lomask, *Vanguard*, 287. On cracking problems in the Sergeant, see Hunley, *Preludes*.

48. Hall, "Vanguard and Explorer," 111; Moore, "Solid Rocket Development," 5–6; Hunley, "Evolution," 27–28 and sources cited there, including Dyer and Sicilia, *Modern Hercules*, 2, 9, 257–58, and Dembrow, OHI. Dembrow, who worked at ABL, said Kincaid and Shuey had relocated from ERL to ABL and done the actual development work there. On this, see also Noyes, *Chemistry*, 17, 26–33, 127–28. My understanding of the cast double-base process has been greatly expanded by e-mails from Ronald L. Simmons, who first became familiar with the propellants used in Hercules' upper stage for Vanguard in 1958. He calls the cast double-base process the "casting powder/solvent process." See esp. his e-mail of July 9, 2002, and résumé, UP.

49. "Project Vanguard Report . . . 1 June 1957," 2–50, NHRC; "Project Vanguard Report No. 9," 7, NHRC; report, A. H. Kitzmiller and E. J. Skurzynski, Hercules Powder Company, to R. Winer, September 17, 1956, subject: MPR—JATO Unit X241 A1 (Project Vanguard) Problem 4–a–81, in NASA, Vanguard Division records, box 6, NA.

50. Hunley, "Evolution," 28; Dyer and Sicilia, *Modern Hercules*, 9, 318–20; B. Wilson, "Composite Motor Case Design"; "Lightweight Pressure Vessels," portion of Ritchey, "Technical Memoir," UP, kindly provided by Ernie Sutton.

51. "Project Vanguard Report . . . 1 June 1957," 2–52, NHRC.

52. For the presence of the 3 percent of aluminum, ABL Report 40, "JATO X248 A2, A Solid Propellant Thrust Unit with High Impulse, High Performance, Wide Applications," November 1958, 7, in NASA, Vanguard Division records, box 6, NA; on the effect of aluminum on combustion instability, see e.g., Povinelli, "Particulate Damping," 1791–96. For the rest of the paragraph, memo, Code 4120 to Code 4100, August 13, 1958, subject: Summary of Conference at ABL and Recommendations, in NASA, Vanguard Division records, box 6, NA; Green and Lomask, *Vanguard*, 287; Holmes, "ABL's Altair," 29. After I wrote this account, Ronald L. Simmons sent me (e-mails of July 10 and 12, 2002, UP) a breakdown of the X248 propellant, which included almost 39 percent nitrocellulose, almost 43 percent nitroglycerine, and almost 12 percent triacetin (a diluting agent) plus the aluminum mentioned in the narrative and lesser ingredients to serve as stabilizers, etc.

53. Attachment to a letter, Director, U.S. Naval Research Laboratory to Commander, Arnold Engineering and Development Center, September 12, 1958, in NASA, Vanguard Division records, box 2, NA; ABL Report 40 (see note 52), which reports the specific impulse in terms of lbf-sec/lbw, the *w* indicating weight rather than mass, but I have substituted the more normal designation lbf-sec/lbm; Stehling, *Project Vanguard*, 128–29n.

Incidentally, Stehling notes that the Stewart Committee had to approve use of the improved X248 motor instead of the X241.

54. The Reynolds number is a nondimensional parameter representing, roughly, the ratio of momentum forces to the viscosity of the fluid through which a body is passing, taking into account representative length; among other uses, the ratio is vital for scale-model testing in wind tunnels, as it provides a basis for extrapolating the test data to full-sized vehicles. For a fuller, more technical definition, see Braslow, *Suction-Type Laminar-Flow Control*, 2; for further discussion, see Hansen, *Engineer in Charge*, 69, 72, 74–76, 318. It was named after Osborne Reynolds (1814–1912) of the University of Manchester who, in Hansen's words, "identified this crucial scaling parameter" (69).

55. "Project Vanguard Report . . . 1 June 1957," 2–13 to 2–15, 2–102, NHRC; Hagen, "Viking and the Vanguard," 130; Martin, "Vanguard," 18, NHRC.

56. "Project Vanguard Report . . . 1 June 1957," 2–103 to 2–104, NHRC.

57. Harwood, *Raise Heaven and Earth*, 254; Furth, "Vanguard," 3.

58. Dominic Edelin, Guidance Control Division, Martin Company, "Separation, Stabilization Final Report X208377," vol. 3, September 1958, 1 (quotation, my italics), and "Systems and Components," vol. 1, August 1, 1958, 1, both in NASA, Vanguard Division records, box 5, NA; Freeman, "Vanguard Control System," 2, 4, NASM.

59. Freeman, "Vanguard Control System," 5–11, NASM.

60. "Project Vanguard Report . . . 1 June 1957," 2–58 to 2–65, 2–69 to 2–75, 2–95, NHRC; Minneapolis-Honeywell Regulator Company, "Vanguard Guidance Sidebar," news release, January 23, 1958, in "Vanguard II (Feb. 17, 1959)," NHRC; Steier, "What Guides the Vanguard," 70, 72; "In the Vanguard." MacKenzie, *Inventing Accuracy*, 182, is the source for the use of strap-down systems in conjunction with small but powerful digital computers; Martin, "Vanguard," 175, NHRC, is the source for their use by 1960 in "many ballistic missile systems." Neither source is specific about which particular rockets or missiles used the strap-down systems. The Minneapolis-Honeywell platform weighed 30 pounds, 70 less than any other system. Yet the Martin report says it provided "accuracy better than required" (29). Steam exhausting from the turbopump in stage 1 provided the thrust in the roll jets for that stage. Stage 2 had no turbopumps, so pressurized gases provided the thrust for its roll jets.

61. Designers for Industry, "Designers for Industry Produced Guidance Subsystem for Vanguard Rocket," news release, n.d., in "Vanguard II (Feb. 17, 1959)," NHRC; "Project Vanguard Report . . . 1 June 1957," 2–15, 2–17, 2–58, 2–63, NHRC; Green and Lomask, *Vanguard*, 59, schematic of Vanguard trajectory; "In the Vanguard"; Steier, "What Guides the Vanguard."

62. "Project Vanguard Report . . . 1 June 1957," 2–77 to 2– 88, NHRC; Stehling, *Project Vanguard*, 266. The integrating accelerometer used the same type of gyro as the Minneapolis-Honeywell platform, but in a pendulous assembly. See Martin, "Vanguard," 34, NHRC.

63. U.S. Congress, "Project Vanguard," 62, 65–66; Green and Lomask, *Vanguard*, 176, 283; Stehling, *Project Vanguard*, 82–83.

64. Green and Lomask, *Vanguard*, 177–82, 283; Stehling, *Project Vanguard*, 106–22,

with many details of problems encountered at Cape Canaveral. For comparative cover-
age of U.S. and Soviet space-launch efforts in this period, see Bille and Lishock, *First
Space Race*; for more detailed coverage of Soviet developments, Siddiqi, *Challenge to
Apollo*.

65. Green and Lomask, *Vanguard*, 196–98; Stehling, *Project Vanguard*, 103, 123;
satellite weight from "Space Activities Summary, Vanguard," December 6, 1957, in "Vanguard Test Vehicle 3," NHRC; substitution of small satellite for instrumented nose cone
from Martin, "Vanguard," 4, NHRC.

66. Green and Lomask, *Vanguard*, 283; "Preliminary Report on TV-3," December
18, 1957, 1 (first quotation), in "Vanguard Test Vehicle 3," NHRC; Stehling, *Project Vanguard*, 24 (second quotation); McDougall, *Heavens and the Earth*, 154 (last two quotations); Heppenheimer, *Countdown*, 127, quoting the *London Daily Herald* for Flopnik
and the *London Daily Express* for Kaputnik. McDougall gives no sources for Stayputnik
or the other two wordplays on Sputnik.

67. Green and Lomask, *Vanguard*, 210; Rosen, comments, May 16, 2002, UP.

68. Green and Lomask, *Vanguard*, 213–14, 217, 283; Stehling, *Project Vanguard*,
157–81, with details of launch efforts leading up to the failed mission.

69. Martin Information Services, "Project Vanguard," [November 23, 1959], in "Vanguard Project, History," NASM; Green and Lomask, *Vanguard*, 204, 219, 285; Stehling,
Project Vanguard, 156, 182–222, 274.

70. NRL, "Transfer," NHRC; Green and Lomask, *Vanguard*, 223; Stehling, *Project
Vanguard*, 237–38; NASA Management Instruction 1052.1 with Attachment A (Executive Order 10783) and Attachment B (Agreement between DoD and NASA), October
1, 1958, effective date October 13, 1959, in "Vanguard Satellite Launching Vehicle 3,"
NASM; Martin, "Vanguard," 96–98, NHRC. Although President Eisenhower's executive
order was dated October 1, NRL notice 5400 says personnel did not transfer to NASA
until November 30. Harry Goett headed the NASA committee, which, according to the
Martin report, made valuable suggestions but found "no major deficiencies in the rocket
design" (98).

71. On these two points, see Rosen, "Viking and Vanguard," 4–5, NHRC.

72. See U.S. Congress, "Project Vanguard," 67–80; Green and Lomask, *Vanguard*,
253–54; Rosen, "What Have We Learned from Vanguard?"

73. Rosen, "Brief History," NHRC; Stehling, *Project Vanguard*, 233–34; Hagen, "Viking and the Vanguard," 139.

74. Rosen, "Brief History," NHRC; Hagen, "Viking and the Vanguard," 139; Green
and Lomask, *Vanguard*, 254–55. For a Silverstein bio, see Levine, *Managing NASA*,
310.

75. Hagen, "Viking and the Vanguard," 140; *CPIA/M1*, unit 412, Minuteman Stage
3 Wing I, January 1964, and unit 427, Altair I, March 1964; Bedard, "Composite Solid
Propellants," 9. Thanks to CPIA for permission to cite the manual and to Tom Moore
of that agency for pointing out the Altair connection. Thanks too to Steve Benson for
pointing out Bedard's article online.

76. "Vanguard, Historical Significance," n.d., in "Vanguard Project, History," NASM.
Cf. Hagen, "Viking and the Vanguard," 132, 135, 139–40.

Chapter 2. The Thor-Delta Family of Space Launch Vehicles, 1958–1990

1. On Thor missiles, see Hunley, *Preludes*, chap. 7.

2. Isakowitz, *Space Launch Systems* (1991), 294–95; Douglas Missile & Space, "Thor History," SMC/HO; Arms, "Thor," 4–2 to 6–8, SMC/HO; "SAMSO Chronology," 387–92, SMC/HO; Spires, *Beyond Horizons*, 71; SMC, "Air Force Satellite Launches," 1–4; GSCF, "Delta Expendable Launch Vehicle," 1, NHRC; U.S. President, *Report . . . 1971*, 116–19; U.S. President, *Report . . . 1974*, 133–35.

3. Meyers, "Delta II," 1–2; U.S. President, *Report . . . 2001*, 140; Boeing, "Delta III"; Boeing, "Delta IV"; Pae, "Delta IV's First Launch."

4. Arms, "Thor," 4–3 to 6–23, SMC/HO; *CPIA/M5*, unit 47, LR101–NA-1000 lb Thrust [vernier for Atlas MA-5 and Thor MB-3-2]; *CPIA/M5*, unit 83, LR79–NA-7, LR79–NA-9, 150,000 lb Thrust [Thor MB-3 Basic and Block I]; *CPIA/M5*, unit 85, LR79–NA-11, 170,000 lb Thrust, Thor IRBM XSM-75 and Booster for Space Vehicles. See also Hunley, *Preludes*, chap. 7, on development of basic engines for Atlas and Thor, the same engines used for these space boosters. An odd and little noticed fact about the MB-3 engines was that they used RJ-1 kerosene, which differed somewhat from the RP-1 used in most kerosene-burning rocket engines. On this, see Meyers, "Delta II," 6–7. RJ-1 is not listed in such standard propellant references as the Liquid Propellant Information Agency's *Liquid Propellant Safety Manual* (esp. chap. 10, Hydrocarbon Fuels, 1) or Kit and Evered, *Rocket Propellant Handbook* (esp. 283–84). According to Butler, "Reliable Delta," 29, RJ-1 was a ramjet fuel having higher density and therefore greater thrust per gallon than RP-1.

5. Rockefeller, "Project Able," 3, SMC/HO; "Air Force Ballistic Missile Division and the Able Programs," 3–4, SMC/HO; J. Powell, "Thor-Able and Atlas-Able," 219; "Aerojet to Build Able Star," for Aerojet's being a subcontractor to Douglas on Able; WDD news release 58–5, "Project Able Fact Sheet," 3–4, in "Re-Entry—(Able-01) (Gen T)" folder, SMC/HO. For Rand Corporation background, see S. Johnson, *Culture of Innovation*, 32. Thanks to Jack Neufeld for a copy of the Johnson history.

6. "Aerojet to Build Able Star," for the propellants used on Able; "Development of Project Able Second Stage, Propulsion System," in "Briefing Charts—Able Projects," SMC/HO; Arms, "Thor," 4–2, 4–3, SMC/HO; letter, R. R. Bennett to L. G. Dunn, March 25, 1958, and "Thor Missile 118 Flight Summary," both in "Re-Entry—(Able-01) (Gen T)" folder, SMC/HO; Stiff, "Storable Liquid Rockets," 8; J. Powell, "Thor-Able and Atlas-Able," 221. See chapter 1 of this book for Vanguard engine development. "Thor Missile 118 Flight Summary" shows Able's vacuum thrust at 7,800 pounds. The 7,500–pound figure comes from Powell, 221.

7. J. Powell, "Thor-Able and Atlas-Able," 219; Arms, "Thor," 4–4, 4–6, B-5, SMC/HO; "Air Force Ballistic Missile Division and the Able Programs," 4, SMC/HO; "Thor Missile 118 Flight Summary," (see note 6). For a different but compatible perspective on these Thor-Able launches, see Marcus, "Pioneer Rocket," 26–30, with promise of a follow-on part 2.

8. Arms, "Thor," 6–8, 6–9, SMC/HO; STL, "Hard Impact Lunar Flight Experiment," SMC/HO; Ezell, *NASA Historical Data Book* 2: 303.

9. Ritland, "Able Program," 2, SMC/HO, which identifies the third stage as ABLE X-248–A3; E. J. Skurzynski, Project Engineer, Allegany Ballistics Laboratory, Development Progress Report, "Design and Testing of JATO X248 A3, 40–DS-3000," November 11, 1958, in NASA, Vanguard Division records, box 6, NA; STL, "Hard Impact Lunar Flight Experiment," 5–6, SMC/HO; STL, "1958 NASA/USAF Space Probes," 59, SMC/HO; Arms, "Thor," 6–8, 6–9, SMC/HO; J. Powell, "Thor-Able and Atlas-Able," 221; *CPIA/M1*, unit 427, 40–DS-3100, X248A5, Altair I.

10. STL, "1958 NASA/USAF Space Probes," 61–63, 82–83, SMC/HO, which I follow for the speeds, converting them from feet to miles per second; Ritland, "Able Program," 2, SMC/HO; Arms, "Thor," 6–10, 6–12, SMC/HO; J. Powell, "Thor-Able and Atlas-Able," 222–23; Ezell, *NASA Historical Data Book* 2: 303–5.

11. STL, "1958 NASA/USAF Space Probes," 62–63, 84, SMC/HO; J. Powell, "Thor-Able and Atlas-Able," 223; Siddiqi, *Deep Space Chronicle*, 18–19.

12. Arms, "Thor," 4–6, 4–7, SMC/HO; J. Powell, "Thor-Able and Atlas-Able," 220–22.

13. ARDC, "Space System Development Plan," III-1 to III-7, III-16 to III-19, AFHRA; "Able 3 Flight Summary" and "Launch Guidance & Tracking Scheme—Able 3 & 4" and "Vehicle Characteristics—Able-3," all in "Briefing Charts—Able Projects," SMC/HO, the last of which shows somewhat lower thrusts for the three stages than shown in the ARDC plan, which I follow since it was revised after the mission and presumably is more accurate than the undated briefing chart; J. Powell, "Thor-Able and Atlas-Able," 224; Arms, "Thor," 6–16, SMC/HO; Ezell, *NASA Historical Data Book* 2: 237.

14. Letters of September 8, October 8, and November 9, 1959, AFBMD/WDPCR to Director, Advanced Research Projects Agency, subject: TRANSIT Program Progress Report for August 1959; . . . for September 1959; . . . for October 1959, all from "Monthly Progress Reports for Transit" folder, SMC/HO; Weil, "Final Report," 5–7, SMC/HO; U.S. President, *Activities . . . 1959*, 24 (quotations); N. Friedman, *Seapower and Space*, 49, with an excellent explanation of the Transit system; J. Powell, "Thor-Able and Atlas-Able," 224.

15. ARDC, "Space System Development Plan," III-46 to III-59, AFHRA; "Launch Guidance & Tracking Scheme" chart (see note 13), which, like the ARDC plan, shows the same guidance and control system for Able 4 as for Able 3; Ezell, *NASA Historical Data Book* 2: 307; J. Powell, "Thor-Able and Atlas-Able," 224, which indicates a new "Burroughs Corp. guidance system" for Able 4, contrary to the first two sources. Powell says the Block II MB-3 propulsion system developed 165,000 pounds, 5,000 less than was normal for that system but significantly higher than the MB-1 on Able 3. AFBMD, "Progress Report Able-4," 3, SMC/HO, written after the launch, states that the "vehicle and payload were similar to those tested in the highly successful ABLE-3 earth satellite (Explorer VI)." Signed by a brigadier general, this official report lists the thrust of stage 1 as 152,957 pounds (less than for Able 3), stage 2 as 7,682 (more than Able 3), and stage 3 as 3,100 pounds (less than Able 3)—see above in the narrative for Able 3 figures. The same report refers to the "STL Advanced Guidance System utilizing a ground based Burroughs Corporation Mod 1 guidance computer" (17) in the second stage.

16. J. Powell, "Thor-Able and Atlas-Able," 224; Siddiqi, *Deep Space Chronicle*, 25; Ezell, *NASA Historical Data Book* 2: 307; U.S. President, *Activities . . . 1960*, vi.

17. J. Powell, "Thor-Able and Atlas-Able," 224; Arms, "Thor," 6–13 to 6–16, SMC/HO; Ezell, *NASA Historical Data Book* 2: 348, 352, 354, and 3: 266; Weil, "Final Report," 9, SMC/HO. Tiros stood for Television Infrared Observation Satellite.

18. J. Powell, "Thor-Able and Atlas-Able," 224; AFBMD, "Proposed AJ10–104 Final Report," SMC/HO; AFBMD, "ABLE-STAR (AJ10–104)," SMC/HO. Of various spellings, including "AbleStar" and "ABLE STAR," I have adopted "Able-Star" for reasons of appearance.

19. Weil, "Final Report," SMC/HO, lists STL contract responsibility for Transit 1–B, the first satellite launched by Thor–Able-Star, as including "design and fabrication of Able-Star second stage" (3) but later states that STL was able to "consign fabrication and test of the Able-Star second stage to Aerojet-General" (53). "Aerojet to Build Able Star" says BMD awarded Aerojet the contract in the fall of 1959 but kept the facts classified until early February 1960. It also says, "Aerojet is program manager of all phases of the *Able Star* vehicle, including control, analysis, design, fabrication, checkout, ground testing and flight testing," but adds that Space Electronics was responsible as subcontractor for "fabrication, checkout and launch services of the electronic system." Stambler, "Simplicity Boosts Able-Star," 59, 63–64, published in August 1961, reveals Space Electronics' "recent" merger with Aerojet's Spacecraft Division and provides details about the guidance and control system, such as that the electronics package included the onboard inertial reference system, programmer, sequencer, and an integrating accelerometer. Stambler says little about the radio guidance system except that it was "designed by Bell Telephone Labs." But since STL designed and built a stage-2 guidance system for Able-Star before October 1960, it seems likely, given the short development time for the upper stage, that this was the same system used with Able.

20. Stambler, "Simplicity Boosts Able-Star," 59; AFBMD, "ABLE-STAR (AJ10–104)," 1–3, SMC/HO; "AbleStar Is Newest Aerojet Triumph," 2 (quotation); Gordon et al., *Aerojet*, III-135.

21. AFBMD, "Proposed AJ10–104 Final Report," SMC/HO, sections entitled "System Requirement," "System Description," "Design Features," "Subsystem Description," and "System Operation," with most of the information corroborated by sources cited below; Stambler, "Simplicity Boosts Able-Star," 59–63; Holmes, "Technological First"; "Aerojet to Build Able Star"; Aerojet, "Model . . . AJ10–104," SMC/HO; *CPIA/M5*, unit 59, AJ10–104, -104B, which provides the thrust and specific impulse for only the 40:1 expansion and, according to Stambler, shows the estimated nominal rather than the rated specific impulse. I give the approximate rated figure.

22. AFBMD, "Proposed AJ10–104 Final Report," SMC/HO, section entitled "System Operation"; Stambler, "Simplicity Boosts Able-Star," 64; Holmes, "Technological First," 44.

23. Letter, BMD/WDPCR to Director, Advanced Research Projects Agency, November 9, 1959, subject: Development of AJ10–104 (ABLE-STAR) Upper Stage Vehicle as of Oct. 31, 1959; letter, BMD/WDLPM-4 to Director, Advanced Research Projects Agency, December 8, 1959, subject: Development of AJ10–104 (ABLE-STAR) Upper Stage Vehicle as of Nov. 30, 1959, both in "Development of AJ10–104," SMC/HO.

24. Weil, "Final Report," 2, 12–25 (first two quotations, 17), SMC/HO; Arms, "Thor,"

6–21, 6–22, 6–26, 6–27, SMC/HO; "Thor Able and Thor Ablestar," 3–4, SMC/HO; AFBMD, "Chronology," 2, SMC/HO; U.S. President, *Activities . . . 1960*, 24–25 (last three quotations); U.S. President, *Activities, 1961*, 92, 94, 99; U.S. President, *Activities, 1962*, 118.

25. Weil, "Final Report," 38–52 (quotation, 45), SMC/HO; U.S. President, *Activities, 1961*, 38–39; AFBMD, "Chronology," 3, SMC/HO; "Thor Able and Thor Ablestar," 4, SMC/HO.

26. U.S. President, *Activities, 1961*, 39; U.S. President, *Activities, 1962*, 121; Weil, "Final Report," 53, SMC/HO; "Thor Able and Thor Ablestar," 4, SMC/HO; N. Friedman, *Seapower and Space*, 49.

27. Weil, "Final Report," 31–37, SMC/HO; U.S. President, *Activities, 1961*, viii, 26; Spires, *Beyond Horizons*, 140; Jaffe, *Communications in Space*, 104–5; Whalen, *Satellite Communications*, 66–67, 111, 160.

28. Arms, "Thor," 6–21, 6–23, SMC/HO; M. Wade, "Thor Able-Star"; U.S. President, *Activities, 1963*, 132, 134; U.S. President, *Activities, 1964*, 141–42, 147; U.S. President, *Activities, 1965*, 145, 150. For all but the last of these launches, the presidential reports to Congress list the satellites as "Defense" and the payload as "Not stated." The final launch included one of these (Transit 5B-7, according to Mark Wade's site) plus a number of payloads related to the Space Surveillance system, which North American Air Defense Command had taken over from the Navy for operational control.

29. See, e.g., "Twentieth Birthday," NHRC; "Thor," in NASA, "500 Thor Launches," LC/MD; Stiff, "Storable Liquid Rockets," 8.

30. Ballistic Missile Organization, "Chronology," 29, 31, UP; Day, "Corona," 10; Bromberg, *Space Industry*, 23–24; Yaffee, "Bell Adapts Hustler," 53; DiFrancesco and Boorady, "Agena," 2, 5; Roach, "Agena," 6, NHRC. Roach tells us (7) that he was one of the engineers who worked on developing Agena. Interestingly, in 1959 he also taught a graduate course in rocketry for the State University of New York at Oswego. On Lockheed and Polaris, see Hunley, *Preludes*, chap. 9.

31. Bromberg, *Space Industry*, 12–13, 23; Roach, "Agena," 6, NHRC; "History," *Rendezvous* (n.d.), 16, and "'55–'65, Textron and Bell . . . and a New Era," *Rendezvous* 9, no. 4 (1970): n.p., in "Bell Rendezvous," NHRC.

32. Roach, "Agena," 6–7, NHRC; Yaffee, "Bell Adapts Hustler," 57; DiFrancesco and Boorady, "Agena," 5–6.

33. Day, "Corona," 15–17, indicates the launch was unsuccessful; Arms, "Thor," 6–28, SMC/HO, says, "Discoverer 1 was successfully placed into a polar orbit which, if not entirely satisfactory from the standpoint of apogee and perigee requirements, was the first such orbit ever effected by a United States spacecraft"; SMC, "Air Force Satellite Launches," 3, lists the launch as successful; McDowell, "US Reconnaissance Satellite Programs," 23–24, has "Agena failed?" in a table but leans the other way in the narrative without giving a definitive answer.

34. Roach, "Agena," 7–8, NHRC; Yaffee, "Bell Adapts Hustler," 58–59; DiFrancesco and Boorady, "Agena," 5–6. There seems to be no information on the total payload capability of Thor–Agena A with either engine, so specification of the 500-pound increase is not very helpful by itself. For information on and an illustration of a venturi tube, see

Mason P. Wilson Jr., "Venturi Tube," in *McGraw-Hill Encyclopedia of Science and Technology*, 19: 217.

35. Yaffee, "Bell Adapts Hustler," 59; McDowell, "US Reconnaissance Satellite Programs," 23–24; SMC, "Air Force Satellite Launches," 3.

36. Yaffee, "Bell Adapts Hustler," 59; DiFrancesco and Boorady, "Agena," 5–6; R. Powell, "Evolution," 121–22; SMC, "Air Force Satellite Launches," 3.

37. Cf. McDowell, "US Reconnaissance Satellite Programs," 24, 26; SMC, "Air Force Satellite Launches," 3–4; McDowell, Agena missions table, 1–2. On the NASA launches, see Ezell, *NASA Historical Data Book* 2: 289, 360–63.

38. Arms, "Thor," 6–39, 6–41, SMC/HO; Isakowitz, *Space Launch Systems* (1991), 295; E. Sutton, *History of Thiokol*, 53, 92, 98 (pages hand-numbered); *CPIA/M1*, unit 237, XM33E5, Castor, Scout-2nd stage; Thiokol, *Rocket Propulsion Data*, TX-33–35; McDowell, Agena missions table, 2; Ezell, *NASA Historical Data Book* 2: 249, 290. Arms, 6–41, and Isakowitz, 295, differ on the thrust-augmented Thor, Isakowitz giving the thrust with strap-on solids as 317,050 pounds, Arms supplying the figure I adopt. The Castor I used on Scout was designated TX-33–35, but CPIA lists its thrust as identical with that shown in Arms for use on the thrust-augmented Thor. See Hunley, *Preludes*, chap. 9, on development of the PBAA binder.

39. Roach, "Agena," 8, NHRC; Yaffee, "Bell Adapts Hustler," 59; DiFrancesco and Boorady, "Agena," 5–6; "Final Report of the Survey Team," JPL; Feld, "Agena Engine," 1, 5–7, 9.

40. R. Powell, "Evolution," 121–29; Lockheed Missiles & Space Company, "Agena: Proven Vehicle for Advanced Space Technology," 8, in folder 010195, "Launch Vehicles: Agena," NHRC; "NASA Projects in Brief—Agena B," NHRC.

41. McDowell, "US Reconnaissance Satellite Programs," 26–28; Arms, "Thor," 6–39, SMC/HO; DiFrancesco and Boorady, "Agena," 5; Ezell, *NASA Historical Data Book* 2: 267, 298, and 3: 173, 283–84, 376.

42. Arms, "Thor," 6–39, 6–41, 6–43, 6–44, SMC/HO; "SAMSO Chronology," 168, SMC/HO; Isakowitz, *Space Launch Systems* (1991), 295; E. Sutton, *History of Thiokol*, 98 (handscribed number, p. 16 of the appendices, which have printed numbers); *CPIA/M1*, unit 237 and unit 582, Castor II; Thiokol, "Catalog," item "TX-354–4, Castor II," NASM. There seems to be no agreement on Thorad specifications. Isakowitz says its tanks were 11 feet longer than the TAT's, increasing burn time from 146 seconds to 167. The SAMSO chronology shows it 14 feet longer and with its burn time extended from 146 seconds to 216.

43. Roach, "Agena," 9, NHRC; DiFrancesco and Boorady, "Agena," 2–3, 5–6; Guill, Sterbentz, and Gordy, "Agena," 10, 13–14; *CPIA/M5*, unit 164, Agena model YLR81–BA-11. Roach gives the dates of the Air Force contract for Bell model 8533, which presumably became 8096–39. At least the details he provides are consistent with the latter model number.

44. "SAMSO Chronology," 367–69, SMC/HO; McDowell, Agena missions table, 4–8; McDowell, "US Reconnaissance Satellite Programs," 28–33.

45. "SAMSO Chronology," 153, SMC/HO; A. Wilson, "Burner 2," 210; *CPIA/M1*, unit 427, X248A5, Altair I. See chapter 1 of this book for treatment of the X248 A2 and this

chapter for coverage of Thor-Able. The configuration that the CPIA depicts for the A5 version conforms well with information in the sources for the A2 version in chapter 1.

46. A. Wilson, "Burner 2," 211; "SAMSO Chronology," 153, 170, 368, SMC/HO; Spires, *Beyond Horizons*, 147, 170; U.S. President, *Activities, 1965*, 131, 136, 142, 151; U.S. President, *Activities, 1966*, 149. Wilson lists two other successful Thor–Burner I launches of meteorological satellites, but the report to Congress for 1965, pages 142 and 151, shows them as having been launched by Thor/FW-4S vehicles. The FW-4S is discussed below in conjunction with its use on Delta. Two of Wilson's dates also differ by one day from those shown in the SAMSO chronology and the report to Congress, which agree, so I use their dates.

47. A. Wilson, "Burner 2," 210; "SAMSO Chronology," 164, SMC/HO.

48. Thiokol, "Catalog," item "TE-M-364–2 & 3 (Star 37B & 37D)," NASM; A. Wilson, "Burner 2," 210. Thiokol lists the binder only as a hydrocarbon, Wilson as a polyhydrocarbon, but the Thiokol catalog shows the same propellant designation as the Star 37E, which, according to *CPIA/M1*, unit 576, used a CTPB binder. Where Wilson's data differ slightly from those of the Thiokol catalog, I follow the catalog.

49. Boeing, "Burner II and Burner IIA," NASM; Boeing, "Burner II for Synchronous Mission Applications," 17–23, NASM; A. Wilson, "Burner 2," 211.

50. A. Wilson, "Burner 2," 211; "SAMSO Chronology," 186–87, 210, 387, 391–93, SMC/HO; Spires, *Beyond Horizons*, 151; U.S. President, *Activities, 1966*, 156; U.S. President, *Activities, 1967*, 125, 128–29. Again, there are some one-day discrepancies among these sources as to launch date, probably due to the difference between Greenwich Mean Time and local time at the launch site, in this case Vandenberg AFB, California.

51. Thiokol, "Catalog," item "TE-M-442–1, Star 26B," NASM; A. Wilson, "Burner 2," 211. Again, where the two sources differ on details, I follow the catalog. See also "SAMSO Chronology," 185–87, 198, SMC/HO. The difference in the propellant between Star 37B and Star 26B was the reduction of ammonium perchlorate by 1 percent and the addition of a catalyst that comprised 1 percent of the propellant (presumably by weight).

52. A. Wilson, "Burner 2," 211; J. Taylor, *Jane's*, 836; "SAMSO Chronology," 246, 368, 387–88, SMC/HO; Boeing, "Burner II for Synchronous Mission Applications," 39, NASM; SMC, "Air Force Satellite Launches," 1–2; Spires, *Beyond Horizons*, 148; U.S. President, *Report . . . 2000*, 26, 100. I am grateful to Frank Winter for pointing out that *Jane's* had information on rocket motors.

53. V. Johnson, "Delta," 51, NHRC; Chien, "Reliable Workhorse"; D. Wade, "Delta Family," 373–76; "Delta II Becomes New Medium Launch Vehicle," AFHRA; Covault, "Boeing," 24; Knauf, Drake, and Portanova, "EELV," 38, 40–42.

54. Rosen, "Brief History," 1, NHRC; Corliss, "Delta," 3–1, NHRC, where Corliss says "Delta's genesis is best described in Rosen's own words" and proceeds to quote Rosen's brief history for two more pages. See also Glennan, *Birth of NASA*, 336–37, 357, for bios and correct titles of Hyatt and Silverstein; Ezell, *NASA Historical Data Book* 2: 611, for a NASA organizational chart at that time.

55. Rosen, "Brief History," 1–2, NHRC; cf. chapter 1 and this chapter, above, for corrections to Rosen's account. Corliss, "Delta," 3–2, 3–3, NHRC, quotes Rosen without qualification.

56. Rosen, "Brief History," 3 (quotations), NHRC; Corliss, "Delta," 3–3, 3–4, NHRC; V. Johnson, "Delta," 51, NHRC.

57. Corliss, "Delta," 2–6, 2–12, 2–13, 3–4, 3–9, NHRC; V. Johnson, "Delta," 54–56, 67–68, NHRC; Arms, "Thor," 6–49, 6–50, SMC/HO. For Schindler's biography, see "William R. Schindler"; Loory, "Quality Control . . . and Success."

58. Corliss, "Delta," 3–4, NHRC; V. Johnson, "Delta," 68, NHRC; *CPIA/M5*, unit 58, Engine AJ10–118.

59. Arms, "Thor," 6–50, SMC/HO; V. Johnson, "Delta," 56, NHRC; *CPIA/M1*, unit 427, 40–DS-3100, X248A5, Altair I. The thrust increased with the ambient temperature. Sources do not always accompany performance data with temperatures, but I have tried to be specific where possible.

60. Ezell, *NASA Historical Data Book* 2: 353–55, 368–72, 375; Arms, "Thor," 6–49, SMC/HO; Rosen, "Brief History," 3, NHRC; V. Johnson, "Delta," 62–64, NHRC; GSCF, "Delta Expendable Launch Vehicle," 4, NHRC.

61. See chart 2 in GSCF, "Delta Expendable Launch Vehicle," 9, NHRC; D. Wade, "Delta Family," 373; Kork and Schindler, "Thor-Delta," 4. Meyers, "Delta II," 1, shows a different sort of chart than the GSFC fact sheet, giving a stronger sense of the rise in payload capability, which was most pronounced between the 6925 and 7925 models, climbing from 3,190 to 4,010 pounds lifted to GTO.

62. NASA, "500 Thor Launches," LC/MD; Meyers, "Delta II," 1–2; Forsyth, "Delta," 117; reliability calculated from GSCF, "Delta Expendable Launch Vehicle," 4–8, NHRC, where data on successes and failures presumably refer only to launches, not satellites themselves, which may have failed after a successful launch.

63. V. Johnson, "Delta," 68–69, NHRC; *CPIA/M5*, unit 58, which gives data for the AJ10–118 and its A, C, and D models without differentiating among them; Kork and Schindler, "Thor-Delta," 3–5.

64. Kork and Schindler, "Thor-Delta," 3–4; *CPIA/M1*, unit 427, X248A5, and unit 415, X258B1, Altair II; GSCF, "Delta Expendable Launch Vehicle," 9, NHRC. Kork and Schindler give the X258 thrust as 6,200 pounds; I follow the CPIA manual's more conservative figure, taken from Hercules/ABL publications and data.

65. Kork and Schindler, "Thor-Delta," 3–4; GSCF, "Delta Expendable Launch Vehicle," 9, NHRC. Kork and Schindler say total liftoff thrust was 325,000 pounds (3), but they give the thrust for the three strap-ons as 53,000 pounds each and for the Thor as 172,000 (4), which add up to the 331,000 pounds I give in the text, not 325,000. Arms, "Thor," 6–60, SMC/HO, shows the combined thrust of the strap-ons as 161,550 (53,850 apiece); with Thor, this yields a total thrust at liftoff of 333,550 pounds.

66. Arms, "Thor," 6–58 to 6–62, SMC/HO; Kork and Schindler, "Thor-Delta," 2–3; Ezell, *NASA Historical Data Book* 2: 378–89.

67. Kork and Schindler, "Thor-Delta," 3–4, 6–7; GSCF, "Delta Expendable Launch Vehicle," 9, NHRC; *CPIA/M1*, unit 480, FW-4. Forsyth, "Delta," 121, says the Delta E incorporated the Castor IIs as "strap-on augmentation," but Kork and Schindler, 4, show the TX-33–52 (Castor I) on the E. Obviously, there are inconsistencies among sources about which models incorporated which improvements. Arms, "Thor," 6–63, SMC/HO, shows the Delta E as using either the Castor I or the TX-354–5 (Castor II).

68. Lindsey, "UTC Solves Scout Stage Spin Problem."

69. Kork and Schindler, "Thor-Delta," 3–5; GSCF, "Delta Expendable Launch Vehicle," 4–5, 9, NHRC; J. Taylor, *Jane's*, 643; Thiokol, "Catalog," item "Star 37B & 37D," NASM; Arms, "Thor," 6–64, 6–66, SMC/HO; Corliss, "Delta," 5–3, NHRC; Bonnett, "Cost History," 21. Despite the title, I found Bonnett's cost data on Thor-Delta not particularly useful, but other information was helpful.

70. Bonnett, "Cost History," 6; Arms, "Thor," 6–66, SMC/HO; Isakowitz, *Space Launch Systems* (1991), 203; *CPIA/M1*, unit 582, Castor II; *CPIA/M5*, unit 194, AJ10–118F, Second Stage Propulsion System. Cf. *CPIA/M5*, unit 122, Titan III Transtage, AJ10–138 (preliminary design completed November 1962); *CPIA/M5*, unit 145, AJ10–118, AJ10–118A/C/D/E; Forsyth, "Delta," 126. On the innovative fiberglass combustion chamber, see chapter 6's discussion of the Titan III transtage.

71. Bonnett, "Cost History," 6, 8, 17, 20; "NASA/Grumman Lunar Module," NASM; information from procurement list for lunar module, by telephone from Joshua Staff, Cradle of Aviation Museum, Garden City, N.Y.; Kurten, "Apollo Experience Report," 9 (also quotations); Gunn, "Delta and Thor/Agena," 7, NASM.

72. Corliss, "Delta," 5–13, 5–14, NHRC; Arms, "Thor," 6–68 to 6–71, SMC/HO; Ezell, *NASA Historical Data Book*, 3: 266–69, 340; Bromberg, *Space Industry*, 13; Library of Congress, *Astronautics and Aeronautics, 1970*, 418; U.S. President, *Report . . . 1972*, 85.

73. Fuller and Minami, "Thor/Delta Booster Engines," 41; Bilstein, *Stages to Saturn*, 97; Ezell, *NASA Historical Data Book* 2: 56; on the X-1 experimental engine, Hunley, *Preludes*, chap. 7. Bilstein, in an uncharacteristic slip of the pen, has NASA awarding the contract for what became the H-1 in September 1958, but NASA became an agency only on October 1.

74. Bilstein, *Stages to Saturn*, 98–104, 414–15.

75. Fuller and Minami, "Thor/Delta Booster Engines," 41, 46–47; *CPIA/M5*, unit 173, H-1 Booster Engines, and unit 196, RS2701A.

76. Bonnett, "Cost History," 6; Arms, "Thor," 6–44, 6–70, 7–3, SMC/HO; Thiokol, "Catalog," items "TE-M-364–2 & 3" and "TE-M-364–4, Star 37E," NASM; GSCF, "Delta Launch History," launch 101, UP; U.S. President, *Report . . . 1974*, 127; GSCF, "Delta Expendable Launch Vehicle," 9, NHRC.

77. Forsyth, "Delta," 127.

78. The enormous difference here seems incredible, but the available data don't allow determination of whether the weights are comparable.

79. J. Taylor, *Jane's*, 837; *CPIA/M5*, unit 194, AJ10–118F-Delta Second Stage Propulsion System, and unit 208, Delta 2nd Stage Engine. I follow the CPIA data, supplied by Aerojet and TRW, where they conflict with *Jane's*, but *Jane's* provides some information not included in the CPIA manual. The designation 2914 was part of a new system instituted in 1972 to delineate the nature of a particular Delta vehicle. Here the "2" indicated that the first stage was the extended long-tank Thor with the RS-27 engine. (A leading "0" would mean the long-tank Thor with the old MB-3 engine, a "1" the extended long-tank Thor with the MB-3.) The second digit gave the number of Castor strap-on motors: 0, 3, 6, or 9. The third digit showed which second stage was used, with "0" for the old 3.63–foot-diameter stage and "1," initially, for the newer 8–foot fairing with the AJ10–118F engine, although by 1974 a "1" meant TRW's TR-201 engine. The fourth digit

identified the third stage: "0" for none, "2" for the UTC FW-4D motor, "3" for the Star 37D, "4" for the newer Star 37E. As this suggests, the four-digit designator constantly changed to keep up with new stages. Corliss, "Delta," 6–13, 6–14, NHRC, and FAS, "Delta," provide keys to the Thor designations at two different times.

80. FAS, "Delta," 4–5; *CPIA/M1*, unit 582, Castor II, and unit 583, TX-526, Castor IV. Thiokol developed the Castor IV initially for the Athena-H advanced ballistic reentry system test vehicle in 1968–69; U.S. President, *Report . . . 1975*, 103; Meyers, "Delta II," 4.

81. Thiokol, *Aerospace Facts*, Spring 1979, 21, 25, NASM. On the Antares III, see *CPIA/M1*, unit 577, Antares IIIA, and J. Taylor, *Jane's*, 837. On Minuteman III, see Hunley, *Preludes*, chap. 9, and *CPIA/M1*, units 457 (Aerojet's version) and 547 (for Thiokol's similar variant). On the development of HTPB, see Hunley, *Preludes*, chap. 9. On carbon-carbon, see Luce, "Composite Technology," 6, NASM; Marshall, *Composite Basics*, 1–4; Jortner, "Transient Thermal Responses," 19.

82. Aerojet, "Delta Second Stage Engine, AJ10–118K," n.d., in "Aerojet AJ10 Series," NASM; FAS, "Delta," 6. Cf. *CPIA/M5*, unit 208, TR-201, and unit 221, AJ10–118 FJI. The latter unit is for the second stage on the Japanese N-II, which the Aerojet fact sheet on the 118K model lists as one of that engine's uses. The FJI engine (and presumably the 118K as well) was ablatively cooled, but the CPIA publication, while showing other features in common with the 118K fact sheet, lists a weight of 1,105 pounds. Perhaps the FJI was an intermediate engine between the 118F and the 118K.

83. Bromberg, *Space Industry*, 113, 130–31; Rumerman, *NASA Historical Data Book* 5: 22, 368; Isakowitz, *Space Launch Systems* (1991), 203; "Delta II Becomes New Medium Launch Vehicle," AFHRA; Zea, "Delta's Dawn," 46; Colucci, "Blue Delta," 42. Navstar stood for Navigation Signal Timing and Ranging.

84. Isakowitz, "Space Launch Systems," (1995), 234–36; Meyers, "Delta II," 3–5; Forsyth, "Delta," 132.

85. Isakowitz, *Space Launch Systems* (1995), 237; FAS, "Delta," 6.

86. Meyers, "Delta II," 7–8; GSCF, "Delta Launch History," launch 184, UP; GSCF, "Delta Expendable Launch Vehicle," 9, NHRC; U.S. President, *Report . . . 1989–1990*, 141; SMC, "Air Force Satellite Launches," 7; Spires, *Beyond Horizons*, 248–49.

87. Fuller and Minami, "Thor/Delta Booster Engines," 47–48; Isakowitz, *Space Launch Systems* (1995), 234–35; Meyers, "Delta II," 3; Colucci, "Blue Delta," 42.

88. Alliant, "Graphite Epoxy Motor," NASM; Isakowitz, *Space Launch Systems* (1995), 234–36; GSCF, "Delta Expendable Launch Vehicle," 9, NHRC.

89. Spires, *Beyond Horizons*, 249–50; GSCF, "Delta Launch History," launch 201, UP; GSCF, "Delta Expendable Launch Vehicle," 8, NHRC; SMC, "Air Force Satellite Launches," 7; U.S. President, *Report . . . 1991*, 73.

90. SMC, "Air Force Satellite Launches," 7, for the GPS launches; for the rest, Gillam, OHI, 10–11, and sources cited in notes 3 and 53.

Chapter 3. The Atlas Space Launch Vehicle and Its Upper Stages, 1958–1990

1. Isakowitz, *Space Launch Systems* (1995), 201–5; U.S. Air Force, "Atlas Space Boosters," NASM; Convair, "Atlas Fact Sheet," 5, NASM; Convair, "Atlas IIB/Centaur," foreword and xi, SMC/HO; General Dynamics/Astronautics, "Atlas ICBM Fact Sheet," 10,

NASM; Spires, *Beyond Horizons*, 71. The Atlas IIB apparently became the Atlas IIAS. See Hunley, *Preludes*, chap. 9, on Atlas's role in Project Mercury.

2. General Dynamics/Astronautics, "Atlas ICBM Fact Sheet," 8, NASM; Missile Test Center, "Atlas Fact Sheet," 3, NASM; Chapman, *Atlas*, 152–65, which has an extended account of the event; Spires, *Beyond Horizons*, 138; Heppenheimer, *Countdown*, 130–31; Crouch, *Aiming for the Stars*, 146. For comparisons of Soviet and American space-related developments, both Heppenheimer and Crouch are useful. Isakowitz, *Space Launch Systems* (1991 and subsequent editions), provides technical data on launch vehicles from around the world.

3. See chapter 2 on development of the Able upper stages from the upper stages used on the Vanguard launch vehicle.

4. "Air Force Ballistic Missile Division and the Able Programs," 7a, SMC/HO; ARDC, "Space System Development Plan," III-2 to III-7, III-64 to III-67, AFHRA; Ritland, "Able Program," 2–3, SMC/HO; AFBMD, "Chronology," 1, SMC/HO; J. Powell, "Thor-Able and Atlas-Able," 224–25; Ezell, *NASA Historical Data Book* 2: 35.

5. ARDC, "Space System Development Plan," III-64 to III-66, AFHRA; "Briefing Charts—Able Projects," p. ABL-9, SMC/HO; J. Powell, "Thor-Able and Atlas-Able," 225.

6. J. Powell, "Thor-Able and Atlas-Able," 225; Ritland, "Able Program," 3, SMC/HO.

7. AFBMD, "Able Program Progress Report," 2–4, SMC/HO; Ritland, "Able Program," 5, SMC/HO; "Program Test Objectives, Atlas/Able 5," NASM; ARDC, "Space System Development Plan," III-83, AFHRA; J. Powell, "Thor-Able and Atlas-Able," 225. Powell claims the third stage never fired, but the 1960 progress report refers to its "40–second propellant burning period."

8. Ritland, "Able Program," 3–6, SMC/HO; J. Powell, "Thor-Able and Atlas-Able," 225, which says the third failure was due to "premature Able stage ignition," but the AFBMD progress report states that the recovered wreckage did not reveal any second-stage combustion. See also Isakowitz, *Space Launch Systems* (1995), 204–5; Ezell, *NASA Historical Data Book* 2: 35.

9. See chapter 2 for development of the Agena upper stage.

10. "SAMSO Chronology" 4, 77, 80, 88, 92, 95, SMC/HO; Canney, "Rockets Used by NASA," 12, NHRC; Hall, "Ranger: A Chronology," 82; Isakowitz, *Space Launch Systems* (1995), 205; Spires, *Beyond Horizons*, 71; Pike, "Atlas," 179; Martin, "Brief History," pt. 2, 40–41; STL, *STL Space Log*, 41–42, NASM; Stone, "U.S. Schedules Additional Samos Launches"; "Air Force Presses"; Peebles, *Corona Project*, 100–101; Peebles, *Guardians*, 62.

11. "SAMSO Chronology," 97, 372, SMC/HO; Spires, *Beyond Horizons*, 156–57; Ezell, *NASA Historical Data Book* 2: 311–13; Hall, "Ranger: A Chronology," 237, 248 (quotation); Siddiqi, *Deep Space Chronicle*, 31; U.S. Congress, "Project Ranger," 3; Hall, *Lunar Impact*, 25–32, 94–104; Baker, *Chronology*, 89. The term "parking orbit" is a bit of a misnomer, since the spacecraft was not stationary with respect to Earth (as a parked car would be) but was circling the planet.

12. Newell to Schriever, JPL. Hall, *Lunar Impact*, 104, says the switch stopped the flow of the oxidizer, but Newell clearly states that the malfunctioning switch failed to start fuel flow.

13. Hall, *Lunar Impact*, 105–9; Newell to Schriever, JPL; Siddiqi, *Deep Space Chronicle*, 31–32; Ezell, *NASA Historical Data Book* 2: 313.

14. Newell to Schriever, JPL; Ezell, *NASA Historical Data Book* 2: 314–15; Siddiqi, *Deep Space Chronicle*, 33–34.

15. See Hunley, *Preludes*, chap. 7, for a discussion of these changes, which were incorporated into the MA-5 system used on the space launch vehicles beginning with the March 1962 launch. For the shift to the baffles and hypergolic ignition, see Martin, "Brief History," pt. 2, 41; for its being *Samos 6*, McDowell, Agena missions table. My thanks to Dr. Harry N. Waldron of SMC/HO for pointing me to this site, which is not definitive but contains much valid information as well as some speculation and some apparently inaccurate information about Agena launchings.

16. Ezell, *NASA Historical Data Book* 2: 335–36; Siddiqi, *Deep Space Chronicle*, 34–35; Office of Space Sciences, "Program Review," 98, NHRC.

17. JPL, *Mariner-Venus 1962*, 97, 101.

18. NASA, Agena Program Presentation, 1–4 to 1–7, JPL; on letter contract, "SAMSO Chronology," 98, SMC/HO.

19. Office of Space Sciences, "Program Review," 98–99, NHRC.

20. Ibid., 107–9; Convair, "Atlas Fact Sheet," 5–6, 10–11, 14, NASM; Hall, *Lunar Impact*, 32, 188–90; Hall, "Ranger: A Chronology," 250; Lewis Public Information Office, news release, "The Agena Project," February 1967, 1–3, in folder 010195, "Launch Vehicles: Agena," NHRC; Dawson and Bowles, *Centaur*, 24–26; Dunar and Waring, *Power to Explore*, 85; Isakowitz, *Space Launch Systems* (1995), 202–4; Powell and Richards, "Atlas E/F," 229. Already in September 1961, Marshall wanted to give up its role providing technical direction for the Agena. Eventually, in January 1963, NASA's Lewis Research Center in Cleveland assumed responsibility for NASA's use of Agenas and their first stages. Apparently Lewis's management helped establish more stringent standards for the Atlas, including the General Electric guidance system.

21. Ezell, *NASA Historical Data Book* 2: 266, 268, 316–18; Siddiqi, *Deep Space Chronicle*, 35–36, 43, 47–48; Isakowitz, *Space Launch Systems* (1995), 205; McDowell, Agena missions table; Dawson and Bowles, *Centaur*, 61.

22. Ezell, *NASA Historical Data Book* 2: 337–38; Siddiqi, *Deep Space Chronicle*, 43–44; JPL, *Mariner-Mars 1964*, esp. 45; "SAMSO Chronology," 8, 125, 128, SMC/HO; McDowell, Agena missions table; Spires, *Beyond Horizons*, 153; "SAMSO Chronology," 385, SMC/HO, which lists all of the Vela launches, four more of them by Atlas-Agena in pairs on July 17, 1964, and July 17, 1965, the rest by Titan IIIC.

23. Ezell, *NASA Historical Data Book* 2: 262, 321–25, 397–99; Siddiqi, *Deep Space Chronicle*, 56, 58, 61, 63, 66; Isakowitz, *Space Launch Systems* (1995), 205; Bullock, "Summary," 16, NASM.

24. H. Taylor, "Atlas Launch Vehicle"; Convair, "Atlas Fact Sheet," 13, NASM; Isakowitz, *Space Launch Systems* (1995), 203; Martin, "Brief History," pt. 2, 41; *CPIA/M5*, unit 181, Atlas MA-5 Booster, and unit 182, Atlas MA-5 Sustainer, which show different propellant flow rates and turbopump pressures for different thrust levels; Convair, "Atlas Family," 26, 68, NHRC; "Agena Project" (see note 20), 3.

25. Martin, "Brief History," pt. 2, 41, 45; Ezell, *NASA Historical Data Book* 2: 270;

McDowell, Agena missions table; Ezell, *NASA Historical Data Book* 3: 344; "Twentieth Birthday," NHRC. McDowell and Martin disagree on the April 1978 launch date and also on the payload; I follow Martin.

26. NASA, "First Launch," NASM; General Dynamics/Astronautics, "Centaur Primer," NHRC. The NASA news release gives the performance gain as 40 percent, "Centaur Primer" as 35 percent.

27. General Dynamics, Space Systems, "Atlas/Centaur," NASM; Ezell, *NASA Historical Data Book* 2: 328–31; 3: 169–70, 305–11; Rumerman, *NASA Historical Data Book* 5: 82; Dawson and Bowles, *Centaur*, 222–24; General Dynamics News, fact sheet "Commercial Atlas/Centaur Program," document III-8 in Geiger and Clear, "History," vol. 3, AFHRA.

28. The point is brought out well in Green and Jones, "Bugs That Live at -423°." I thank Dwayne Day for calling to my attention this excellent article in an unusual publication.

29. Thiel, "Practical Possibilities," 4 (quotation), NASM; on Goddard and Oberth's comments, Hunley, *Preludes*, chap. 1. See also Sloop, *Liquid Hydrogen*, 236, 258–64.

30. Sloop, *Liquid Hydrogen*, 13–14, 20–26, 37–38, 49–58; Osborne et al., "Liquid-Hydrogen Rocket," 279–318.

31. Sloop, *Liquid Hydrogen*, 81–93, 102–12.

32. Ibid., 113, 141–66; Dawson and Bowles, *Centaur*, 17–18; Tucker, "RL10," 125.

33. Sloop, *Liquid Hydrogen*, 191–94; Newlon, "Krafft Ehricke"; Ehricke, OHI, 32.

34. Ehricke, OHI, 33, 51–53; Chapman, *Atlas*, 153; Sloop, *Liquid Hydrogen*, 194; Dawson and Bowles, *Centaur*, 1–12.

35. Dawson and Bowles, *Centaur*, 13, 18–19; Sloop, *Liquid Hydrogen*, 178–79, 194–95; "SAMSO Chronology," 56, SMC/HO; Spires, *Beyond Horizons*, 57–58. Following General Dynamics' acquisition of Convair in 1954, the company established Astronautics as "a separate operating division of Convair" in 1955–56. For the quotation, see Convair, "Atlas Fact Sheet," 3, NASM.

36. Sloop, *Liquid Hydrogen*, 200–201; "Statement of Krafft A. Ehricke, Director, Advanced Studies, General Dynamics/Astronautics," in U.S. Congress, "Centaur Program," 5, 63–66.

37. Ehricke, OHI, 57–59; Ezell, *NASA Historical Data Book* 2: 45; U.S. Congress, "Centaur Program," 6.

38. Ehricke, OHI, 59; Heald, "LH_2 Technology," 207.

39. U.S. Congress, "Centaur Program," 4–5, 105–6.

40. Green and Jones, "Bugs," 14, 19; Martin, "Brief History," pt. 2, 43; Heald, "LH_2 Technology," 209–10; Richards and Powell, "Centaur Vehicle," 99. Green and Jones interchange the terms "concave" and "convex," as is clear from Heald, 209, diagram.

41. U.S. Congress, "Centaur Program," 9, 51, 97; Dawson and Bowles, *Centaur*, 19–20, 34, 51, 74; Stambler, "Centaur," 74; John L. Sloop, memo to Director of Space Sciences, December 18, 1961, in "Centaur Management & Development," NHRC; Schubert, "Centaur," 173–75, NHRC.

42. U.S. Congress, "Centaur Program," 7, 33, 47, 66; Hans Herbert Hueter bio, MSFC fiche 1067, copy in file 001055, NHRC; Schubert, "Centaur," 121–80, NHRC.

43. Sloop, December 18 memo (see note 41), 1–2 (first two quotations); Sloop, memo to Director of Space Sciences, December 20, 1961, 3 (third quotation), in "Centaur Management & Development," NHRC; Sloop, *Liquid Hydrogen*, 323, for his bio.

44. Stambler, "Centaur," 73–75; Schubert, "Centaur," 175, NHRC (quotation); Dawson and Bowles, *Centaur*, 74.

45. Tucker, "RL10," 126–28 (quotations, 126–27).

46. Ibid., 128–29.

47. Ibid., 130–31.

48. Richards and Powell, "Centaur Vehicle," 100; Green and Jones, "Bugs," 21–22.

49. Tucker, "RL10," 132–37; U.S. Congress, "Centaur Program," 115–16; Schubert, "Centaur," 131, NHRC.

50. U.S. Congress, "Centaur Program," 116; Tucker, "RL10," 139.

51. Richards and Powell, "Centaur Vehicle," 101; Heald, "LH$_2$ Technology," 213.

52. Richards and Powell, "Centaur Vehicle," 102–3; Isakowitz, *Space Launch Systems* (1995), 205.

53. Richards and Powell, "Centaur Vehicle," 103; Davis, Correspondence, 17; Heald, "LH$_2$ Technology," 206; U.S. Congress, "Centaur Program," 12–27.

54. Sloop, December 20 memo (see note 43) (quotations, 2); letter, Hans Hueter, Director, Light and Medium Vehicles Office [MSFC], to J. R. Dempsey, President, General Dynamics/Astronautics, January 4, 1962, in "Centaur Management & Development," NHRC.

55. Hueter to Dempsey, 2.

56. U.S. Congress, "Centaur Program," 9, 61–62; Dawson and Bowles, *Centaur*, 35–36, citing a Convair interview by Sloop on April 29, 1974, and quoting Grant Hansen on what Dempsey told him about the company's thinking. On Hansen's background, see letter, William H. Pickering to Norman L. Baker, President, National Space Club, December 19, 1963, enclosure 1, in William Pickering Publications Collection, 1932–1971, box 3, folder 29, JPL.

57. Geddes, "Centaur," 25, 28–29; U.S. Congress, "Centaur Program," 2, 9, 37, 66, 104.

58. U.S. Congress, "Centaur Program," 5, 89; Martin, "Brief History," pt. 2, 43; Schubert, "Centaur," 136, NHRC; General Dynamics/Astronautics, "Centaur Primer," 17, 31, NHRC; General Dynamics/Astronautics, "A Primer," 16, NHRC.

59. Martin, "Brief History," pt. 2, 43; General Dynamics/Astronautics, "Centaur Primer," 18–21, NHRC; Librascope Division, "General Precision/Librascope Computer Guides Centaur," n.d., in "Centaur General," NHRC, which locates Librascope in Glendale, Calif.; Air Force Missile Development Center, news release 64–152–R, "AFSC News," October 1, 1964, microfilm roll 32,273, frame 215, AFHRA; Central Inertial Guidance Test Facility, Final Report, Centaur (-1) MGS [Missile Guidance System] Sled Test, vol. 1 (Operational Hardware Evaluation), September 1964, ix, 1, A-1, microfilm roll 32,273, frame 217, AFHRA; Central Inertial Guidance Test Facility, Final Report, Centaur-1 Missile Guidance System Sled Test, vol. 2: Accuracy Evaluation, November 1964, xiii, 5, 48, 54, 62–63, microfilm roll 32,273, frame 6, AFHRA; U.S. Congress, "Centaur Program," 2; NASA, "First Launch," 2–3, 2–4, NASM. On Librascope's larger

and earlier drum computer, the LGP-30, see Ceruzzi, *Modern Computing*, 42. As early as 1965, NASA was referring to the airborne digital computer on Centaur as "built by the Kearfott Division of General Precision, Inc.," but presumably this only reflected an organizational reshuffling within General Precision. See NASA, "Centaur Test," NASM.

60. U.S. Congress, "Centaur Program"; U.S. Congress, "Centaur Launch Vehicle," 11 and passim.

61. U.S. Congress, "Centaur Launch Vehicle," 12.

62. Deane Davis to Eugene Emme (NASA historian), February 6, 1983, with attached account (Davis quotations, 4), in folder 010190, "Launch Vehicles: Atlas," NHRC; Sloop, *Liquid Hydrogen*, 208 (Ehricke quotation). Davis's account is published as "Seeing Is Believing," 196–98 (quotations, 198).

63. U.S. Congress, "Centaur Program," 59 (quotations); Dawson and Bowles, *Centaur*, 52, for von Braun's attempt to have Centaur cancelled.

64. "History of the George C. Marshall Space Flight Center," July 1–December 31, 1962, 85, in "Centaur General," NHRC, for date of transfer; Sloop quotations from his *Liquid Hydrogen*, 183; Davis quotations from his letter to Emme (see note 62); final quotations from Geddes, "Centaur," 28, reflecting Hansen's views; Silverstein bio from Dawson, *Engines and Innovation*, 169–70, 177–78, and Glennan, *Birth of NASA*, 357.

65. NASA news release 62–209, "Liquid Hydrogen Program Stepped Up," September 30, 1962, in "Centaur General," NHRC; "History" (see note 64), 85.

66. U.S. Congress, "Centaur Program," 66–68; General Dynamics/Astronautics, "Centaur Primer," 28–29, 41–42, NHRC; Dawson and Bowles, *Centaur*, 15–17, 27.

67. Dawson and Bowles, *Centaur*, 59–66; Geddes, "Centaur," 27, 29.

68. Dawson and Bowles, *Centaur*, 62–69; Geddes, "Centaur," 27–28; Sloop, December 20 memo (see note 43), 2, for Rovenger having only three others working for him at GD/A at that time.

69. Green and Jones, "Bugs," 26–34; Richards and Powell, "Centaur Vehicle," 104–6; *CPIA/M5*, unit 115, RL10A-3–1, Centaur and Saturn S-IV Propulsion, and unit 116, Centaur.

70. Green and Jones, "Bugs," 30–38; Richards and Powell, "Centaur Vehicle," 106–7; Launius, *NASA*, 83–84; Isakowitz, *Space Launch Systems* (1995), 203, 205; Convair, "Atlas Fact Sheet," 12, 14, NASM; Ezell, *NASA Historical Data Book* 2: 263, 328–31, 399; Siddiqi, *Deep Space Chronicle*, 55, 57, 61, 63, 65, 67–69; U.S. President, *Aeronautics and Space Activities, 1965*, 161; U.S. President, *Aeronautics and Space Activities, 1968*, 108. The 1968 report does not specify which Atlas was used with the Centaur, but it shows SLV-3A with the Agena and shows the same data for the two Atlas vehicles. The thrust figures in the two reports do not agree with those I have given (under Agena), but both reports indicate "Definitive data are difficult to compile." The payload figures appear correct in both. Besides the last three Surveyors, SLV-3C launched some Applications Technology Satellites, Orbiting Astrophysical Observatories, some of the Mariner and the Pioneer missions, and several Intelsat satellites.

71. Ezell, *NASA Historical Data Book* 3: 222; Pioneer 10 and 11 Mission Descriptions; Dawson and Bowles, *Centaur*, 123–25; Richards and Powell, "Centaur Vehicle," 109–10; Siddiqi, *Deep Space Chronicle*, 93–97; *CPIA/M1*, unit 576, Star 37E. On the use of CTPB

on Minuteman II stage 2 and on development of the binder, see Hunley, *Preludes*, chap. 9. "Star" stands for Spherical Thiokol Apogee Rocket; the number indicates the diameter in inches; see Powell and Richards, "Atlas E/F," 232.

72. Dawson and Bowles, *Centaur*, 125–31; Space and Missile Systems Organization, "Centaur D-1 Payload Users Guide," 3–1, 3–30 to 3–37, NASM; Richards and Powell, "Centaur Vehicle," 110–11; "Atlas Launch Vehicle History"; Capt. Nicholas C. Belmont, "Laboratory Tests of Advanced Centaur Components," January 30, 1968, microfilm roll 32,273, frame 334, AFHRA; Honeywell, "Centaur Inertial Navigation Unit," March 1992, in "Centaur General," NHRC. I have found no trustworthy source for the length of SLV-3C, but General Dynamics/Astronautics, "A Primer," 14, NHRC, puts what apparently was LV-3C at 113 feet long, including 30 (in some sources 31) feet for the Centaur. The diameter was 10 feet. Convair, "Atlas Family," NHRC, says SLV-3C was 51 inches longer than LV-3C (listed as just SLV), which would make SLV-3C some 117 feet long. SLV-3D was listed in "Atlas Launch Vehicle Family" as 132 feet long, but Isakowitz, *Space Launch Systems* (1995), 203, gives the length as 131 feet including payload fairing, which would make the Centaur plus the payload fairing 61.5 feet. "Atlas Family," 39, gives the length of the Centaur D-1 as 31 feet. Richards and Powell say the thrust for SLV-3D was somewhat higher than for SLV-3C, but "Atlas Launch Vehicle History" shows it to have been the same as for SLV-3A, which would have been the same as for SLV-3C. Isakowitz (1995), 203, also says SLV-3D had the same thrust as SLV-3C. Centaur converted to strapdown guidance for Atlas II, which first flew in 1992.

73. Dawson and Bowles, *Centaur*, 125–31; Richards and Powell, "Centaur Vehicle," 111; Ezell, *NASA Historical Data Book* 3: 223; Thiokol, *Aerospace Facts*, July–September 1973, 12, NASM; Walker, *Atlas*, 235.

74. Dawson and Bowles, *Centaur*, 130–31; Ezell, *NASA Historical Data Book* 3: 223; Siddiqi, *Deep Space Chronicle*, 102–3.

75. Dawson and Bowles, *Centaur*, 131–33; Ezell, *NASA Historical Data Book* 3: 304–5; Siddiqi, *Deep Space Chronicle*, 105–6.

76. Isakowitz, *Space Launch Systems* (1995), 203, 205, with the success rate calculated from the SLV-3D entries on page 205, including entries for failures and their basic causes; Richards and Powell, "Centaur Vehicle," 100, 113–14, including longer descriptions of the Atlas failures and information about the upgrades to Centaur and the engine (without explaining how it was "adjusted" to increase thrust, a technique that may have been proprietary and not revealed to outsiders by Pratt & Whitney); Dawson and Bowles, *Centaur*, 222–24, especially for information on the Intelsat satellites but also details on the D-1AR, used in conjunction with Richards and Powell; Ezell, *NASA Historical Data Book* 3: 290, 293–94, 304–11, and Rumerman, *NASA Historical Data Book* 5: 82, for information about satellites. Richards and Powell give the thrust for the "adjusted" engine as 73 kilonewtons, which converts at the rate of 224.8 kN/lb to 16,400 lbs. However, a manufacturer's brochure on the RL10 gives a thrust of 16,500 lbs; see Pratt & Whitney, "RL10 Engine," NHRC. For the RL10A-3–3A, which Richards and Powell say also had a thrust of 73 kN, I use the 16,500 figure.

77. Richards and Powell, "Centaur Vehicle," 114; Tucker, "RL10," 147; Dawson and Bowles, *Centaur*, 242–43; General Dynamics, Space Systems, "Atlas/Centaur," NASM.

78. Dawson and Bowles, *Centaur*, 224, 232, 234–35; Isakowitz, *Space Launch Systems* (1995), 205; U.S. President, *Report . . . 1989–1990*, 143; Richards and Powell, "Centaur Vehicle," 114–15; Martin, "Steel Balloon," 12.

79. Bromberg, *Space Industry*, 100–101, 149–59; Martin, "Brief History," pt. 3, 46; Dawson and Bowles, *Centaur*, 229–42, 245–46. A useful, brief discussion of the history and technology of Ariane is in Verger, Sourbès-Verger, and Ghirardi, *Cambridge Encyclopedia of Space*, 145–48.

80. Martin, "Brief History," pt. 3, 46–47.

81. Ibid., 47 (also quotation); Martin, "Brief History," pt. 2, 44; Tucker, "RL10," 147; Heald, "LH$_2$ Technology," 211, all contributing to the explanation of the two failures; Isakowitz, *Space Launch Systems* (1995), 205; U.S. President, *Report . . . 1989–1990*, 18, 147, on the Combined Release and Radiation Effects Satellite; U.S. Congress, "Recent Launch Failures," 21, for the number of parts on, in this case, an SLV-3C–Centaur.

82. Robertson, "Centaur Canters On," which provides the statistics for Centaur to that time, including seven flights of "Atlas/Centaur D-1T," which presumably were actually *Titan*-Centaur launches; General Dynamics, "Mission Planner's Guide," 1–1, 1–2, A-1 to A-6, NASM; Heald, "LH$_2$ Technology," 220; Dawson and Bowles, *Centaur*, iv–vi; Covault, "Atlas V Soars," 22–24. See Martin, "Brief History," pt. 3, 47–51, for an overview of Atlas versions flown during the 1990s, which are beyond the scope of this book.

83. Powell and Richards, "Atlas E/F," 229–31; Martin, "Brief History," pt. 2, 44.

84. Powell and Richards, "Atlas E/F," 230–31; "ABRES Research Really Not a Breeze," *AFSC Newsreview* 9, no. 2 (July 1965): 16, microfilm roll 31,696, AFHRA; Martin, "Brief History," pt. 1, 60–61; Powell and Richards, "Orbiting Vehicle Series of Satellites," 417–21, 426; *CPIA/M1*, unit 480, FW-4; Convair, "Atlas Family," 42, NHRC. The FW-4S was used as Scout's fourth stage, while an FW-4D served as the third stage on the Thor-Delta launch vehicle. The four failures to orbit OV-1s stemmed from a separation thruster that did not fire, an Atlas that exploded, a collision between the satellite and a portion of the Atlas, and a propulsion module malfunction. See chapter 4 for the Scout family of launch vehicles, chapter 6 for Titan III, and chapter 2 for a description of the FW-4.

85. Powell and Richards, "Atlas E/F," 231–33; U.S. President, *Report . . . 1974*, 44. On the significance of GPS, see N. Friedman, *Seapower and Space*, 228–29, 242, 266–82. For a failed Atlas mission with Burner II, see A. Wilson, "Burner 2," 211.

86. Powell and Richards, "Atlas E/F," 233; U.S. President, *Report . . . 1977*, 41; U.S. President, *Report . . . 1980*, 39. For generic descriptions with diagrams of the Fairchild upper stages, see Porcelli and Vogtel, "Modular, Spin-Stabilized, Tandem Solid Rocket Upper Stage."

87. Powell and Richards, "Atlas E/F," 231, 233–34; Thiokol, *Aerospace Facts*, Spring 1979, 25, NASM. On the Antares III, see *CPIA/M1*, unit 577, Antares IIIA. On HTPB development, see Hunley, *Preludes*, chap. 8. On carbon-carbon, see Luce, "Composite Technology," 6, NASM; Marshall, *Composite Basics*, 1–4; Jortner, "Transient Thermal Responses," 19.

88. Powell and Richards, "Atlas E/F," 231, 234–37; M. Wade, "Star 37S," citing Thiokol's site www.thiokol.com in May 1999 for the information, which appears no longer to be there; J. Taylor, *Jane's*, 836; U.S. President, *Report . . . 1981*, 38.

89. SMC, "Air Force Satellite Launches," 1–2; USAF, "Defense Meteorological Satel-

lite Program," 1–2; NASA, "Defense Meteorological Satellites Program," 1–2; U.S. President, *Report . . . 1982 Activities*, 87; U.S. President, *Report . . . 1979 Activities*, 89; U.S. President, *Report . . . 1991 Activities*, 160.

90. Powell and Richards, "Atlas E/F," 231, 234, 239; FAS, "White Cloud Naval Ocean Surveillance System."

91. Isakowitz, *Space Launch Systems* (1995), 204–5; Powell and Richards, "Atlas E/F," 240. One Atlas E mission launched during 1990 and not covered in the narrative was the so-called Stacksat, consisting of three payloads by DoD: the Polar Orbiting Geomagnetic Survey (POGS) and Selective Communications Experiment (SCE) for the Navy and the Transceiver EXperiment (TEX) for the Air Force. According to the Internet, at least, the upper stage used to launch these three spacecraft was a Thiokol Star 20, which usually served as a Scout fourth stage. On the Star 20, see chapter 4 of this book. For the information on the spacecraft and Star 20 related above, see U.S. President, *Report . . . 1989–1990*, 145–46; McDowell, "Star 20"; "Stacksat (P87–2)." This last has more on the three spacecraft, but note the sponsor's warning, "Information in The Mission and Spacecraft Library is provided without warranty or guarantee. USE AT YOUR OWN RISK."

92. Quotation from Martin, "Steel Balloon," 15, where Martin predicted accurately, "The gleaming stainless steel tanks that are a trademark of the Atlas series of vehicles have proven to be easily adaptable through many vehicle changes, including four new versions currently in development which are likely to still be launching payloads to space in the 21st century."

93. This was the burden of the testimony of NASA's deputy administrator George M. Low and Vincent L. Johnson, associate administrator for space science, among others, before a House subcommittee on NASA oversight in June 1971 about the failure of an integrated circuit on the attempted launch of Mariner 8 on May 8, 1971. The circuit was part of the pitch-gyro preamplifier in the Centaur autopilot. The autopilot had functioned normally during its checkout 40 minutes before launch. See U.S. Congress, "Recent Launch Failures," esp. 3–5, 14–15, 20–25, 81–89.

Chapter 4. The Scout Family of Space Launch Vehicles, 1958–1990

1. On the heritage of Langley under the National Advisory Committee for Aeronautics, see Hansen, *Engineer in Charge*. On the center's contributions to space, see especially Hansen, *Spaceflight Revolution*, in which chapter 7 is devoted to Scout. A shorter version of that chapter with some differences is Hansen, "Learning through Failure." Leiss, "Scout," xxxiii, provides some background on Scout by a veteran of the program. Bille et al., "Small Launch Vehicles," 204–13, also deals with Scout.

2. Hansen, *Spaceflight Revolution*, 214–17 (quotation, 215); Hansen, "Learning through Failure," 22–23; Bille et al., "Small Launch Vehicles," 210–11, which tabulates successes and failures and dates the last launch to 1994. On successful and failed launches, see also the incomplete list in GSCF, "NASA's Scout Launch Vehicle," NHRC. For a discussion of many of the Scout missions, see McDowell, "Scout Launch Vehicle," 102–7. McDowell, 101, lists three missions with imperfect orbits as failures, whereas Goddard counted them as successes. I count partial successes as failures in computing the success rate. McDowell also lists three Blue Scout failures, but since there seems to be no complete

list of Blue Scout launches, the Air Force vehicle is not necessarily included in the 118 Scout launches. See below in the narrative for such information as is available about Blue Scout. Many of the recorded 118 regular Scout launches carried Air Force and Navy payloads or performed missions for the armed services and were listed in the National Aeronautics and Space Council's reports to Congress as Blue Scout missions.

3. Hansen, *Spaceflight Revolution*, 197–200; Shortal, *A New Dimension*, vii and passim. On the history of Wallops, see also H. Wallace, *Wallops Station*, although it has far less material on Scout than Shortal's older study. On Faget, see the biographical sketch in Glennan, *Birth of NASA*, 328. On Thibodaux, see chapter 2 of this book and Shortal, 533–34. On Piland, see Swenson, Grimwood, and Alexander, *This New Ocean*, 65. Hansen discusses Stoney in *Spaceflight Revolution*, 199–200.

4. Shortal, *A New Dimension*, 706–7; *CPIA/M1*, unit 228, Sergeant.

5. Shortal, *A New Dimension*, 707–8; Hansen, *Spaceflight Revolution*, 199–200.

6. Shortal, *A New Dimension*, 708–9; *CPIA/M1*, unit 427, X248A5, Altair I, and unit 248, X254A1 Antares.

7. Shortal, *A New Dimension*, 709; *CPIA/M1*, unit 277, Algol I, Aerojet Senior.

8. Shortal, *A New Dimension*, 708–9; *CPIA/M1*, unit 237, Castor, and unit 228, Sergeant; Thiokol, *Rocket Propulsion Data*, TX-12 (Sergeant) and TX-33–35 (Scout second stage). I use Thiokol performance figures because *CPIA/M1* provides data for Castor at sea level.

9. Leiss, "Scout," 26, 31–32, 84–88; LTV, "Scout," 3–10 to 3–15, SMC/HO; "Press Release Information, Solar Radiation Scout Program," attachment to letter, Space Systems Division/SSVXO (Maj. Reed) to DCEP (Maj. Hinds), April 6, 1962, in "Hyper Environment Test System (TS 609A)" folder, SMC/HO; Vought, "Scout User's Manual," NASM; Bille et al., "Small Launch Vehicles," 206; A. Wilson, "Scout," 457. None of these sources indicates whether the guidance and control computer was analog or digital or even whether Minneapolis-Honeywell provided it or procured it from another company. On all of these initial NASA contracts, see also "NASA Awards Scout Contracts."

10. Leiss, "Scout," 128–33; Shortal, *A New Dimension*, 710–11; GSCF, "NASA's Scout Launch Vehicle," NHRC; L. Wallace, *Dreams, Hopes, Realities*, 41, which provides the year when Wallops became part of Goddard.

11. "Blue Scout Chronology," September 9 and December 7, 1959, and March 9 and June 30, 1960, SMC/HO; Shortal, *A New Dimension*, 717.

12. Hansen, *Spaceflight Revolution*, 205; McDowell, "Scout Launch Vehicle," 100; Shortal, *A New Dimension*, 717–18. Hansen, citing Shortal and two Newport News *Daily Press* articles, places the problem with the fourth-stage heat shield, but Shortal consistently says the third-stage shield experienced the problem.

13. Shortal, *A New Dimension*, 717, 720; Leiss, "Scout," 31–32.

14. Shortal, *A New Dimension*, 715–16; Leiss, "Scout," 31–32.

15. Shortal, *A New Dimension*, 719–20; Hansen, *Spaceflight Revolution*, 207–9; "Blue Scout Chronology," July 1, 1960, SMC/HO; Bille et al., "Small Launch Vehicles," 208. McDowell, "Scout Launch Vehicle," 101, lists this as a failure, but Leiss, "Scout," 437, shows it as a success.

16. A. Wilson, "Scout," 449–50; Hansen, *Spaceflight Revolution*, 209–10; Shortal, *A New Dimension*, 720; U.S. President, *Aeronautics and Space Activities, 1961*, 91.

17. A. Wilson, "Scout," 451; Leiss, "Scout," 32, 90 U.S. President, *Aeronautics and Space Activities, 1961*, 9.

18. A. Wilson, "Scout," 451; Leiss, "Scout," 2, 437, 447. Wilson treats the NOTS-17 as a fifth stage, but Leiss states that only one Scout, vehicle S-191C, which flew on June 3, 1974, had five stages. He adds, "Five previous Scout vehicles had had a NOTS-17 payload kick stage not part of the booster" (2). NASA, Office of Programs, "History of Vehicles Launched," 6, NHRC, is unusual in listing this mission as a failure.

19. Leiss, "Scout," 53–54, 65, 131 (quotation, 53); A. Wilson, "Scout," 451; *CPIA/M1*, unit 428, X254A1, and unit 421, X259A2, Antares II.

20. A. Wilson, "Scout," 452; GSCF, "NASA's Scout Launch Vehicle," NHRC; Leiss, "Scout," 56; *CPIA/M1*, unit 415, X258B1, Altair II, and unit 427, Altair I. The improvements in Altair II resulted in at least three versions of the X258, the A1, A2, and B1. On Altair II, see also Hunley, *Preludes*, chap. 9.

21. Shortal, *A New Dimension*, 708; "Blue Scout Chronology," August 1958 and October 14, 1958, SMC/HO. An entry for December 18, 1958, refers to Scout as System 609A and discusses the Hyper Environment Test System (HETS).

22. Shortal, *A New Dimension*, 708, 712, 716; "Blue Scout Chronology," February 3 and 24, May 8, May 28–29, and December 2, 1959, and March 9 and September 13, 1960, SMC/HO.

23. Krebs, "Blue Scout Junior," 1; Lethbridge, "Blue Scout Junior Fact Sheet," 1; M. Wade, "Blue Scout Junior," 1; Shortal, *A New Dimension*, 705; Winter and James, "50 Years of Aerojet," 690. Krebs lists the Alcor as the AJ10–41, but that may be an error since the AJ10–41 was one of the Able liquid-propellant engines. However, the "Blue Scout Chronology" entry for October 1960 also uses the AJ10–41 designation.

24. On NOTS, see Babcock's *Magnificent Mavericks* and other sources cited in Hunley, *Preludes*, chaps. 5 and 9; also Westrum, *Sidewinder*, esp. chaps. 3, 17, 18. For the fact that AFSWC asked NOTS to design the model 100A motor, see "Solid Propulsion Systems," NAWCWD. For more on propulsion development at NOTS, see Robbins and Feist, "China Lake Propulsion Laboratories," 2–10, NAWCWD. Thanks to China Lake historian Leroy Doig for arranging access to these and other un- or declassified documents. On AFSWC's involvement with Wallops and Blue Scout Junior, see Shortal, *A New Dimension*, 484–573, 702–6; "Blue Scout Chronology," September 21, 1960, and "Blue Scout History," January 1 through June 20, 1962, SMC/HO. The exact nature of AFSWC involvement may still be classified.

25. J. Powell, "NOTS Air-Launched Satellite," 433–36, with many specifications of the rockets obtained from Cdr. William W. West, USN Ret., pilot of the F4D-1; Babcock, *Magnificent Mavericks*, chap. 24; Bille and Lishock, "NOTSNIK," 1–5; Bille and Lishock, *First Space Race*, 140–44. Thanks to Tom Moore of CPIA for a copy of the Notsnik paper. In their book, Bille and Lishock do not consider the F4D-1 a stage and state that Notsnik was the first all-solid-propellant launch vehicle, but since the aircraft carried the vehicle to 41,000 feet, I argue that the F4D-1 operated as a first stage.

26. J. Powell, "NOTS Air-Launched Satellite," 436–39; Babcock, *Magnificent Mavericks*, chap. 24; Bille and Lishock, "NOTSNIK," 7–8.

27. Unclassified extracts from a 1959 NOTS progress report kindly provided by Leroy Doig, China Lake historian, NAWCWD.

28. *CPIA/M1*, unit 327, NOTS Model 100A; "Solid Propulsion Systems," NAWCWD; G. Sutton, *Rocket Propulsion Elements*, 436–37 (also quotation). In all of its editions, Sutton's book was a standard reference work. Sutton had worked at Rocketdyne and elsewhere and had become Hunsaker Professor of Aeronautical Engineering at MIT by 1992. Notice his subtitle's reference to rocket engineering rather than science.

29. M. Wade, "Blue Scout Junior," 1; Krebs, "Blue Scout Junior," 1–2, according to which both of these launches used the NOTS 100A; "Blue Scout Chronology," September 21 (quotation) and November 8, 1960, SMC/HO.

30. M. Wade, "Blue Scout Junior," 1–2; Krebs, "Blue Scout Junior," 1–2; "Blue Scout Chronology," August 17 and December 4, 1961, and July 24, November 21, and December 18, 1962, SMC/HO; "Blue Scout History," SMC/HO. The emergency rocket communications system (ERCS) was a Strategic Air Command (SAC) requirement for reliable command, control, and communications of its missile forces.

31. M. Wade, "Blue Scout 1," 1–2; Krebs, "Blue Scout-1," 1–2; McDowell, "Scout Launch Vehicle," 101; "Blue Scout Chronology," October 1960 and January 7, May 9, May 18, and June 6, 1961 (quotations from last three), SMC/HO. Wade lists the Castor 2 as the second-stage motor for Blue Scout 1, but since it was not introduced until 1965, this is evidently wrong.

32. M. Wade, "Blue Scout 1," 1–2; Krebs, "Blue Scout-1," 1–2; "Blue Scout Chronology," September 1961, SMC/HO; "Blue Scout History" (quotations), SMC/HO.

33. Krebs, "Blue Scout-2," 1–2; M. Wade, "Blue Scout 2," 1–2, which both list only the three launches mentioned in the narrative; Convair, "Summary," AFHRA; U.S. President, *Activities, 1967*, 127; GSCF, "NASA's Scout Launch Vehicle," NHRC; McDowell, "Scout Launch Vehicle," 101–3 (quotation, 103), 105–6, which lists Transit and other DoD launches in the regular sequence of NASA launches, although it lists flights D4, D5, and D8 as Blue Scout 2 launches.

34. Leiss, "Scout," 129; "Blue Scout Chronology," April 1 and October 2, 1961, and December 18, 1962, SMC/HO; "SAMSO Chronology," 201, SMC/HO.

35. "Blue Scout Chronology," March 3, April 12 and 28, and May 25, 1961, SMC/HO; McDowell, "Scout Launch Vehicle," 101.

36. "Blue Scout Chronology," November 1, 1961, SMC/HO; Swenson, Grimwood, Alexander, *This New Ocean*, 392–97. Neither source reveals why Mercury used a Blue Scout or where the backup Scout came from. Sources differ on whether the D designations (like D8) should be rendered with or without a hyphen.

37. Hansen, *Spaceflight Revolution*, 210; GSCF, "NASA's Scout Launch Vehicle," NHRC, in which the listing of Scout launches and failures provides dates to go with Hansen's generalizations; "Blue Scout Chronology," November 30, 1962, SMC/HO, for Stine's retirement in office; narrative above for the Blue Scout failures.

38. "Blue Scout History," SMC/HO; Leiss, "Scout," 437, which shows who procured the vehicles and lists this flight in the regular Scout sequence; McDowell, "Scout Launch Vehicle," 103. The fact that this mission is discussed in "Blue Scout History" does not mean this was a Blue Scout vehicle, because many NASA launches are discussed in "Blue Scout Chronology." I am grateful to Matt Bille for pointing out the secret electronic intelligence payload. On this, see Darling, "GRAB"; Krebs, "USA Military Satellites."

39. "Blue Scout Chronology," August 22, 1962, SMC/HO; "Blue Scout History" (first

quotation), SMC/HO; U.S. President, *Activities, 1962*, 115, which reports that the payload and even S-112's satellite's weight were "not stated" and the objective was "development of space flight techniques and technology." Launch date discrepancies presumably stem from the difference between Greenwich Mean Time and local time on the California coast. On the self-destruction of S-112 and the problems and results of S-114's failure, see A. Wilson, "Scout," 452 (also second quotation); Hansen, *Spaceflight Revolution*, 210.

40. A. Wilson, "Scout," 452; U.S. President, *Activities, 1962*, 121, and *Activities, 1963*, 123; GSCF, "NASA's Scout Launch Vehicle," NHRC.

41. Hansen, *Spaceflight Revolution*, 210–11 (quotations, 211); A. Wilson, "Scout," 451; Leiss, "Scout," 2–3.

42. A. Wilson, "Scout," 451; *CPIA/M1*, unit 277, Algol I, and unit 363, Algol IIA.

43. A. Wilson, "Scout," 452; Hansen, *Spaceflight Revolution*, 211–12; Leiss, "Scout," 50.

44. Leiss, "Scout," 50 (see also 62 for Algol IIC data, unfortunately not suitable for comparison with Algol IIA data in this text); McDowell, "Scout Launch Vehicle," 101; U.S. President, *Activities, 1968*, 96. Leiss, the only source I have found that describes the Algol IIC nozzle, provides no further details.

45. A. Wilson, "Scout," 452; Leiss, "Scout," 441.

46. A. Wilson, "Scout," 452; Hansen, *Spaceflight Revolution*, 212–14 (also quotations).

47. A. Wilson, "Scout," 452–53; Hansen, *Spaceflight Revolution*, 210–15.

48. Hansen, *Spaceflight Revolution*, 214–16; A. Wilson, "Scout," 453–55; U.S. President, *Activities, 1963*, 135; GSFC, "NASA's Scout Launch Vehicle," NHRC. Not all Scouts launched in this period were recertified. Many were new vehicles that were simply certified as having been checked out for reliability.

49. A. Wilson, "Scout," 454–55; Leiss, "Scout," 53; U.S. President, *Activities, 1965*, 149.

50. *CPIA/M1*, unit 237, Castor, 415, Altair II, and unit 480, FW-4; Thiokol, "Catalog," item "Castor II (TX-354–4)," NASM; Leiss, "Scout," 56; A. Wilson, "Scout," 449, 454–55. The designation Altair III is confusing because it covers both the UTC FW-4S and the Thiokol TE-M-640 (Star 20), discussed below. Leiss distinguishes between the two by calling the first Altair III and the second Altair IIIA. Wilson calls them both Altair IIIA, with FW4S or TEM640 in parentheses.

51. *CPIA/M1*, units 363, Algol IIA, and 523, Algol III; A. Wilson, "Scout," 449, 456, which shows different lengths for the two motors, but the CPIA manual shows them as the same; Leiss, "Scout," 63. The CPIA manual gives performance at sea level for Algol II and under vacuum conditions for Algol III. Wilson shows a longer burn time for Algol II, but his figures are not compatible with the other two sources, and CPIA shows a longer burn time for Algol III. The 404–pound payload capability was for a launch in an easterly direction.

52. A. Wilson, "Scout," 453–54, 456, 459; Bille et al., "Small Launch Vehicles," 208–9; Leiss, "Scout" 63; U.S. President, *Report . . . 1972*, 89. A table in Wilson, 459, places the first D-1 launch (with Algol III) on August 13, 1972, from Wallops.

53. A. Wilson, "Scout," 457; Leiss, "Scout," 2, 57; *CPIA/M1*, unit 535, BE-3–9A, Alcyone, Scout 5th Stage; McDowell, "BE-3"; Krebs, "Scout-E1"; U.S. President, *Report . . . 1974*, 128 (quotation).

54. Leiss, "Scout," 56–57, 447 (quotations, 57).

55. *CPIA/M1*, unit 579, TE-M-640, Star 20, and unit 480, FW-4; Leiss, "Scout," 57, which states without further information about the propellant or motor that "Thiokol successfully demonstrated an advanced high energy propellant in Altair IIIA configuration in 1974"; Thiokol, "Thiokol's Solid Propellant Upper and Lower Stages," which gives a slightly lower specific impulse (expressed in newtons but converting to almost 292 lbf-sec/lbm) than *CPIA/M1*, unit 579; M. Wade, "Star 20A," which gives the specific impulse as 292 sec. and states "Flown: 2" without any indication of where, when, or even on what vehicle, although presumably it was a Scout.

56. U.S. President, *Report . . . 1974*, 129 (quotation); Ezell, *NASA Historical Data Book* 3: 183; Leiss, "Scout," 57, 447, which lists the launch as a success; McDowell, "Scout Launch Vehicle," 101, which counts it as a failure.

57. Leiss, "Scout," 54; *CPIA/M1*, unit 538, X259B4, Antares IIB, and unit 577, TE-M-762, Star 31, Antares IIIA; B. Wilson, "Composite Motor Case Design," slides "The 70's—Kevlar Is King" and "Composite Motor Cases by Thiokol"; Zimmerman, Linsk, and Grunwald, "Solid Rocket Technology for the Eighties," 9; "Bidirectional Woven Kevlar"; "Aramid Fiber."

58. Leiss, "Scout," 54, 448; U.S. President, *Report . . . 1979*, 90; A. Wilson, "Scout," 449; Krebs, "Scout," 2, 5. This account follows Krebs and his description of the G-1, with which Wilson agrees. Bille et al., "Small Launch Vehicles," 210–11, lists the same configuration as the X-5C and lists the G-1 as having the Algol 2B as its first stage and flying just once, in 1968.

59. Spires, *Beyond Horizons*, 267; McDowell, "Scout Launch Vehicle," 107; Krebs, "Scout," 5; U.S. President, *Report . . . 1989–1990*, 146; U.S. President, *Report . . . 1994*, 24, 72, which see for further details about MSTI; Darling, "MACSAT." The sources I have found are curiously silent about why NASA did not continue to produce and use Scout. The presidential *Aeronautics and Space Report* for 1994 mentions the last launch but says nothing about why Scout was phased out. The same is true of NASA, "Scout Launch Vehicle to Retire."

60. Isakowitz, *Space Launch Systems* (1995), 256; Bromberg, *Space Industry*, 119–20; U.S. President, *Report . . . 1993*, 23, 96; Curry, Mendenhall, and Moulton, "Air-Launched Space Booster," 2; Kadish, memo, AFHRA; Covault, "Commercial Winged Booster."

61. Isakowitz, *Space Launch Systems* (1995), 256; Bille et al., "Small Launch Vehicles," 214–15; SMC, "Pegasus Launch Vehicle," 3.

62. Isakowitz, *Space Launch Systems* (1995), 261–62, 264; SMC, "Pegasus Launch Vehicle," 1–2 (quotations, 2); Curry, Mendenhall, and Moulton, "Air-Launched Space Booster," 3. See chapter 7 of this book for a discussion of the Flexseal nozzles.

63. Curry, Mendenhall, and Moulton, "Air-Launched Space Booster," 2; Bille et al., "Small Launch Vehicles," 214; Covault, "Commercial Winged Booster"; Isakowitz, *Space Launch Systems* (1995), 261 (quotations). Above Mach 8, presumably the Pegasus was high enough that there was no longer any aerodynamic heating.

64. Orbital Sciences Corporation, "Pegasus Air-Launched Space Booster Payload Users Guide," December 1988, 3, document V-18 in Geiger and Clear, "History," vol. 5, AFHRA; Isakowitz, *Space Launch Systems* (1995), 261; U.S. President, *Report . . . 1993*,

96; Bille et al., "Small Launch Vehicles," 216–17. Each of these sources gives slightly different specifics about the three motors. I follow the payload users guide.

65. "Pegasus Air-Launched Space Booster" (see note 64), 2; Isakowitz, *Space Launch Systems* (1995), 261; Bille et al., "Small Launch Vehicles," 216. Bille et al. is the source for the 241-second burn time, which seemed long to one of the coauthors, Pat Johnson, who kindly reviewed a draft of this chapter and commented on the burn time specifically. The payload users guide does not cover the fourth stage, but no other stage burned that long.

66. Isakowitz, *Space Launch Systems* (1995), 256–60; ATK Thiokol Propulsion, "Castor 120"; Orbital Sciences Corporation, "Pegasus Mission History."

67. SMC, "Pegasus Launch Vehicle," 3.

Chapter 5. Saturn I through Saturn V, 1958–1975

1. See, e.g., Williamson, "Biggest of Them All," 313, 320–21; NASA, "Saturn V News Reference," iv, 1–2, 9–1; E. Clark, "Brain Trust," 33; Isakowitz, *Space Launch Systems* (1991), 292; Bilstein, "Saturn Launch Vehicle Family," 186; Bilstein, *Stages to Saturn*, 369.

2. Williamson, "Biggest of Them All," esp. 304–5; Bilstein, *Stages to Saturn*, esp. xv–xvi, 15–16, 36–38, 46–48, 91–92, 95–96, 134, 140–47, 150, 191; Bilstein, "Saturn Launch Vehicle Family," 116; Murray and Cox, *Apollo*, 149–51, 180, 313–14; MSFC, "Saturn Illustrated Chronology," 1–2, 9, NASM; NASA, "Saturn IB News Reference," 1–1, 2–2. The Saturn news references are available in various archives including NHRC as well as on the Internet.

3. Bilstein, *Stages to Saturn*, 25–28; MSFC, "Saturn Illustrated Chronology," 1–5, NASM; Murray and Cox, *Apollo*, 54, which points out in a footnote that Saturn was the next planet out from Jupiter, a previous von Braun–group rocket. Since both were also ancient Roman gods (not Greek, as Murray and Cox say), when Abe Silverstein named the spacecraft they would launch, he appropriately chose Apollo (who bore the same name in Greek and Latin).

4. In Hunley, *Preludes*, chap. 7, and this book, chap. 2.

5. Bilstein, *Stages to Saturn*, 28–29; MSFC, "Saturn Illustrated Chronology," 2–3, NASM.

6. Bilstein, *Stages to Saturn*, 30–31; MSFC, "Saturn Illustrated Chronology," 2–4, NASM.

7. Murray and Cox, *Apollo*, 156; Dunar and Waring, *Power to Explore*, 43; Bilstein, *Stages to Saturn*, 337; and see Hunley, *Preludes*, chaps. 2 and 6, and this book, chap. 4.

8. Dunar and Waring, *Power to Explore*, 19, 39–46; and see Hunley, *Preludes*, chap. 8, and this book, chap. 2.

9. Dunar and Waring, *Power to Explore*, 51; Bilstein, *Stages to Saturn*, 263; S. Johnson, "Phillips," 691; Stuhlinger, OHI, 48–49. Stuhlinger said the notes were sent to von Braun's secretary on Friday, but they were referred to as "Monday" or "weekly notes." For a recent perspective on the weekly notes and von Braun in general, see Dunar, "Wernher von Braun."

10. Bilstein, *Stages to Saturn*, 41, 44–46; MSFC, "Saturn Illustrated Chronology," 5–8, NASM.

11. MSFC, "Saturn Illustrated Chronology," 8–10, 12–13, 17–20, 22, NASM; Ezell, *NASA Historical Data Book* 2: 56–57.

12. Bilstein, *Stages to Saturn*, 97–99; McCool and Chandler, "Development Trends," 296; *CPIA/M5*, unit 84, H-1, April 1965.

13. Bilstein, *Stages to Saturn*, 97, 99; McCool and Chandler, "Development Trends," 296; *CPIA/M5*, unit 84, H-1, and unit 85, LR79–NA-11 Thor, May 1965. See also Hunley, *Preludes*, chap. 7, on MA-3.

14. MSFC, "Saturn Illustrated Chronology," 6, 11–12, NASM; Bilstein, *Stages to Saturn*, 97–98. Herring, *Way Station to Space*, 104, 200, 347, for the name changes of the Mississippi facility.

15. Bilstein, *Stages to Saturn*, 98–102; *CPIA/M5*, unit 84, H-1, 2, which shows oxidizer and fuel flow rates increasing between the 165,000– and the 188,000–pound engines. Bilstein, 101, mentions the "Thor-Atlas injectors, uprated to . . . 188,000 pounds . . . of thrust," suggesting that changes in the injectors increased the thrust.

16. For above two paragraphs, mainly Bilstein, *Stages to Saturn*, 188–89; some details from Heusinger, "Saturn Propulsion Improvements," 25; Ezell, *NASA Historical Data Book* 2: 56–58; U.S. President, *Activities, 1964*, 128.

17. For above two paragraphs, mainly "Final Report S-IV," LC/MD (quotations, 2–3); also Bilstein, *Stages to Saturn*, 184–85. The report says the helium shutoff valve was "changed to another design for flight of S-IV-5." I interpret this to mean the S-IV stage on vehicle SA-5, since SA-6 did not fly until May 28, 1965, after the report was written.

18. Bilstein, *Stages to Saturn*, 241–45, 477; NASA, "Saturn IB News Reference," 1–4; Douglas, "Payload Planner's Guide," 38–39, NASM.

19. MSFC, "Saturn Illustrated Chronology," 6, 25, NASM.

20. Bilstein, *Stages to Saturn*, 77–78, 324–25; Heusinger, "Saturn Propulsion Improvements," 25.

21. Bilstein, *Stages to Saturn*, 78–81, 325; MSFC, "Saturn Illustrated Chronology," 17, NASM; and see Hunley, *Preludes*, chap. 9, for the problem of base heating on Polaris.

22. Bilstein, *Stages to Saturn*, 81.

23. For above three paragraphs, ibid., 324–37; MSFC, "Saturn Illustrated Chronology," 46–47, NASM; NASA, "Saturn IB News Reference," 12–2.

24. NASA, "Saturn IB News Reference," 12–2; U.S. President, *Aeronautics and Space Activities, 1964*, 128; same report, *1965*, 148, and *1966*, 161; Ezell, *NASA Historical Data Book* 2: 56. These sources differ slightly in their data, so all figures in the narrative should be taken as approximations.

25. Roger D. Launius, introduction to Glennan, *Birth of NASA*, xxi–xxii, xxv, plus Glennan, 312–16; MSFC, "Saturn Illustrated Chronology," 8, 14, NASM; Levine, *Managing NASA*, 16, 317; and see below in the narrative.

26. Lambright, *Powering Apollo*, esp. xi, 2, 100–105, 108–9; Levine, *Managing NASA*, 5, 19; on Kennedy's decision to land a man on the Moon, see the classic account, Logsdon, *Decision*, esp. 91–94, 97, 105, 109, 111, 122–23.

27. Lambright, *Powering Apollo*, 114–16; Levine, *Managing NASA*, 19 (quotation); Murray and Cox, *Apollo*, 152.

28. Lambright, *Powering Apollo*, 116–17 (quotation, 117); bio of Mueller in Levine, *Managing NASA*, 308.

29. Murray and Cox, *Apollo*, 152–54 (headline, 152; quotation, 153; story of investigation; 153–54); Lambright, *Powering Apollo*, 117, which characterizes Mueller and relates Webb's concession to him; Ezell, *NASA Historical Data Book* 2: 6, 625, 642, for full names and positions of Disher and Tischler.

30. Bilstein, *Stages to Saturn*, 349; excerpted interview of Mueller by Robert Sherrod, April 21, 1971, in Swanson, *Before This Decade Is Out*, 108. On the conception of all-up testing on Minuteman, see Hunley, *Preludes*, chap. 9.

31. Levine, *Managing NASA*, 6 (citing source for first quotation); Bilstein, *Stages to Saturn*, 349 (next two quotations); Mueller, memo to Webb, with enclosed press release, LC/MD. Also in LC/MD, "Apollo Program—All-Up Concept" claims hardware savings from all-up testing totaled $705 million.

32. Bilstein, *Stages to Saturn*, 349–51 (first quotation, 349; fourth and fifth, 351); Dunar and Waring, *Power to Explore*, 94–95 (second and third quotations, 94).

33. Van Nimmen and Bruno, *NASA Historical Data Book* 1: 611, organizational chart, 1 November 1963; Levine, *Managing NASA*, 5–6, 175; Lambright, *Powering Apollo*, 118; Bilstein, *Stages to Saturn*, 269; Dunar and Waring, *Power to Explore*, 66–67 (quotation, 66); MSFC, "Apollo Program Management," LC/MD, esp. 4–1, which shows how the IO organization fit with concepts introduced under Mueller's aegis.

34. Dunar and Waring, *Power to Explore*, 67; Lambright, *Powering Apollo*, 118; S. Johnson, "Phillips," 694–95, 700; Phillips, OHIs by Ray, September 25, 1970, 22, and July 22, 1970, 8, 12–13; Phillips bio in Levine, *Managing NASA*, 309.

35. S. Johnson, "Phillips," 697, 700–703 (quotations); E. Clark, "Brain Trust."

36. S. Johnson, "Phillips," 700–704; MSFC, "Apollo Program Management," 4–29, LC/MD; Levine, *Managing NASA*, 156–57; U.S. Navy, briefing, LC/MD; Swanson, *Before This Decade Is Out*, 101–2, 108. For further insights into Phillips's management approach, see Phillips, memo to Mueller (1964), LC/MD.

37. Phillips, OHI by Ordway, with extracts from Phillips's Wernher von Braun Memorial Lecture, NASM, January 28, 1988, 1, 3–4, 7–8.

38. Phillips, OHI by Sherrod (first three quotations); Murray and Cox, *Apollo*, 159–60 (remaining quotations).

39. NASA, "Saturn IB News Reference," 1–3 to 2–5; MSFC, "Saturn Illustrated Chronology," 50, 58, 60, NASM.

40. Bilstein, *Stages to Saturn*, 83; NASA, "Saturn IB News Reference," 3–1, 8–1, and S-IB Stage Fact Sheet.

41. Cf. *CPIA/M5*, unit 84, H-1 with 165,000- and 188,000–pound thrust, and unit 173, H-1 with 205,000–pound thrust; NASA, "Saturn IB News Reference," H-1 Engine Fact Sheet, for 200,000–pound thrust; Bilstein, *Stages to Saturn*, 97, for the association of engines with flight numbers but not the changes in the H-1s. Reportedly, between the 200,000- and the 205,000–pound H-1s, the fuel flow rates declined slightly while those of the oxidizer continued to go up, but this seems inconsistent with the statement that the mixture ratio remained unchanged.

42. Bilstein, *Stages to Saturn*, 140–41; Glennan, *Birth of NASA*, 149–50, where note 6 summarizes a May 31 Glennan memo providing pricing information; extracts from

"The Apollo Spacecraft: A Chronology," 1: 53–54, in "Propulsion J-2," NHRC; memo, NASA/MLP (Tischler) to NASA/ML (Rosen), July 8, 1962, subject: AGC Proposal, in "Propulsion J-2," NHRC. Tischler does not specifically mention the contract but says the J-2 "was deliberately specified to require conservative design in order to meet NASA's use schedules."

43. Bilstein, *Stages to Saturn*, 141–43; MSFC, "Saturn Illustrated Chronology," 46, 50, NASM; [Rockwell International], "Data Sheet, J-2 Rocket Engine," June 3, 1975, in "Propulsion J-2," NHRC; memo, Eldon W. Hall, NASA/MEE, to Dr. Joseph F. Shea, NASA/ME, July 30, 1962, subject: Transfer of Engine Development Responsibility, in "Propulsion J-2," NHRC.

44. Bilstein, *Stages to Saturn*, 138, 144–45; memo, Tischler to Rosen (see note 42).

45. Bilstein, *Stages to Saturn*, 145–46; *CPIA/M5*, unit 87, engine J-2, 200,000 lb thrust, April 1965, 4.

46. *CPIA/M5*, unit 87, 3, compared with unit 116, RL10A-3–3, 5; NASA, "Saturn IB News Reference," 6–2 to 6–4; Brennan, "Milestones," 8; Strangeland, "Turbopumps," 38.

47. Bilstein, *Stages to Saturn*, 150–52; Strangeland, "Turbopumps," 42.

48. *CPIA/M5*, unit 87, 2, 4; unit 125, J-2, 225K, 1; unit 174, J-2, 230K, 1, 3; NASA, "Saturn IB News Reference," 6–1 to 6–5; NASA, "Saturn V News Reference," 6–4. Sources differ about which J-2 engines flew on each Saturn IB and Saturn V flight, but interestingly, on March 4, 1966, Rocketdyne was still testing 200,000–pound engines at Santa Susana, while on May 6 the first of the uprated 230,000-pound J-2s had arrived at MSFC for static testing. See Library of Congress, *Astronautics and Aeronautics, 1966*, 85, 168–69.

49. For above two paragraphs, Bilstein, *Stages to Saturn*, 166–78; NASA, "Saturn IB News Reference," S-IVB Stage Fact Sheet; MSFC, "Saturn Illustrated Chronology," 50, NASM; Ezell, *NASA Historical Data Book*, 2: 56; Douglas, "Payload Planner's Guide," 36, 38, NASM.

50. Bilstein, *Stages to Saturn*, 182–83, 429; Douglas, "Payload Planner's Guide," 38, NASM; NASA, "Saturn IB News Reference," B-5/6, NASM. On page 182, Bilstein refers to Moog Industries as the developer of the actuators, but on 429 in the list of subcontractors, he gives the name as Moog Servo Controls, Inc. Perhaps the company changed its name at some point during the 1960s.

51. Bilstein, *Stages to Saturn*, 245; Haeussermann, "Guidance and Control," 232–35; NASA, "Saturn IB News Reference," 1–4, 7–1, 7–3, 7–5, B-2, NASM.

52. For above two paragraphs, AFMDC news release 64–152–R, October 1, 1964, on microfilm roll 32,273, frames 215–16, AFHRA; also from AFHRA, on microfilm roll 32,270, these eight documents: AFMDC news release 62–349–R, n.d., 1–3, frames 498–500; AFMDC/MDOT progress report MR-3, "Saturn Guidance Sled Test Program," March 12, 1963, 1–3 (first two quotations, 2), frames 404–6; AFMDC/CIGTF, "Final Report, Saturn ST-124 Inertial Platform Test Program, Phase IIB Summary," January 1965, vii, 1–2, 40, 53, frames 127–87; AFMDC/CIGTF, "Final Report, Saturn Gas Bearing and Accelerometer Components Sled Test, Phase IIC Summary," vii (third quotation), 1–2, 28, frames 201–36; AFMDC/CIGTF, "Interim Report for the ST-124 (Saturn Inertial Platform Rocket Sled Test Program)," 1: 1–2, 4–6 (fourth quotation, 6), 11–12, frames 293–311; AFMDC/CIGTF, "Final Report, Rocket Sled Test Program of the Saturn

ST124M Inertial Guidance Platform, Phase 3," 2: ii, 1, 37–38, frames 510–54; AFMDC/ CIGTF, "Final Report," Appendix B: System Sled Run Velocity Profiles, 1: v (last quotation), frames 581, 586; AFMDC/MDSO, "Saturn Guidance Sled Test Program," August 17, 1966, 1–2, frames 506–7.

53. NASA, "Saturn IB News Reference," 7–3 to 7–7, B-2; Douglas, "Payload Planner's Guide," 38, NASM; MSFC, "Astrionics System Handbook, Saturn," 4–29, 4–33, 4–35, NASM; Bilstein, *Stages to Saturn*, 243–52.

54. MSFC, "Astrionics System Handbook, Saturn," 3–14 to 3–20, 4–125, 4–127, NASM; Seltzer, "Saturn IB/V Astrionics System," 5, NASM.

55. NASA, "Saturn IB News Reference," 7–5, 7–6, and IU Fact Sheet; Douglas, "Payload Planner's Guide," 39, NASM; Bilstein, *Stages to Saturn*, 251.

56. NASA, "Saturn IB News Reference," 7–6, 7–7.

57. Bilstein, *Stages to Saturn*, 245.

58. NASA, "Saturn IB News Reference," Saturn IB Fact Sheets. There is a generic fact sheet, with separate ones for AS-203, AS-204, and AS-205. The last two used 225,000– pound J-2s. Apparently none of the Saturn IBs used 205,000–pound engines.

59. NASA, "Saturn IB News Reference," 12–1 to 12–3; U.S. President, *Activities, 1966*, 148.

60. For above two paragraphs, NASA, "Saturn IB News Reference," 12–2 to 12–4; U.S. President, *Activities, 1966*, 153; Bilstein, *Stages to Saturn*, 339.

61. For above two paragraphs, NASA, "Saturn IB News Reference," 12–2 to 12–4; U.S. President, *Activities, 1966*, 148, 155; Bilstein, *Stages to Saturn*, 340; NSSDC, "Spacecraft AS-202"; Brooks, Grimwood, and Swenson, *Chariots for Apollo*, chap. 8, "Qualifying Missions" section.

62. Bilstein, *Stages to Saturn*, 340; Lambright, *Powering Apollo*, 143–45 (quotation, 145). For a more extensive account of the fire, see, e.g., Murray and Cox, *Apollo*, 184– 225.

63. Bilstein, *Stages to Saturn*, 340–43; NASA, "Saturn IB News Reference," 12–5, 12–6, and Saturn IB Fact Sheet for AS-204; U.S. President, *Activities, 1968*, 95; Swanson, *Before This Decade Is Out*, 102. The reason for the Apollo 7 designation was that Apollo 4 and 6 in Saturn V preceded the AS-205 mission; see Bilstein, 414, 416.

64. Bilstein, *Stages to Saturn*, 343–44, 414–16; Library of Congress, *Astronautics and Aeronautics, 1968*, 252; U.S. President, *Activities, 1968*, 99; Orloff, *Apollo by the Numbers*, 14–23; Ezell, *NASA Historical Data Book*, 3: 104–5, 111; NASA, "Saturn IB News Reference," Saturn IB Fact Sheet for AS-205.

65. MSFC, "Saturn Illustrated Chronology," 4, 13–14, NASM; Bilstein, *Stages to Saturn*, 58–60; Williamson, "Biggest of Them All," 315; Ezell, *NASA Historical Data Book* 2: 59.

66. For above two paragraphs, Bilstein, *Stages to Saturn*, 104–7 (first quotation, 105); Kraemer, *Rocketdyne*, 161–63; NASA, "Saturn V News Reference," 1–1 (second quotation); MSFC, "Saturn Illustrated Chronology," 4, NASM; Glennan, *Birth of NASA*, 13, 15. On some details, Bilstein and Kraemer do not agree. On the test stands at Edwards, see Stiffler and Eppley, "Flight Test Center, 1 Jan.–30 June 1960," 69, 180, 183, and "Flight Test Center, 1 Jan.–30 June 1961," 58–59, 71–72, appendix K, both AFFTC/HO; "Test Stand 1B," AFFTC/HO. Bilstein, 124, says that at the peak of its development, the F-1

used five different test stands at the Rocket Engine Test Site, but he does not identify the other two.

67. Bilstein, *Stages to Saturn*, 107–8 (first quotation, 108); Brennan, "Milestones," 8 (second quotation); *CPIA/M5*, unit 173, H-1 Booster Engines, 4, and unit 175, F-1, Saturn V Booster Engine, 5.

68. NASA, "Saturn V News Reference," 3–1 to 3–5 and F-1 Engine Fact Sheet; Brennan, "Milestones," 8–9; *CPIA/M5*, unit 173, H-1 Booster Engines, 4, and unit 175, F-1, Saturn V Booster Engine, 1–6; Bilstein, *Stages to Saturn*, 109–10. As Bilstein shows on page 416, vehicle 504 flew on Apollo 9 on March 3, 1969, and a table he reproduces on page 110 shows that the 1,522,000–pound engine weighed 20,180 pounds at burnout compared to 20,096 for the 1,500,000–pound version.

69. Jaqua and Ferrenberg, "Art of Injector Design," 4, 6, 9.

70. Bilstein, *Stages to Saturn*, 109–13; Murray and Cox, *Apollo*, 145, 147–48.

71. Murray and Cox, *Apollo*, 148–49 (quotations, 149); Bilstein, *Stages to Saturn*, 113; "Experimental Engines Group," 21.

72. Murray and Cox, *Apollo*, 150–51, 179–80; Bilstein, *Stages to Saturn*, 115–16; Brennan, "Milestones," 9. For simultaneous and parallel efforts at the Air Force Rocket Propulsion Laboratory at Edwards AFB to solve combustion stability problems, see Herzberg, "Rocket Propulsion Laboratory," 1: II-108 to II-158, AFFTC/HO, where there is considerable emphasis on trial-and-error methodologies. For the PFRT date, see *CPIA/M5*, unit 175.

73. Murray and Cox, *Apollo*, 146–47; Bilstein, *Stages to Saturn*, 109; NASA, "Saturn V News Reference," 3–1 to 3–4; *CPIA/M5*, unit 175. I follow the news reference for the liquid-oxygen flow, Murray and Cox apparently having interposed two numbers. For more on achieving combustion stability in the F-1, see Fred E. C. Culick and Vigor Yang, "Overview of Combustion Instabilities in Liquid-Propellant Rocket Engines," in Yang and Anderson, *Liquid Rocket Engine Combustion Instability*, 8–9 and sources cited; Kraemer, *Rocketdyne*, 166.

74. Description of development from Bilstein, *Stages to Saturn*, 116–19; information on René 41 from Herzberg, "Rocket Propulsion Laboratory," 1: IV-346, AFFTC/HO. Available only through the Air Force, Herzberg's history is a wonderful source for such arcane information.

75. Bilstein, *Stages to Saturn*, 191; NASA, "Saturn IB News Reference," Saturn IB Fact Sheet; NASA, "Saturn V News Reference," 1–6, 2–1, and Saturn V Fact Sheet; Dunar and Waring, *Power to Explore*, 87.

76. Barton Hacker and E. M. Emme, notes on interview with Rosen at NASA HQ, November 14, 1969, in "Rosen, Milton W.," NHRC; Bilstein, *Stages to Saturn*, 192–93; Ezell, *NASA Historical Data Book* 2: 4–5. Bilstein presents the same basic story narrated here but has apparently gotten some of the dates wrong, having the committee, for example, present its report to Holmes on March 20, 1961, when he did not join NASA until October.

77. Bilstein, *Stages to Saturn*, 193–95; Dunar and Waring, *Power to Explore*, 86–87; NASA, "Saturn IB News Reference," 8–1; NASA, "Saturn V News Reference," 8–1.

78. For above three paragraphs, Bilstein, *Stages to Saturn*, 195–207; Herring, *Way Station to Space*, 14, 45, 51, 55–56, 93, 96–97, 125–29.

79. Bilstein, *Stages to Saturn*, 207–9; NASA, "Saturn V News Reference," 1–2, 1–7, and F-1 Engine Fact Sheet; Herring, *Way Station to Space*, 126–28.

80. NASA, "Saturn V News Reference," 1–6; Bromberg, *Space Industry*, 56–57; for characterization of Storms, Heppenheimer, *Countdown*, 207–8 (first quotation, 207), and Gray, *Angle of Attack*, 20, 22–34, 65–72 (Crossfield quotation, 66).

81. Bilstein, *Stages to Saturn*, 209–11; Gray, *Angle of Attack*, 89; NASA, "Saturn V News Reference," Second Stage Fact Sheet.

82. Bilstein, *Stages to Saturn*, 211–14; NASA, "Saturn V News Reference," Second Stage Fact Sheet, Third Stage Fact Sheet, 1–6, 8–2. Bilstein, 211, says the government built the Seal Beach facilities; "Saturn V News Reference," 8–2, refers to North American–owned administrative buildings. So apparently what the government built were the manufacturing facilities.

83. Bilstein, *Stages to Saturn*, 214–16. I have greatly simplified the detailed descriptions in this excellent source.

84. Gray, *Angle of Attack*, 153–56 (quotation, 154); Bilstein, *Stages to Saturn*, 217–22. Bilstein also uses the phrase "ballooned out of shape" (219).

85. Bilstein, *Stages to Saturn*, 222–23; Gray, *Angle of Attack*, 159–61; Dunar and Waring, *Power to Explore*, 88–89. Not surprisingly, Gray tends to blame the animosity between Marshall and North American on Rees, while Dunar and Waring tend to emphasize North American's failings.

86. Gray, *Angle of Attack*, 196–98; Bilstein, *Stages to Saturn*, 222–24; Dunar and Waring, *Power to Explore*, 90.

87. Gray, *Angle of Attack*, 198; Bilstein, *Stages to Saturn*, 224–25, 228, 269.

88. Gray, *Angle of Attack*, 198–99; Bilstein, *Stages to Saturn*, 225–26; Murray and Cox, *Apollo*, 166–71, 183; Lambright, *Powering Apollo*, 107–8; Bromberg, *Space Industry*, 58–59.

89. Mueller, letter to Atwood (quotations, 1, 3), LC/MD.

90. Phillips, memo to Mueller (1965), LC/MD (quotations, 1, 6); Bilstein, *Stages to Saturn*, 227–28; Gray, *Angle of Attack*, 202, 208.

91. Bilstein, *Stages to Saturn*, 224–33; Gray, *Angle of Attack*, 209, 253–55; Murray and Cox, *Apollo*, 231–36; Bromberg, *Space Industry*, 70–71; Lambright, *Powering Apollo*, 174–75.

92. Bilstein, *Stages to Saturn*, 179, 185–86; NASA, "Saturn V News Reference," 1–3 and Third Stage Fact Sheet; NASA, "Saturn IB News Reference," S-IVB Stage Fact Sheet.

93. NASA, "Saturn V News Reference," 5–10 to 5–11, A-4.

94. Bilstein, *Stages to Saturn*, 186, 353–60.

95. For above two paragraphs, NASA, "Saturn V News Reference," 7–1 to 7–5 (quotations, 7–4) and unpaginated Instrument Unit Fact Sheet; Seltzer, "Saturn IB/V Astrionics System," 5–10, NASM.

96. NASA, "Saturn V News Reference," 12–1 to 12–2; U.S. President, *Activities, 1967*, 131; Senate, "Summary," 6.

97. Bilstein, *Stages to Saturn*, 360–61; NASA, "Saturn V News Reference," 12–2; Senate, "Summary," 5, 16, 27–28; Murray and Cox, *Apollo*, 312.

98. NASA, "Saturn V News Reference," 12–2 to 12–4 (first quotation, 12–2); Sen-

ate, "Summary," 360–61 (von Braun quotation, 361). Despite what the narrative states above, the official report, U.S. President, *United States Aeronautics and Space Activities . . . 1968*, 96, counts the mission a success.

99. Senate, "Summary," 16–17, quotations from both pages of Phillips's testimony.

100. Bilstein, *Stages to Saturn*, 362–63; Senate, "Summary," 6, 16–18; Phillips, draft article, LC/MD; NASA, "Saturn V Tests Completed," NHRC.

101. Bilstein, *Stages to Saturn*, 362–63; Senate, "Summary," 18; NASA, "Saturn V Tests Completed," 2–4, 5A, NHRC; "Minutes," 3, LC/MD; Dunar and Waring, *Power to Explore*, 97.

102. Bilstein, *Stages to Saturn*, 361–63; Murray and Cox, *Apollo*, 313–14; "Minutes," 3, LC/MD; Senate, "Summary," 22–27, testimony of George H. Hage, Phillips's deputy director of Apollo.

103. "Minutes," 5, LC/MD; Senate, "Summary," 28, 34 (Mueller quotations, 34). Still another problem arose on AS-502 when a section of the spacecraft containing the lunar module ripped soon after the pogo event. The problem, which proved unrelated, was diagnosed with some difficulty, then solved by modifications. See Murray and Cox, *Apollo*, 313n.

104. Bilstein, *Stages to Saturn*, 366–67; Lambright, *Powering Apollo*, 197–98; Orloff, *Apollo by the Numbers*, 32–33.

105. Bilstein, *Stages to Saturn*, 366–68; Orloff, *Apollo by the Numbers*, 32–35, 40–43, 275–76; U.S. President, *United States Aeronautics and Space Activities . . . 1968*, 101. Orloff, 275–76, is the source for the rated thrusts of F-1 and J-2 engines used on this and subsequent Apollo missions. Other sources differ, but as Orloff used Saturn launch vehicle flight evaluation reports, I accept his data here.

106. Library of Congress, *Astronautics and Aeronautics, 1969*, 434–35; Orloff, *Apollo by the Numbers*, 52–59, 275–76; Bilstein, *Stages to Saturn*, 368.

107. Orloff, *Apollo by the Numbers*, 54, 59–62.

108. Bilstein, *Stages to Saturn*, 368–72; Orloff, *Apollo by the Numbers*, 72–81, 275–76.

109. Orloff, *Apollo by the Numbers*, 137–57, esp. 137; Bilstein, *Stages to Saturn*, 372–77 (quotation, 374). For an extended account of all the Apollo missions, see, e.g., Murray and Cox, *Apollo*, 237–458.

110. NASA, "Apollo 14 Saturn Modified," NHRC; North American Rockwell, "Modifications," NHRC; Orloff, *Apollo by the Numbers*, 171–72, 186, 200, 213–14 , 228, 241–42, 254.

111. See the tally in Bilstein, *Stages to Saturn*, 422. Conversion to 2006 dollars, with 1965 as the base year, uses S. M. Friedman, "Inflation Calculator."

Chapter 6. Titan Space Launch Vehicles, 1961–1990

1. Launius, "Titan," 166; Isakowitz, *Space Launch Systems* (1995), 292.

2. Launius, "Titan," 166–67; Richards and Powell, "Titan 3 and Titan 4," 123; Piper, "Titan III," 18, SMC/HO. Copy of Piper kindly provided by SMC History Office.

3. There is, of course, an enormous literature on Kennedy and McNamara, but for the developments described here, see Launius, *NASA*, 55–56; J. Neufeld, *Ballistic Missiles*, 4, 212–13, 215; S. Johnson, *Culture of Innovation*, 198–201.

4. S. Johnson, *Culture of Innovation*, 198–201.

5. Ibid., 204–5 (quotation, 205); Shapley, *Promise and Power*, 100–101.

6. Piper, "Titan III," 17–19, SMC/HO. Under PPB, McNamara assigned the task of improving research-and-development management to Rubel, but not until September 1961, as Stephen Johnson points out in *Culture of Innovation*, 206.

7. Piper, "Titan III," 20–30, SMC/HO; Launius, "Titan," 166 (quoting the Planning Group); Spires, *Beyond Horizons*, 81; Levine, *Managing NASA*, 225–26. For a brief biographical sketch of Golovin, see Glennan, *Birth of NASA*, 331.

8. Piper, "Titan III," 31–37 (quotations, 32, 32n, 37, 31, 37), SMC/HO; S. Johnson, *Culture of Innovation*, 206–9.

9. For above two paragraphs (bracketing the heading), Piper, "Titan III," 101–3, SMC/HO; "SAMSO Chronology," 110, SMC/HO; Harwood, *Raise Heaven and Earth*, 356; [Space Systems Division], "SLV-5 Program," 9 (quotation), SMC/HO.

10. For above three paragraphs, "Brainpower First" (quotation, 139, which also pictures Adelman next to taller Putt); Lindsey, "Tighter Rocket Market," 22; "Highlights," 35–42, CSD, and a narrative, chronology, and biographic materials kindly provided by CSD librarian Karen Schaffer; Adelman, OHI; Altman, OHI; Gorn, *Harnessing the Genie*, 48–49, 89–93; Gorn, *Universal Man*, 90–101, 136–37, 150–54. The solid-propellant portion of this chapter is also indebted to Hunley, "Minuteman."

11. Gordon et al., *Aerojet*, IV-38; Klager, "Segmented Rocket Demonstration" (diagram of the joint, 168); Andrepont and Felix, "Motor Development," 2; B. Ross Felix, former vice president of engineering and technology, CSD, comments, August 22, 2000, UP.

12. Klager, "Segmented Rocket Demonstration," 163–87; Gordon et al., *Aerojet*, IV-38 to IV-39; Andrepont and Felix, "Motor Development," 2, 19. For details see these sources (which sometimes differ) and Hunley, "Minuteman," 269–72.

13. For above two paragraphs, mainly Andrepont and Felix, "Motor Development," 3, 7, 19. Where there is internal disagreement on particulars, I follow page 3. See also Hawkes, "Larger Booster Capabilities," esp. 57; "Solid-Fueled Rocket Nears Crucial Test," esp. 49; UTC, "120–Inch," esp. 5, JPL; and details later in this chapter. On Adelman and Minuteman, see Hunley, *Preludes*, chap. 9.

14. UTC, "120–Inch," 16, JPL; Andrepont and Felix, "Motor Development," 7; "Most Reliable Booster," section "CSD Facilities Growth through Titan III," CSD.

15. These insights are from Thackwell, "Application of Solid Propellants," UP. As Thackwell remarked in a note to (unnumbered) page 21, "the solid propellant rocket's thrust is proportional to the product of the *burning rate* of the propellant and the *burning area* of the grain, both of which can be readily varied over a wide range." Further, Thackwell pointed out in his "Large Solid Rocket Programs," 10, that shorter burning times and lower gravity losses could be achieved "by varying the oxidizer particle size distribution and employing a nozzle [with] a larger size throat and exit cone."

16. UTC, "120–Inch," 16, JPL; Andrepont and Felix, "Motor Development," 7.

17. Piper, "Titan III," 130–31, 136, SMC/HO; letter, Brig. Gen. J. S. Bleymaier, System Program Director, SSD/SSB, to Lt. Gen. Donald L. Putt (Ret.), United Aircraft, July 23, 1963, supporting document 24 to Piper's history; UTC, "120–Inch," 13, JPL.

18. Hunley, "Evolution," 30–31 and sources cited there, esp. *CPIA/M1*, unit 559, Titan

III and Titan IIIC Zero Stage; Weyland, Willoughby, Backlund, and Hill, OHI, with information on the toughness of UTC's PBAN from Weyland, then UTC's senior research propellant chemist; Andrepont and Felix, "Motor Development," 7.

19. "Most Reliable Booster," section "Development of Major Components," CSD; UTC, "120–Inch," 3, 5, JPL; Richards and Powell, "Titan 3 and Titan 4," 125; Andrepont and Felix, "Motor Development," 7; Adelman, OHI; Felix, comments, UP and PI. As Felix explained, on both the Titan III and shuttle solid-rocket-motor nozzle throats, "phenolic resin glues together layers of fabric that have either been graphitized or carbonized in a furnace. This avoids the thermal stress problems that cause cracking on large diameter monolithic graphite throats." He characterized the "throat package" as "graphite phenolic in rosette construction" but said "carbon phenolic" was a simplified term for it. Other sections of this narrative about Titan IIIC are also indebted to Felix's judicious and invaluable comments.

20. "Most Reliable Booster," section "Development of Major Components," CSD; Felix, comments, August 22, 2000, UP; Andrepont and Felix, "Motor Development," 7; Richards and Powell, "Titan 3 and Titan 4," 125; "Titan 34D Recovery Program," 26, 35, LC/MD; Ballistic Systems Division, "Independent Panel Reviews," August 1986, 1 of 3, n.p., LC/MD. The last two sources, which provide the clearest illustrations I have found of the Titan III clevis joint, document the investigation of the Titan 34D-9 mishap on April 18, 1986. Apparently a weak case bond allowed hot gases to reach the case, causing it to fracture in flight. The investigation revealed no basic design deficiencies in the motor, and the clevis joint design in particular was not implicated. Corrective action was to improve the bonding process and the methods of detecting inadequate bonds. For this description, besides the sources from the Phillips Collection just cited, see "Titan 34D Flight Readiness Meeting," 10, 12, LC/MD.

21. "Most Reliable Booster," section "Development of Major Components," CSD; Andrepont and Felix, "Motor Development," 7; Judge, "Westinghouse Know-How." The Ladish rolled-ring forging had been developed by the date of Judge's article.

22. The Blue Scout preceded Titan IIIC, but it was much smaller and partly developed by NASA.

23. *CPIA/M1*, unit 559; "Highlights," 42, CSD; "SAMSO Chronology," 159, SMC/HO.

24. Richards and Powell, "Titan 3 and Titan 4," 126, 129; Piper, "Titan III," 104, 120, SMC/HO; Chulick et al., "Titan Engines," 31–32; *CPIA/M5*, units 79–81, Engines LR91–AJ-5, -7, and -9, as well as units 92–94, LR87–AJ-5, YLR87–AJ-7, and YLR87–AJ-9, respectively. Martin performed the structural modifications. For other details of the -9 engines, see Stiff, "Storable Liquid Rockets," 9 and figures 35 and 36.

25. "SAMSO Chronology," 131, 133, 139, SMC/HO; Piper, "Titan III," 104–5, SMC/HO.

26. For the two engines' design similarity, Richards and Powell, "Titan 3 and Titan 4," 126; *CPIA/M5*, units 146, AJ10–137, and 154, Titan III Transtage. Gibson and Wood, "Apollo Experience Report," 7, says that "the subcontractor" (Aerojet, not named) began design and development for the Service Propulsion Subsystem engine on April 9, 1962. My sources provide no firm start date for Transtage development. Thanks to Jane Odom, NHRC archivist, for locating the Gibson and Wood report in the NHRC and transmitting it to me.

27. Huzel and Huang, *Design* (1967), 118.

28. *CPIA/M5*, units 146, AJ10–137, and 154, Titan III Transtage.

29. Piper, "Titan III," 91, 121, 125–26 (quotation, 126), SMC/HO; *CPIA/M5*, unit 154, Titan III Transtage; Hatton, "Titan Transtage Rockets"; Huzel and Huang, *Modern Engineering*, 378.

30. For above three paragraphs, mainly Ross, "Life at Aerojet University," 5, 47, 74–75 (quotations, 74–75), UP. Date of actual delivery of engines is from Stiff, "Storable Liquid Rockets," 9.

31. Stiff, "Storable Liquid Rockets," 9. The presumption is the author's, but Stiff's tone suggests that the temperature of 350°F on the back side of the injector was a problem.

32. Gordon et al., *Aerojet*, III-146 (first and fourth quotations); *CPIA/M5*, unit 154, Titan III, Transtage (other quotations); Richards and Powell, "Titan 3 and Titan 4," 126, for the fuel cooling of the baffle. Sources differ on Transtage's specific impulse, but this figure is considerably higher than most.

33. *CPIA/M5*, unit 154, Titan III, Transtage, 1–3; "Titanium Tanks Machined for Titan 3 Transtage"; Stiff, "Storable Liquid Rockets," figures 37 and 38.

34. Piper, "Titan III," 50, 66, 68–70, 110, 121, SMC/HO; Richards and Powell, "Titan 3 and Titan 4," 126. For the disappointment of the program office in SSD, see Bleymaier, "Approval of Titan III," 10–18, 23–25, SMC/HO. Bleymaier argued that delaying development of a more sophisticated guidance and control system would cost more in the long run than doing it together with the rest of Titan III, as additional test vehicles and facilities would be needed.

35. For above two paragraphs, Richards and Powell, "Titan 3 and Titan 4," 127; Piper, "Titan III," 131, 136–37, SMC/HO.

36. Richards and Powell, "Titan 3 and Titan 4," 130; "SAMSO Chronology," 144, 146, SMC/HO.

37. Richards and Powell, "Titan 3 and Titan 4," 130, which says separation could not be confirmed; U.S. President, *Activities, 1964*, 147, which claims separation did occur.

38. Richards and Powell, "Titan 3 and Titan 4," 130; "SAMSO Chronology," 154, SMC/HO; U.S. President, *Activities, 1965*, 133; Krebs, "Transtage."

39. Richards and Powell, "Titan 3 and Titan 4," 130; "SAMSO Chronology," 158, SMC/HO; U.S. President, *Activities, 1965*, 141; Krebs, "Transtage."

40. Richards and Powell, "Titan 3 and Titan 4," 131; "SAMSO Chronology," 159, SMC/HO; U.S. President, *Activities, 1965*, 145.

41. Richards and Powell, "Titan 3 and Titan 4," 130–31; "SAMSO Chronology," 164, SMC/HO; U.S. President, *Activities, 1965*, 153.

42. Richards and Powell, "Titan 3 and Titan 4," 131; "SAMSO Chronology," 166–67, SMC/HO; U.S. President, *Activities, 1965*, 157–58.

43. Spires, *Beyond Horizons*, 139–40; Richards and Powell, "Titan 3 and Titan 4," 131; "SAMSO Chronology," 173–74, SMC/HO; U.S. President, *Activities, 1966*, 152–53.

44. Richards and Powell, "Titan 3 and Titan 4," 130–32; "SAMSO Chronology," 176, SMC/HO.

45. Richards and Powell, "Titan 3 and Titan 4," 134–35; Launius, "Titan," 167; Chulick et al., "Titan Engines," 32; Martin Marietta, "Titan 23C-4," 4–4, AFHRA, reporting on the second Titan 23C flight; "Most Reliable Booster," section "Development of Major Components," CSD.

46. Chulick et al., "Titan Engines," 33–34; Richards and Powell, "Titan 3 and Titan 4," 134–35. Specific impulse figures vary among sources. These are from Chulick et al.

47. Chulick et al., "Titan Engines," 33, 35; *CPIA/M5*, units 150, YLR91–AJ-9, and 222, LR91–AJ-11. Curiously, although Chulick et al. discuss improved specific impulse in their narrative, in figure 12 on page 33 they show identical thrust (100,000 pounds) and specific impulse (316 lbf-sec/lbm) for the -9 and -11 engines, both in vacuum. The CPIA manual provides specific figures that are slightly different.

48. Delco Systems, "Inertial Guidance System Components," LC/MD; Ballistic Systems Division, "Independent Panel Reviews," LC/MD; 6555th Aerospace Test Group, "Preliminary Report," 2, and "Launch Evaluation Report," 21, both AFHRA; Richards and Powell, "Titan 3 and Titan 4," 135, which identifies the computer as the Magic 351 (perhaps an earlier variant of the one actually used on the Titan 23C launches, as indicated by the launch evaluation report just cited). I have written "apparently" because the Delco Systems document cited, although in a Titan III folder, makes no specific reference to a particular Titan III, but it fits with information from other sources and includes the Magic 352 computer as part of the inertial guidance system components. The permutations of AC Spark Plug, AC Delco, and Delco Electronics are confusing. On September 4, 1970, B. P. Blasingame, formerly the guidance expert at Western Development Division, signed a mass-mailing letter as Manager, Milwaukee Operations, Delco Electronics Division of General Motors stating that on September 1, AC Electronics and Delco Radio Divisions of General Motors had consolidated to form Delco Electronics Division. However, an Internet site, "About ACDelco," lists a 1974 merger of AC Spark Plug Sales Division with United Delco Division to create AC-Delco Division. This site also says that in 1971 United Motor Services changed its name to United Delco Division. In any event, Delco Electronics in 1971 prepared the report "Program 624A-Titan III Inertial Guidance System," AFHRA.

49. Richards and Powell, "Titan 3 and Titan 4," 134–35; Isakowitz, *Space Launch Systems* (1995), 293; "Most Reliable Booster," section "Highlights from the Past," table "Notable Firsts," CSD; Cleary, "Cape: Military Space Operations," comment about December 13, 1973, launch specifically from <www.globalsecurity.org/space/library/report/1994/cape/cape2–5.htm>; Spires, *Beyond Horizons*, 160; "SAMSO Chronology," 209 (quotation), SMC/HO; Library of Congress, *Astronautics and Aeronautics, 1970*, 433, which provides the apogee and perigee of the satellite; Martin Marietta, "Titan 23C-4," 4–1, 4–2, AFHRA, reporting on the first Titan 23C mission, including the fact that this mission (Titan IIIC-19) used the LR87–AJ-11 engines in stage 1 but the LR91–AJ-9 engine in stage 2.

50. Richards and Powell, "Titan 3 and Titan 4," 130, 135; Isakowitz, *Space Launch Systems* (1995), 293; Wilman, "Space Division," 64, 101, SMC/HO, copy generously provided to author.

51. "Space Systems Summaries: Titan"; Richards and Powell, "Titan 3 and Titan 4," 133; "SAMSO Chronology," 152–53, 361, SMC/HO; Wilman, "Space Division," 101, SMC/HO; Fink, "Titan 34D Booster," 48.

52. Richards and Powell, "Titan 3 and Titan 4," 133–34, 137; McDowell, "US Reconnaissance Satellite Programs," 30–33; McDowell, Agena missions table; Guill, Sterbentz,

and Gordy, "Agena," 20–21, which has information on Ascent Agena's guidance system; Turco, "Titan Family Guidance," 162, and microfiche supplement, 20, which shows the Titan 34B guidance and control system including an HDC (Honeywell Digital Computer, Highly Distributed Control?) 501 missile guidance computer, an AGS (Agena Guidance System?) strapdown inertial measurement unit H448, a three-axis reference system, a lateral acceleration sensing system, and other features. This information is not comparable to that provided by Guill, Sterbentz, and Gordy, so it is uncertain whether Turco's information applies to Ascent Agena or the basic Agena D, which had an older guidance and control system. "Titan Launch History Summary" lists Titan 34B separately, but taking Titan IIIB and 34B together it shows only one failure, evidently not counting the Agenas, which technically were part of the payload except for the Ascent Agenas, but this seems an artificial distinction. Agena was hardly just a kick motor.

53. "SAMSO Chronology," 5, 189, SMC/HO; Richards and Powell, "Titan 3 and Titan 4," 136; "Space Systems Summaries: Titan." Turco, "Titan Family Guidance," microfiche, 9, shows a diagram of the radio guidance for at least the 23D and 24B versions of Titan III in which ground radar measured the position and velocity of the Titan, sending the data to a ground computer that calculated acceleration and velocity and computed the vehicle orientation needed to reach the proper position in space, then transmitting the results of these calculations to the vehicle. As shown in separate diagrams on pages 13 and 16, the vehicle's computer then commanded adjustments to the thrust vector control systems in the various stages.

54. Richards and Powell, "Titan 3 and Titan 4," 136–37; McDowell, "US Reconnaissance Satellite Programs," 31–32. Heyman, "Directory," app. 3, p. 6, states that there were 24 Titan IIID launches through August 28, 1985, of which one failed. But McDowell, 32, lists the August 28, 1985, launch, which did fail, and one on December 4, 1984, as Titan 34D launches. Since it had the same payload as the previous Titan IIID launches through November 17, 1982, Heyman may have confused the launchers. The Web site "Titan Launch History Summary" also shows 22 successful launches and no failures.

55. Ezell, *NASA Historical Data Book* 3: 38, 41–42; "SAMSO Chronology," 225, SMC/HO.

56. For above three paragraphs, Dawson and Bowles, *Centaur*, 141–45; quotation from author interview with Lewis engineer Joe Nieberding.

57. For above three paragraphs, Dawson and Bowles, *Centaur*, 145–47; Ezell, *NASA Historical Data Book* 3: 42.

58. Dawson and Bowles, *Centaur*, 149, 159–60; Ezell, *NASA Historical Data Book* 3: 42; Siddiqi, *Deep Space Chronicle*, 108, 110–12, 115, 117–22; Launius, *NASA*, 101–4. See chapter 2 of this book on development of the Delta solid-propellant stage used on the Helios missions and the sources cited above for details of the scientific returns from the Titan-Centaur missions.

59. Fink, "Titan 34D Booster," 48; Richards and Powell, "Titan 3 and Titan 4," 138; "SAMSO Chronology," 280, 423, SMC/HO; U.S. President, *Report . . . 1982*, 87. In the Fink article, a quotation from Lt. Col. Gilbert F. Kelley, chief of the Titan Integration Division at SAMSO, also supported the view that the 5.5–segment SRM was a secondary feature to the long-tank first stage.

60. "SAMSO Chronology," 286, 307, SMC/HO; Richards and Powell, "Titan 3 and Titan 4," 138; "Most Reliable Booster," section "CSD Facilities Growth through Titan III," CSD.

61. *CPIA/M1*, units 559, Titan III and Titan IIIC Zero Stage, and 578, Titan 34D Stage Zero SRM; "Space Systems Summaries: Titan," 76; U.S. President, *Report . . . 1986*, 144.

62. Aerojet, "Propulsion Characteristics Summary," in "Aerojet AJ10 Series," NASM; *CPIA/M5*, units 122 and 154, AJ10–138; Krebs, "Titan-34D Transtage," which lists the seven Transtage launches with Titan 34D and shows that they were AJ-10–138A versions; Richards and Powell, "Titan 3 and Titan 4," 140.

63. Statement of Maj. Gen. William R. Yost, director of Space Systems and Command, Control, Communications, HQ USAF, to House of Representatives, Committee on Armed Services, March 1980, supporting document Conclusion-1 to Waldron, "Inertial Upper Stage," SMC/HO. Quotations from Yost, 3–5. Waldron's conclusion, 213–19, contains a sophisticated discussion of reasons for the "shortcomings of the IUS development program" (213). On uses of the IUS, see FAS, "Inertial Upper Stage," which, however, attributes a launch on September 4, 1989, to IUS that according to other sources involved a Transtage.

64. Dunn, "Inertial Upper Stage," 1–2 (first quotation, 1); Yost, statement (see note 63), 5–7 (other quotations); Dooling, "Third Stage for Space Shuttle," 559; Boeing Aerospace Company, "Interim Upper Stage (IUS)," January 13, 1977, supporting document III-17 in Waldron, "Inertial Upper Stage," SMC/HO; Waldron, "Inertial Upper Stage," 209, SMC/HO; Jenkins, *Space Shuttle*, 244. My thanks to Jenkins for a copy of this book.

65. Boeing, "IUS" (see note 64), 2; Waldron, "Inertial Upper Stage," 95, SMC/HO. See Hunley, *Preludes*, chap. 9, for the development of HTPB.

66. Waldron, "Inertial Upper Stage," 53–54, 95–96, SMC/HO.

67. Waldron, "Inertial Upper Stage," 96–112, SMC/HO; Dunn, "Inertial Upper Stage," 2–3; Chase, "IUS," 13–15, 18.

68. Waldron, "Inertial Upper Stage," 56, 61, SMC/HO; Dumoulin, "Payloads," 6–7, 9–10; Boeing, "IUS" (see note 64), 2–3; Richards and Powell, "Titan 3 and Titan 4," 138.

69. Waldron, "Inertial Upper Stage," esp. 208–9, SMC/HO; Yost, statement (see note 63), esp. 8.

70. Richards and Powell, "Titan 3 and Titan 4," 138–39; Dunn, "Inertial Upper Stage," 3; U.S. President, *Report . . . 1982*, 86–87.

71. Wilman, "Space Division," 1, SMC/HO.

72. Chase, "IUS," 2–5; FAS, "Inertial Upper Stage"; U.S. President, *Report . . . 1986*, 144; "SAMSO Chronology," 364, SMC/HO.

73. Richards and Powell, "Titan 3 and Titan 4," 139–40; Martin Marietta, "Titan IV User's Handbook," I-7, SMC/HO. On the solid rocket motor failure, see also note 20. "Titan Launch History Summary" on the Internet agrees with the 80 percent success rate (12 of 15 launches) for Titan 34D.

74. Bromberg, *Space Industry*, 130, 153; Launius, "Titan," 170; NASA, "NASA and Martin Marietta Sign Commercial Launch Agreement"; Lethbridge, "Commercial Titan III Fact Sheet"; Isakowitz, *Space Launch Systems* (1995), 292, 294–99.

75. Lethbridge, "Commercial Titan III Fact Sheet"; Isakowitz, *Space Launch Systems* (1995), 298; Launius, "Titan," 170; McDowell, "The Present Day."

76. U.S. President, *Report . . . 1989–1990*, 144–45; U.S. President, *Report . . . 1992*, 9; Jenkins, "Broken in Midstride," 406–07; Krebs, "Intelsat-6"; M. Wade, "HS 393"; FAS, "Skynet 4"; Krebs, "Skynet." Sources differ about whether the launch took place December 31 or January 1, probably depending on whether they report GMT or local time at the launch site.

77. U.S. President, *Report . . . 1992*, 9, 81; U.S. President, *Report . . . 1993*, 17; Siddiqi, *Deep Space Chronicle*, 153; Krebs, "Commercial Titan-3"; Launius, "Titan," 170.

78. Isakowitz, *Space Launch Systems* (1995), 295; Launius, "Titan," 170; Jenkins, "Broken in Midstride," 409; Jenkins, *Space Shuttle*, 289; Heppenheimer, *Countdown*, 342.

79. Powell and Caldwell, "New Space Careers for Former Military Missiles," 124–25; Stumpf, *Titan II*, 265–71; Martin Marietta Astronautics, "Titan II," NASM; Capt. Jeffrey J. Finan, Titan II 23G-1 Launch Controller, memo for record, "Titan II 23G-1 Processing History," November 21, 1988, document III-9 in Geiger and Clear, "History," vol. 3, AFHRA.

80. Stumpf, *Titan II*, 270, which lists launches through part of 1999; U.S. President, *Report . . . 1989–90*, 160, for payload capacities; U.S. President, *Report . . . 1999*, 14–15, for the nature of a later mission; U.S. President, *Report . . . 2000*, 26, 100, for the launching of Titan II's second DMSP satellite. "Titan Launch History Summary" shows twelve total launches by (what appears to be) early 2003, all successful. This apparently does not include a final successful Titan II launch in October 2003. "Space Launch Report, 2003 World Space Launch Log," January 25, 2004, <www.geocities.com/launchreport/log2003.html>, shows two successful Titan II (23G) launches in 2003. However, a suborbital Titan II mission with a Star 37 solid rocket motor failed in 1993 because of the Star 37 stage, marring Titan II's perfect record as a launch vehicle.

81. Richards and Powell, "Titan 3 and Titan 4," 140; Lockheed Martin, "Titan IV," SMC/HO.

82. Martin Marietta, "Fact Sheet, Titan IV," n.d., document III-28 in Geiger and Clear, "History," vol. 3, AFHRA; *CPIA/MI*, unit 578, Titan 34D, Stage Zero SRM, compared with unit 629, Titan IV SRM, July 1996; Richards and Powell, "Titan 3 and Titan 4," 139–41.

83. Richards and Powell, "Titan 3 and Titan 4," 140–41; Air Force fact sheet "Titan IVB," January 29, 1997, document II- 180 in Carlin et al., "Air Force Materiel Command," vol. 5, AFHRA; Hunley, "Minuteman," 281–83, for further details on the SRMU.

84. Martin Marietta, "[Titan IV] System," XII-5, XII-6, SMC/HO; Richards and Powell, "Titan 3 and Titan 4," 140.

85. Lethbridge, "Titan IVA Fact Sheet"; Lockheed Martin, "Titan IV," SMC/HO; U.S. President, *Report . . . 1991*, 179; U.S. President, *Aeronautics and Space . . . 1992*, 10; U.S. President, *Report . . . 1993*, 13, 23; U.S. President, *Report . . . 1994*, 11.

86. Martin Marietta, "Fact Sheet, Titan IV" (see note 82); Lockheed Martin, "Titan IV," SMC/HO; "Titan IV"; Isakowitz, *Space Launch Systems* (1995), 299.

87. [Delco Electronics], "The Hemispherical Resonator Gyro," in "Delco Systems Operations 1992," LC/MD, itself part of a file entitled "Delco Electronics Carousel 400 INS for Widebody Aircraft, Delco's new family of inertial systems." The Carousel 400 was

indeed designed for aircraft but, like the Transtage system, it could have been adapted for launch vehicles. George Hartley Bryan was a professor of mathematics at the University College of North Wales, Bangor, from 1896 to 1926. He published his discoveries about the rotation-sensing properties of a vibrating wineglass in 1890 and also cowrote a book on mathematical astronomy and a paper titled "The Longitudinal Stability of Aerial Gliders." See Jones, "Astronomical Research"; Malm, "For Want of a Nail"; Tomayko, *Computers Take Flight*, 3.

88. "Honeywell Hexad Fault Tolerant Guidance System" (first quotation) in "Component Research: Guidance," NHRC; Broad, "New Age of Precision" (second quotation).

89. Honeywell, "Centaur Inertial Navigation Unit," in "Centaur General," NHRC; "Titan IV Guidance Control Unit," in folder 010805, "Honeywell," NHRC. Except for their titles and photographs (two different views, one with a man and one with a woman looking at the units), these documents are identical. Internal evidence suggests they date from about 1988. See also Honeywell, "Guidance and Control for America's Future," in "Component Research: Guidance."

90. Griffiths and Angus, "Low Cost Guidance."

91. Bowles, "Eclipsed by Tragedy," 425, 432–33; quotation from Martin Marietta, "Fact Sheet, Titan IV" (see note 82).

92. Robertson, "Centaur Canters On," 24; Pratt & Whitney, "RL10 Engine," NHRC.

93. Richards and Powell, "Titan 3 and Titan 4," 141–42; U.S. President, *Report ... 1989–1990*, 142.

94. "NOSS"; Thomson, "Titan Launch Dispenser"; U.S. President, *Report ... 1989–1990*, 147; Richards and Powell, "Titan 3 and Titan 4," 142.

95. FAS, "Titan-4 Launch History"; "Titan Launch History Summary"; U.S. President, *Report ... 1991*, 159–60; U.S. President, *Report ... 1992*, 75; U.S. President, *Report ... 1994*, 11, 70; U.S. President, *Report ... 2000*, 27; Spires, *Beyond Horizons*, 213, 266–67; "Lockheed Martin's Last Titan IV Delivers National Security Payload to Space," *Space Daily*, October 19, 2005, <www.spacedaily.com/news/spysat-051.html>; Aerospace Corporation, "Launch," <www.aero.org/programs/launch.html>.

96. Harry Waldron, "Last of the Titans ... SMC, Titan Vehicles Share Lifelong History," *Astro News*, October 28, 2005 <www.aerotechnews.com/Astro/Astro_102805dbl.pdf>. Dr. Waldron talks about Titans providing access to space for fifty years as of 2005, but this includes the history of Titans I and II as missiles, not launch vehicles. Even so, it is generous, since the first "operational" flight of Titan I was in 1960.

97. For above two paragraphs, Knauf, Drake, and Portanova, "EELV," 38, 40; Anselmo, "EELV Finalists." See also Launius, "Titan," esp. 181.

98. Covault, "Atlas V Soars," 22–23; "Military EELV Launches" (quotations), which also says the Air Force estimated it could launch a heavy payload on an EELV booster for $150 million, compared with $400 million on a Titan IVB; Boeing, "Successful First Launch;" Tod Flemming, "Air Force Launches Final Titan IV Rocket," *Astro News*, October 28, 2005 <www.aerotechnews.com/Astro/Astro_102805dbl.pdf>. For a later estimate of EELV costs, still much lower than those for Titan IVB, see chapter 8, note 20.

Chapter 7. The Space Shuttle, 1972–1990

1. The existence of two technical histories of the space shuttle makes extensive coverage less necessary than for many other subjects of this book, on which information is much more scattered and less readily available. See Jenkins, *Space Shuttle*, and Heppenheimer, *Development of the Space Shuttle*.

2. On precursors to the shuttle, see Heppenheimer, *Decision*, 12–14, 49–54, 73–84; Jenkins, *Space Shuttle*, 15–101; Jenkins, "Broken in Midstride," 358–63; Guilmartin and Mauer, *Shuttle Chronology*, esp. 1: I-1 to I-102. On budgetary and other issues, besides Jenkins, "Broken in Midstride," 369–71, see esp. Launius, "Decision," 17–32. The literature on Nixon, the 1960s and early 1970s, and the Vietnam conflict is far too extensive to cite here. I thank Nadine Andreassen of the NASA History Office, who was kind enough to have the *Shuttle Chronology* copied for me.

3. Jenkins, "Broken in Midstride," 358–71; Bromberg, *Space Industry*, 77–92; and for actual NASA budgets, U.S. President, *Report . . . 2001*, 142–43. For detailed accounts of these developments, see Heppenheimer, *Decision*, 245–90, 331–88; Jenkins, *Space Shuttle*, 139–52; Guilmartin and Mauer, *Shuttle Chronology*, esp. 4: V-5 to V-18 and V-23 to V-135. For inflation-adjusted dollars I use S. M. Friedman, "Inflation Calculator."

4. This account compresses an enormously complex series of developments, based on Jenkins, "Broken in Midstride," 371–75 (Nixon quotation, 375); Bromberg, *Space Industry*, 91–93; Guilmartin and Mauer, *Shuttle Chronology*, esp. 4: V-71, V-72, V-99 to V-102, V-193, V-194, V-237 to V-240, and 5: VI-45, VI-46; Jenkins, *Space Shuttle*, 139–51, 167–73; Heppenheimer, *Decision*, 357–422. Although Jenkins says ("Broken in Midstride," 371; *Space Shuttle*, 139–40) that NASA had decided on some version of an external tank or stage-and-a-half design in June 1971, the other accounts suggest a more complex reality, with a longer decision period.

5. Guilmartin and Mauer, *Shuttle Chronology*, 4: V-55, V-56 (quotation).

6. Heppenheimer, *Decision*, 179; Bromberg, *Space Industry*, 88–89 (quotation, 88); Dunar and Waring, *Power to Explore*, 152–67, 622–24.

7. Jenkins, *Space Shuttle*, 109–10; Heppenheimer, *Decision*, 237–39; Brennan, "Milestones," 14–15; Wiswell and Huggins, "Astronautics Laboratory"; *CPIA/M5*, unit 174, J-2 Engine, and unit 184, J-2S Engine Propulsion System. NASA had terminated J-2S development in March 1970. The preliminary design had been completed but there had been no qualification date. Incidentally, the Astronautics Laboratory was a later name of the Air Force's Rocket Propulsion Laboratory.

8. Heppenheimer, *Decision*, 240–42; Jenkins, *Space Shuttle*, 110; Heppenheimer, *Development*, 134. Heppenheimer has Anderson as president of Rockwell International, but as Bromberg states in *NASA and the Space Industry*, 97, North American Rockwell did not become Rockwell International until early 1973. See chapter 5 in this book for Castenholz's role in F-1 problem solving.

9. Jenkins, *Space Shuttle*, 110, 184; Heppenheimer, *Decision*, 242–44; Guilmartin and Mauer, *Shuttle Chronology*, 4: V-34, V-78; Bromberg, *Space Industry*, 98–99; Dunar and Waring, *Power to Explore*, 288–89 (quotation, 288).

10. Heppenheimer, *Development*, 126; Jenkins, *Space Shuttle*, 224; Biggs, "Space Shut-

tle Main Engine," 75–76; Martinez, "Power Cycles," 17. For some interesting perspective on staged combustion and the alternative aerospike concept, see Kraemer, *Rocketdyne*, 187–217. For Russian engines using a staged combustion cycle, see Kraemer, 220–21.

11. Biggs, "Space Shuttle Main Engine," 71 (first quotation), 81; Heppenheimer, *Decision*, 240 (second quotation); Heppenheimer, *Development*, 136; *CPIA/M5*, unit 174, J-2 Engine, and unit 195, Space Shuttle Main Engine.

12. Heppenheimer, *Development*, 127–28, 133–34; Wiswell and Huggins, "Astronautics Laboratory."

13. Heppenheimer, *Development*, 134–35; Dunar and Waring, *Power to Explore*, 296–97, 625; Jenkins, *Space Shuttle*, 225; Biggs, "Space Shuttle Main Engine," 77.

14. Biggs, "Space Shuttle Main Engine," 79–80; Heppenheimer, *Development*, 136–37.

15. For above three paragraphs, Biggs, "Space Shuttle Main Engine," 80–87. For clarity and simplicity I have chosen slightly inexact terminology. Where I speak of the arrival of liquid oxygen, Biggs speaks of "priming": "An oxidizer system is said to be 'primed' when it is filled with liquid down to the combustor such that the flow rate entering the injector is equal to the flow rate leaving the injector to be burned in the combustor" (84).

16. For above three paragraphs, Biggs, "Space Shuttle Main Engine," 88–91; Heppenheimer, *Development*, 140–42; Martinez, "Power Cycles," 22. Heppenheimer, 141, says rotordynamics experts came from both the United States and Britain, but he names only their leaders, Matthew Ek, chief engineer at Rocketdyne, and Otto Goetz, leading turbomachinery expert at Marshall Space Flight Center.

17. For above two paragraphs, Biggs, "Space Shuttle Main Engine," 92–96; Heppenheimer, *Development*, 144–47, 429n53.

18. For last two paragraphs, Biggs, "Space Shuttle Main Engine," 87–118; Heppenheimer, *Development*, 148–71; Jenkins, *Space Shuttle*, 225–27.

19. Aeronautics and Space Engineering Board, *Liquid Rocket Propulsion Technology*, 16.

20. Rockwell, "Press Information," 9, 51–53; Jenkins, *Space Shuttle*, 225–27.

21. *CPIA/M5*, unit 174, J-2 Engine, unit 175, F-1 Saturn V Booster Engine, and unit 195, Space Shuttle Main Engine.

22. Rockwell, "Press Information," 21.

23. Dunar and Waring, *Power to Explore*, 286–90; Joliff, "Pershing Weapon System," 23, 25, 28–35, 56, 113; Pershing System Description, 1: 3–4, 39–40, RATL. Joliff's Pershing history is still classified. I thank the Army Missile Command History Office for providing unclassified portions.

24. Jenkins, *Space Shuttle*, 184–85; Heppenheimer, *Development*, 71–78; Logsdon et al., *Exploring the Unknown*, vol. 4, Document II-17, "The Comptroller General of the United States, Decision in the Matter of Protest by Lockheed Propulsion Company, June 24, 1974," 269. See also chapter 6 of this book for Aerojet's involvement with segmented solid-rocket motors.

25. Jenkins, *Space Shuttle*, 185–86; Bromberg, *Space Industry*, 100; Heppenheimer, *Development*, 71–78; "Comptroller General" (see note 24), 271 (quotation).

26. The above two paragraphs summarize a longer treatment in Hunley, "Minute-

man," 271–78, which draws heavily upon Andrepont and Felix, "Motor Development," 4–6, 9–14, 17, 20, plus Gordon et al., *Aerojet*, IV-42 and IV-43, and E. Sutton, "Polymers to Propellants," 17. One technology developed in the 623A program that did not become part of the shuttle was use of a new case material called maraging steel. This was a tough, strong steel that was low in carbon and high in nickel. It formed hardening precipitates as it aged. The primary focus of the testing was maraging steel with 18 percent nickel, which demonstrated greater fracture toughness than conventional steels with the same strength levels, allowing thinner and lighter cases. Aerojet featured this steel in its proposal for the shuttle boosters, but Thiokol used an older steel in its design for the shuttle. See Defense Metals Information Center, "Third Maraging Steel Project Review," esp. 1–3, A-17, A-44; Andrepont and Felix, 4–6; Jenkins, *Space Shuttle*, 153–54, 185–86.

27. Jenkins, *Space Shuttle*, 186; Hunley, "Minuteman," 278–80; Rockwell, "Press Information," 21–22; *CPIA/M1*, unit 556, Space Shuttle Booster; Heppenheimer, *Development*, 175.

28. Heppenheimer, *Development*, 174.

29. "UTC Prepares for Shuttle Booster Role," 15 (quotation); Jenkins, *Space Shuttle*, 186; Heppenheimer, *Development*, 174–75. Accounts of USBI tend to be confused about names. At the time of contract award, the parent firm of UTC was named United Aircraft Corporation. In 1975 it changed its name to United Technologies Corporation, and United Technology Center (UTC) then became Chemical Systems Division (CSD).

30. Heppenheimer, *Development*, 175–76 (Hardy quotation, 175; Thirkill, 176). For Thiokol's development of PBAN and its low cost, see above and Hunley, *Preludes*, chap. 9; for the importance of cost control in shuttle development, which Heppenheimer does not mention in this connection, although well aware of it, see the sources in note 3.

31. Heppenheimer, *Development*, 176; Rockwell, "Press Information," 9, 21–22; *CPIA/ M1*, unit 613, Redesigned Space Shuttle Booster, June 1993. The last source gives the thrust-time curve for the SRB on the redesigned booster. Since the propellant configuration "was recontoured to reduce the stress fields between the star and cylinder portion of the propellant grain" in the redesigned version, the basic curve could have changed somewhat, but the description in the narrative seems consistent with evidence available for this chapter on the original SRBs. See NASA, *National Space Transportation System Reference*, 33a, 33d (quotation), 35; cf. the 1984 counterpart in Rockwell, 22.

32. Heppenheimer, *Development*, 176–77.

33. Above three paragraphs draw on many sources. For the Titan motor, see chapter 6 and its sources, esp. Felix, comments on manuscript, August 22, 2000, UP; also Hunley, "Minuteman," 278–80, for both the Titan and the shuttle; Andrepont and Felix, "Motor Development," 7. For the shuttle SRB, Heppenheimer, *Development*, 177–78; Hunley, "Evolution," 31, 37, and the sources cited there, esp. Adelman, OHI; Andrepont and Felix, 14; NASA, *National Space Transportation System Reference*, 33a–33c. Heppenheimer, 178–79, says the shuttle had 180 pins, but both Jenkins, *Space Shuttle*, 186, and Andrepont and Felix, 14, say 177, so I follow them. That the orientation of the tang and clevis and the heating strip were significant is suggested by the fact that after the *Challenger* disaster, the redesigned field joints incorporated external heaters with weather seals to keep the joint and O-rings at a minimum temperature of 75°F and to waterproof

the joint. The number of pins may also have been significant, since after the *Challenger* disaster the mating pins were lengthened to improve shear strength, a change that might have been unnecessary with more pins. See NASA, *National Space Transportation System Reference*, 33b, and below in this chapter.

34. Heppenheimer, *Development*, 178–79; Felix, comments, August 22, 2000, UP; Rockwell, "Press Information," 21–22, where on page 22 the nozzle expansion ratio is listed as 7:79, presumably a misprint for 7.79; *CPIA/M1*, unit 556, which gives the 1979 initial expansion ratio as 7.16, and unit 613, which gives the expansion ratio of the redesigned SRB as 7.72; e-mail from Dennis R. Jenkins, September 17, 2003, which confirms that for the first seven shuttle missions the expansion ratio was 7.16:1, changing to 7.72:1 from the eighth mission to the present.

35. Hunley, "Minuteman," 280; Jenkins, *Space Shuttle*, 429; Rockwell, "Press Information," 22; Isakowitz, *Space Launch Systems* (1991), 252; Heppenheimer, *Development*, 179–81. Heppenheimer has a lot of good information on the nozzle and its functioning but gives its amount of swivel as 7.1 degrees. Both Jenkins and Rockwell say it was 8 degrees, which I have accepted, although Heppenheimer cites AIAA papers on the SRB nozzles and may be correct.

36. Jenkins, *Space Shuttle*, 425–26; Rockwell, "Press Information," 21. Thiokol had set up its Wasatch Division in Utah in 1956–57; this is where it developed the Minuteman first-stage motor as well as the 156–inch motors and the shuttle motors. See E. Sutton, "Polymers to Propellants," 14–16; "Spacecraft Propulsion," in Thiokol, *Aerospace Facts*, Spring 1983, 4–5, NASM.

37. Jenkins, *Space Shuttle*, 227; Heppenheimer, *Development*, 178, 183.

38. For above two paragraphs, Heppenheimer, *Development*, 186–89; Jenkins, *Space Shuttle*, 227–28.

39. Jenkins, *Space Shuttle*, 186–87; Rockwell, "Press Information," 46; Heppenheimer, *Development*, 68–69 (quotation, 68); Dunar and Waring, *Power to Explore*, 293.

40. Dunar and Waring, *Power to Explore*, 292 (first quotation), 294, 301 (second quotation), 323; Jenkins, *Space Shuttle*, 186; CAIB, *Report*, 9 (third and fourth quotations). Even before the *Columbia* accident, Jenkins had recognized that the ET was "a fairly advanced piece of engineering" (186), and at the outset Eberhard Rees at Marshall had stated that the tank was "very challenging to work on, . . . very complex and difficult" (Dunar and Waring, 292).

41. Jenkins, *Space Shuttle*, 186, 231; Heppenheimer, *Development*, 191–92; Dunar and Waring, *Power to Explore*, 292, where they quote James Kingsbury characterizing the ET as the "structural backbone of the [shuttle] stack."

42. Heppenheimer, *Development*, 190–91; Jenkins, *Space Shuttle*, 229.

43. Jenkins, *Space Shuttle*, 230–31, 422; Rockwell, "Press Information," 46; Dunar and Waring, *Power to Explore*, 304.

44. Rockwell, "Press Information," 47–49; Jenkins, *Space Shuttle*, 421; Heppenheimer, *Development*, 194.

45. Heppenheimer, *Development*, 191–92; Dunar and Waring, *Power to Explore*, 304–5 (quotation, 304); Jenkins, *Space Shuttle*, 231, 421; CAIB, *Report*, 14.

46. CAIB, *Report*, 52. For coverage of other aspects of ET testing, see Heppenheimer, *Development*, 193–98; Jenkins, *Space Shuttle*, 229–31.

47. CAIB, *Report*, 52, 53, 122 (quotations, 52, 53).

48. Ibid., 9, 52–53 (quotations, 52–53).

49. Ibid., 97 (first quotation), 121 (last quotation); Vartabedian, "Foam Woes" (other quotations).

50. CAIB, *Report*, 54 (second, fourth, and fifth quotations); Vartabedian, "Foam Woes" (third quotation, paraphrase of Osheroff). Osheroff's quotations from his separate essay in the CAIB report (54), entitled "Foam Fracture under Hydrostatic Pressure." He does not emphasize the negative in that commentary, ending, "it is the simple studies that so far have most contributed to our understanding of foam failure modes."

51. Rockwell, "Press Information," 11, 22, 47.

52. For other features of the shuttle, see Heppenheimer, *Development*; Jenkins, *Space Shuttle*; Iliff and Peebles, *From Runway to Orbit*.

53. On these, see especially Jenkins, *Space Shuttle*, 365–411; normal altitudes of shuttle orbits from Jenkins, 260.

54. Jenkins, *Space Shuttle*, 169, 173, 182–83; Jenkins, "Broken in Midstride," 363–64.

55. Heppenheimer, *Development*, 22–27; Jenkins, *Space Shuttle*, 189–92, 387.

56. See Heppenheimer, *Development*, 246–50; Iliff and Peebles, *From Runway to Orbit*, 221, 237, 243. I discuss some of these issues in a general way in "NASA Dryden," esp. 197, 200–201, 203.

57. Heppenheimer, *Development*, 246–50; Jenkins, "Broken in Midstride," 388; Jenkins, *Space Shuttle*, 211; Iliff and Peebles, *From Runway to Orbit*, 157–65.

58. For technical discussions of parameter identification (also called parameter estimation), see Maine and Iliff, *Applications of Parameter Estimation*; Iliff, "Parameter Estimation for Flight Vehicles." For explanations in lay terms, see L. Wallace, *Flights of Discovery*, 56–57, photo caption; Gorn, *Expanding the Envelope*, 339–42; Iliff and Peebles, *From Runway to Orbit*, x, xi, 10, 54–55, 94, 105, 158–59, 221–23, 236–37, which discusses the use of this tool on shuttle flights and from which much of this paragraph is derived. For management structure in the shuttle program, see esp. Heppenheimer, *Development*, 33–42; Dethloff, *Suddenly, Tomorrow Came*, 228–34. For the various names of Dryden Flight Research Center since its founding in 1946, see Wallace, 21, 23–27. In 2007 there were reports that Congress planned to rename the center one more time, after research pilot/astronaut Neil Armstrong.

59. Jenkins, *Space Shuttle*, 365–84; Rockwell, "Press Information," 5, 109–48; Jenkins, "Broken in Midstride," 364.

60. Rockwell, "Press Information," 109, 148–52; Jenkins, *Space Shuttle*, 385–90; Rumerman, "Fairchild Aviation Corporation," for the date Fairchild Republic went out of business on Long Island.

61. Heppenheimer, *Development*, 198–202; Gordon et al., *Aerojet*, III-169 to III-171; *CPIA/M5*, unit 211, AJ10–190.

62. Heppenheimer, *Development*, 202–3; Jenkins, *Space Shuttle*, 391; *CPIA/M5*, units 209, R-1E-3, September 1982, and 210, R-40–A, September 1982.

63. Heppenheimer, *Development*, 211–13, 221; Jenkins, *Space Shuttle*, 237–38.

64. Jenkins, *Space Shuttle*, 238–39, 395–400; Heppenheimer, *Development*, 213–14.

65. Heppenheimer, *Development*, 223–27; Bromberg, *Space Industry*, 100; Jenkins, *Space Shuttle*, 238.

66. Jenkins, *Space Shuttle*, 240; Hallion and Gorn, *On the Frontier*, 260–61; Heppenheimer, *Development*, 229–30. My thanks to Mike Gorn for copies of *On the Frontier* (a revision and updating of Hallion's original 1984 *On the Frontier*, NASA SP-4303) and his *Expanding the Envelope*, cited earlier.

67. Heppenheimer, *Development*, 225, 232–44 (quotation, 242); Jenkins, *Space Shuttle*, 239–40. The two engineer-historians, who provide much detail on this complicated story, differ on some particulars. For example, Heppenheimer, 225, places the number of tiles at 34,000, while Jenkins, 239, puts it at slightly more than 30,000.

68. Jenkins, *Space Shuttle*, 395–401; Iliff and Peebles, *From Runway to Orbit*, 243, 256, 261.

69. Heppenheimer, *Development*, 215–16, 224; Jenkins, *Space Shuttle*, 395, 398; CAIB, *Report*, 55. As discussed in chapter 4 of this book, Vought was at times part of Ling-Temko-Vought and LTV Aerospace Corporation among other names, at other times known as Vought Corporation in a bewildering succession of corporate name changes. By 2003 its Missiles Division had become part of Lockheed Martin Missiles and Fire Control.

70. CAIB, *Report*, 9 (third quotation), 122 (first quotation); Meyer and Barneburg, "In-Flight Rain Damage Tests," 1–5 (second quotation, 5), 9–11. Italics on "may" added, but the report does emphasize that the conclusion is provisional.

71. These comments summarize much more detailed and specific information in Rockwell, "Press Information," 438–72; Jenkins, *Space Shuttle*, 231–35, 405–8; Heppenheimer, *Development*, 287–322; Tomayko, *Computers in Spaceflight*, 85–133.

72. Jenkins, *Space Shuttle*, 231–33 (quotation, 233); Heppenheimer, *Development*, 290, 296–97, 302; Tomayko, *Computers Take Flight*, 90–91; Tomayko, *Computers in Spaceflight*, 91, 112–13, 116. Tomayko, *Spaceflight*, 108, says the September contract was preceded by a "first," presumably preliminary, contract in March 1973.

73. Heppenheimer, *Development*, 299, 302–3 (first quotation, 302); Jenkins, *Space Shuttle*, 233; Tomayko, *Computers in Spaceflight*, 91–92, 109 (Parten quotation, 109).

74. Tomayko, *Computers in Spaceflight*, 102–4; Jenkins, *Space Shuttle*, 233–35; Heppenheimer, *Development*, 290, 300–301.

75. Tomayko, *Computers Take Flight*, esp. 89, 93–95, 107–8, 125. See also Jenkins, *Space Shuttle*, 233–35, and Tomayko, *Computers in Spaceflight*, 102, where Tomayko wrote that the DFBW "flight program did much to convince NASA of the viability of the synchronization and redundancy management schemes developed for the Shuttle." Another critical contribution of the DFBW project and Dryden Flight Research Center to the shuttle occurred after astronaut Fred Haise experienced a pilot-induced oscillation (PIO) on the fifth landing of the prototype *Enterprise* on October 26, 1977. A slight delay (a fifth to a quarter of a second) in the fly-by-wire flight control system's response to his pilot inputs and his overreaction to what appeared to be a nonresponsive vehicle caused the orbiter to bounce dangerously on landing. After months of sorting out what had happened, Dryden engineers designed a PIO-suppression filter. Tested in the F-8 and installed in the orbiter, it did not entirely eliminate the tendency toward PIO, but the pilot had to make larger movements of the flight controls for a longer time before it would occur. The filter plus practice in simulators reduced the likelihood of a PIO like Haise's. Indeed, his experience may well have saved future shuttle pilots from a

recurrence. On this, see Tomayko, *Computers Take Flight*, 111–14, and especially Iliff's illuminating comments in *From Runway to Orbit*, 182–87.

76. Tomayko, *Computers in Spaceflight*, 96, 99, 115.

77. Heppenheimer, *Development*, 292, 305–6; Tomayko, *Computers in Spaceflight*, 112–13; Jenkins, *Space Shuttle*, 234–35.

78. Rockwell, "Press Information," 23, 36, 54, 75, 439–62; Jenkins, *Space Shuttle*, 406; Heppenheimer, *Development*, 313 (quotation); Tomayko, *Computers in Spaceflight*, 128–30.

79. Rockwell, "Press Information," 441, 452–54; Jenkins, *Space Shuttle*, 402; NASA, "NSTS 1988 News Reference Manual."

80. Rockwell, "Press Information," 36, 449–51; Jenkins, *Space Shuttle*, 402–3.

81. CAIB, *Report*, 14; Isakowitz, *Space Launch Systems* (1991), 252–53.

82. U.S. President, *Report . . . 2001*, 121–29; Jenkins, *Space Shuttle*, 266–75, 294–301, 328–30; Siddiqi, *Deep Space Chronicle*, 143–45; Rumerman, *NASA Historical Data Book* 5: 171. For the development of the IUS, see chapter 6 of this book.

83. Siddiqi, *Deep Space Chronicle*, 145–51; U.S. President, *Report . . . 1995*, 18; Jenkins, *Space Shuttle*, 326. For the development of the Payload Assist Module, see chapter 2 of this book. On its 23rd flight, the shuttle *Atlantis* used a PAM-D2 (for Delta) to launch a communications satellite.

84. Launius, *NASA*, 125–27; Crouch, *Aiming for the Stars*, 278–80; U.S. President, *Report . . . 1989–1990*, 146; U.S. President, *Report . . . 1993*, 1; U.S. President, *Report . . . 1994*, 12; U.S. President, *Report . . . 1995*, 17.

85. Logsdon et al., *Exploring the Unknown* 4: "Presidential Commission on the Space Shuttle *Challenger* Accident, 'Report at a Glance,' June 6, 1986," 358–59, 363; Jenkins, *Space Shuttle*, 267, 277–81; Launius, *NASA*, 250–51, extracts from the Presidential Report on the Challenger Accident; Crouch, *Aiming for the Stars*, 260; Waring, "Accident and Anachronism," 13–14, copy kindly provided by the author; and for Morton Salt's takeover of Thiokol, E. Sutton, "Polymers to Propellants," 22. Thiokol again became a separate corporation in 1989.

86. "Report at a Glance" (see note 85), 363 (quotation); Waring, "Accident and Anachronism," 9–12, 14–15; Jenkins, *Space Shuttle*, 279, 281; Launius, *NASA*, 115–17; CAIB, *Report*, 122–23, for the view that "NASA and contractor personnel came to view foam strikes not as a safety of flight issue, but rather a simple maintenance . . . issue" but not the comparison with *Challenger*.

87. Waring, "Accident and Anachronism," 18, for the argument (by Marshall managers) that the assembly errors caused the accident; "Report at a Glance" (see note 85), 386 (quotation).

88. See, e.g., "Report at a Glance" (see note 85, above), 366–68. But see also Allan J. McDonald and James R. Hansen, "Truth, Lies, and O-Rings: The Untold Story Behind the Challenger Accident: An Excerpt from the Forthcoming Book," *Quest: The History of Spaceflight Quarterly*, 14, no. 3 (2007): 6–11, where McDonald, director of the Space Shuttle Solid Rocket Motor Project in 1986 when the Challenger accident occurred, states his vehement opposition to the launch at the time and accuses NASA of a cover-up.

89. NASA, *National Space Transportation System Reference*, 1: 33a–33b (quotations,

33a). For other changes to the shuttle following the *Challenger* accident, ibid., 27–33, 33c–33h.

90. Jenkins, *Space Shuttle*, 288; Spires, *Beyond Horizons*, 221–23, 228–29; Jenkins, "Broken in Midstride," 374–75, 408–9; Launius, "Titan," 178; U.S. President, *Report . . . 2001*, 127–29.

91. Launius, "Decision," 34. In this connection, see also Ray A. Williamson, "Developing the Space Shuttle," in Logsdon et al., *Exploring the Unknown* 4: 161–91, esp. 182, which notes, "The Shuttle was developed in part to serve DOD needs, which led to higher operations costs than NASA had anticipated."

92. CAIB, *Report*, 19.

93. "Report at a Glance" (see note 85), 371 (quotation).

Chapter 8. The Art of Rocket Engineering

1. For a recent, short, and incisive discussion of these issues, see Butrica, *Single Stage to Orbit*, 47–52.

2. Estimate and general information from Schwiebert, *Air Force Ballistic Missiles*, 139; I have used S. M. Friedman, "Inflation Calculator," to convert from dollars of 1961, slightly past the middle of the period Schwiebert's estimate covers (1955–64), since costs would have climbed significantly after the early years.

3. Bilstein, *Stages to Saturn*, 422; S. M. Friedman, "Inflation Calculator," based on 1965.

4. Of course, science itself has been taken off its pedestal in recent decades by scholars from a variety of disciplines in what have been called the science wars. For a judicious discussion of this complex historiographical issue, see Jacob, "Science Studies."

5. See introduction, note 2, for works of Layton, Vincenti, and Ferguson.

6. Memo, Tischler to Rosen (see chapter 5, note 42).

7. As late as November 2005, shuttle officials still were not certain why foam separated from the external tank, but in early December NASA administrator Michael Griffin said engineers were finally pretty sure that radical temperature changes when liquid oxygen and liquid nitrogen were loaded into the tank caused the foam to crack and be subject to breaking off during launch. See J. Johnson, "Likely Cause."

8. Hale, "Accepting Risk." My thanks to Mr. Hale for his e-mail of November 25, 2004, granting permission to quote him.

9. See Hunley, *Preludes*, chap. 9, note 10, and the sources cited there. Henderson said he had shared his recollections with Rumbel and incorporated the latter's comments before sending them to me.

10. Ibid., notes 78 and 83, plus the narrative coverage documented there.

11. Glasser, OHI, 73, 76 (quotations, 76 and 73).

12. Aitken, *Continuous Wave*, 547 (quotation), but see also 15–16, 522–25, 536–52.

13. On combustion instability, there is an enormous literature. Here I can only cite Price, "Solid Rocket Combustion Instability," and Yang and Anderson, *Liquid Rocket Engine Combustion Instability*, plus the sources cited in both works.

14. The classic account is Armacost, *The Politics of Weapons Innovation: The Thor-Jupiter Controversy*.

15. See, e.g., the comments of Glennan in *Birth of NASA*, 9–12, 19–20, 22–23, 111n4, 258.

16. See sources in chapter 1, note 46.

17. Bio of Adelman provided by Karen Schaffer of CSD; Adelman, OHI; Kennedy, Kovacic, and Rea, "Solid Rocket History at TRW," 13, for Adelman's role on Minuteman. For his contributions to Titan III, see chapter 6.

18. See Hunley, *Preludes*, chap. 3.

19. Knauf, Drake, and Portanova, "EELV," 38, 40; Anselmo, "EELV Finalists."

20. Covault, "Atlas V Soars," 22–23; "Military EELV Launches"; Boeing, "Successful First Launch"; "EELV Evolved Expendable Launch Vehicle." Incidentally, according to Boeing, News Release, "Boeing and Lockheed Martin Complete United Launch Alliance Transaction," <http://www.boeing.com/ids/news/2006/q4/061201a_nr.html>, Boeing and Lockheed Martin formed a joint venture called United Launch Alliance on December 1, 2006, to "combine the production, engineering, test and launch operations" of the two firms for the Delta and Atlas vehicles. Whereas "Military EELV Launches" had put the cost of an EELV launch at $150 million (compared with $400 million for a Titan IVB), a NASA Kennedy Space Center site stated that a fiscal year 2006 budget request put the cost of an average EELV launch, including associated services, at about $170 million, still much cheaper than a Titan IVB launch. See <http://science.ksc.nasa.gov/shuttle/nexgen/CapeMap/cx41.htm>, Note 4. On the RD-180, see G. Sutton, *Liquid Propellant Rocket Engines*, 505–8, 590. Sutton's book has an extensive treatment of Russian, Ukrainian, and Soviet engines (531–736) and also covers German, French, Japanese, British, Chinese, and Indian engines (737–801).

21. See, e.g., Butrica, "Quest for Reusability," 463.

22. For above two paragraphs, NASA, "Constellation Program," and associated links; J. Johnson, "Likely Cause"; "NASA Weighing 2 Rocketdyne Engines." As the last article states, in 2005 the Pratt & Whitney Division of United Technologies had acquired Rocketdyne, previously part of Boeing. On the names Ares I and Ares V, see also Reddy, "NASA Christens Its New Rockets."

23. First paragraph, Doug Richardson, "Rocket Engineers Explore Burning Issues," *Jane's International Defense Review* (1, 3), posted August 14, 2007 on a subscriber website accessed through <http://www.janes.com/news/defence/systems/idr/idr070815_1_n.shtml> and e-mailed to the author by Jane's Information Group on August 17, 2007; second paragraph, CAIB, *Report*, 19. See Hunley, *Preludes*, chap. 7, for the success rate of early Atlas flights.

Glossary of Terms, Acronyms, and Abbreviations

The definitions of terms below are intended primarily for the general reader. As such, they are deliberately couched in simple, lay language and will not satisfy the rigor that the scientific and engineering communities expect. Unfortunately, many more precise definitions are not comprehensible to nontechnical audiences. These definitions thus constitute a compromise between exactness and comprehensibility. The technical community will probably find this glossary useful mostly for the spelling out of acronyms and abbreviations that may be unfamiliar or forgotten.

AAS – American Astronautical Society
ABL – Allegany Ballistics Laboratory
ablation – The vaporizing of a substance to dissipate or carry away heat—used for nose cones of reentry bodies, combustion chambers, and nozzles or nozzle throats where heating was intense
ABMA – Army Ballistic Missile Agency
AC – Atlas-Centaur
accelerometer – An instrument for measuring acceleration
ADDJUST – Automatic Determination and Dissemination of Just Updated Steering Terms
AFB – Air Force Base
AFBMD – *See* BMD
AFHRA – Air Force Historical Research Agency
AFMDC – Air Force Missile Development Center
AFRSI – Advanced flexible reusable surface insulation, one type of insulation on the shuttle orbiters
AFSC – Air Force Systems Command
AFSWC – Air Force Special Weapons Center

AGC – Aerojet General Corporation

AIAA – American Institute of Aeronautics and Astronautics

AJ – Aerojet

AMC – Air Materiel Command; Army Missile Command

ANNA – Army, Navy, NASA, Air Force

apogee – The farthest point from Earth's center of a satellite orbiting Earth

ARDC – Air Research and Development Command, a precursor of Air Force Systems Command

ARPA – Advanced Research Projects Agency

ATS – Applications Technology Satellites

attitude – The position of a rocket or spacecraft in relation to its axes and an external data point such as the horizon or a star

axial – Circling around a cylinder or other elongated body in a direction essentially perpendicular to the length of that body.

baffle – A device to prevent sloshing in liquid-propellant tanks or combustion instability in liquid-propellant combustion chambers

Bell Labs – Bell Telephone Laboratories

BMD – [Air Force] Ballistic Missile Division (formerly WDD)

booster – A somewhat ambiguous term applied to any kind of rocket that adds to or provides lift. Most specifically, it refers to (1) strap-on or other rocket motors or engines, especially solid, that augment lift capability during launch and the early part of flight. More generally, it can mean (2) an entire launch vehicle, (3) the first stage of a multistage launch vehicle, or even (4) a rocket that lifts a satellite from one orbit to a higher one, as in "Agena booster," which refers to an Agena upper stage used in this way.

BSD – [Air Force] Ballistics Systems Division, one of two divisions that split from BMD on April 1, 1961

CAIB – Columbia Accident Investigation Board

Caltech – California Institute of Technology

castable – Of a fluid including a binder, oxidizer, and fuel, able to be poured into a case and cured into a solid propellant

cavitation – The formation of bubbles in the "plumbing" of a liquid-propellant rocket engine, interfering with the flow of propellants or other liquids

CFD – Computational fluid dynamics

CIGTF – Central Inertial Guidance Test Facility, at AFMDC

combustion instability – Oscillations in an operating combustion chamber (whether liquid- or solid-propellant) that can be so great as to destroy the engine or motor

composite – A type of solid propellant consisting of separate particles of oxidizer, possibly a separate fuel, and other substances dispersed in an elastic matrix that serves as a binder and also a fuel

control – To provide commands to rocket or missile actuators that will cause the vehicle to follow a desired trajectory, usually with provision for feedback to a guidance-and-control system; as a noun, the provision of such commands

Convair – Consolidated Vultee Aircraft Corporation

CPIA – Chemical Propulsion Information Agency

cryogenic – Extremely cold

CSD – Chemical Systems Division (of United Technologies Corporation)

CSM – [Apollo] Command and Service Module, which actually consisted of separate command and service modules

CTPB – Carboxy-terminated polybutadiene, a solid-propellant binder that was used somewhat sparingly because of its high cost compared with PBAN

cut-and-try engineering – Cutting or constructing parts and trying them out to fix undiagnosed or at least unpredicted problems; if they worked, the fabricator still might not understand fully the nature of the underlying problem, but their success or failure added to the engineering data base for future design; in this book, roughly synonymous with "trial-and-error" and "empirical"

delta guidance – A system of guidance in which a rocket's velocity and position at the end of powered flight are calculated in advance but achieved in a flexible manner, not simply returned to a preplanned trajectory whenever a deviation occurs

DFBW – Digital fly-by-wire

DMSP – Defense Meteorological Satellite Program

DoD – Department of Defense

Doppler effect – A change in the frequency with which (radio) waves from a moving object reach a receiver, depending on speed; used in control devices for missiles and rockets to track trajectory and for related purposes

drag – A retarding force from the atmosphere operating on a body passing through that atmosphere

DSCS – Defense Satellite Communications System

DSP – Defense Support Program

EELV – Evolved Expendable Launch Vehicle

ET – External tank

exhaust velocity – The speed of the gases expelled from a rocket nozzle; a measure of effectiveness extrusion – A method of producing a propellant grain by forcing it through dies, either with the propellants suspended in a solvent or in a dry (solventless) process

F – Fahrenheit

film cooling – Introducing a flow (film) of fuel down the inside wall of a liquid-propellant combustion chamber and/or exhaust nozzle to protect it from the heat of combustion

FLTSATCOM – Fleet Satellite Communications

G – Acceleration equal to the force of gravity at sea level

gas generator – A device for starting operation of turbopumps that uses the same propellants as the combustion chamber, avoiding the need in earlier liquid-propellant rockets for a separate system with its own propellants

GD/A – General Dynamics/Astronautics

GE – General Electric Company

GEM – Graphite epoxy motor

geostationary – In an orbit above the equator at an altitude of 22,300 miles and traveling at a speed such that the object remains above a fixed point on Earth

geosynchronous – Geostationary

gimbal – A device that permits a body (such as an engine, nozzle, or guidance-and-control device) to rotate in any direction or that suspends it so that it remains stable when the larger structure, such as a rocket, changes its attitude; as a verb, to "steer" by rotating an engine or nozzle using a gimbal

GN&C – Guidance, navigation, and control

GPO – Government Printing Office

GPS – [Navstar] Global Positioning System

grain – A mass of propellant, usually configured to provide a predetermined thrust time curve (a graph of the amount of thrust over time as a propellant burns)

GSFC – [NASA] Goddard Space Flight Center

GTO – Geostationary transfer orbit, an orbit from which a satellite can move to a geosynchronous orbit

guidance – Determining a trajectory and velocity for a missile or rocket to reach a desired position from another position or location

gyroscope – A device that rotates about an axis like a children's top. It responds to a disturbing angular force by moving slowly (precessing) in a direction at right angles to that of the force. The precession is predictable, so by mounting gyroscopes in such a way that they respond in only one or two directions, they can be used to indicate acceleration or angular velocity. These indications together with electrical pickoff devices can then be used by a guidance and control system to direct servomechanisms to vanes, gimbals, or other devices to adjust the attitude of a rocket and thus, in effect, steer it.

HAL/S – A computer language perhaps named in honor of Hal Laning

heterogeneous engineering – Social engineering that involves winning support for a project or goal, as distinguished from designing or manipulating objects in the physical world

HMX – A high explosive used in some rocket propellants

HTPB – Hydroxy-terminated polybutadiene, a solid-propellant binder that is superior to CTPB and also cheaper but not as low in cost as PBAN

hypergolic – Of propellants, igniting upon contact with each other without need of an igniter

IAA – International Academy of Astronautics

IAF – International Astronautics Federation

IBM – International Business Machines Corporation

ICBM – Intercontinental ballistic missile, usually defined as having a range of at least 5,000 miles

IDCSP – Initial Defense Communication Satellite Program

IGY – International Geophysical Year, July 1, 1957, to December 31, 1958

IMU – Inertial measurement unit

inertial guidance – Determining a trajectory and velocity for a missile or rocket so it will reach a desired position from another position or location, using self-contained, automatic devices such as gyroscopes and accelerometers that respond to inertial forces (changes of direction and/or speed) and feed information to a computer

injector – A device in a liquid-propellant engine that atomizes and mixes propellants as it introduces them to the combustion chamber for ignition

Intelsat – International Telecommunications Satellite Consortium

internal cavity – A hollow area in a propellant grain where the burning of propellants occurs; also called a perforation

IO – Industrial Operations, function at MSFC

IOP – Input-output processor

IRBM – Intermediate range ballistic missile, usually defined as having a range of at least 1,500 miles but less than 5,000 miles

IRFNA – Inhibited red fuming nitric acid, a liquid oxidizer

ISS – Integrated Spacecraft System

IU – Instrument Unit, component of the Saturn rockets

IUS – Inertial Upper Stage (originally, Interim Upper Stage)

IWFNA – Inhibited white fuming nitric acid, a liquid oxidizer

JANNAF – Joint Army Navy NASA Air Force

JATO – Jet-assisted takeoff

JP – [Kerosene] jet fuel, as in JP-4

JPL – Jet Propulsion Laboratory, Caltech

JSC – [NASA] Johnson Space Center

KSC – [NASA] Kennedy Space Center

lbf-sec/lbm – Pounds of thrust per pound of propellant burned per second, a measure of specific impulse

LES – Lincoln Experimental Satellite

LOC – [NASA] Launch Operations Center (later KSC)

longeron – A lengthwise framing member of an airplane or rocket structure

LTV – A shortened name for Ling-Temco-Vought

LV – Launch vehicle

MA – Mercury-Atlas, as in MA-1, MA-2

Mach number – Speed in relation to that of sound

mass fraction – The mass of the propellant in a rocket stage divided by the total mass of the stage; the higher the mass fraction, the more efficient the stage, other parameters being equal

Milstar – Military Strategic and Tactical Relay System

MIT – Massachusetts Institute of Technology

MSC – [NASA] Manned Spacecraft Center (later JSC)

MSFC – [NASA] Marshall Space Flight Center

MSTI – Miniature Sensor Technology Integration

NA – National Archives

NAA – North American Aviation

NACA – National Advisory Committee for Aeronautics, predecessor of NASA

NASA – National Aeronautics and Space Administration

NASM – National Air and Space Museum, Smithsonian Institution

NDS – Navigation Development System

NOAA – National Oceanic and Atmospheric Administration

NOSS – Naval Ocean Surveillance System

NOTS – Naval Ordnance Test Station

nozzle – A device at the end (initially the bottom) of a rocket engine or motor that accelerates the expanding gases from the combustion chamber to increase thrust

NRL – Naval Research Laboratory

OGO – Orbiting Geophysical Observatory

OHI – Oral history interview

OMSF – [NASA] Office of Manned Space Flight

OSC – Orbital Sciences Corporation

OV – Orbiting vehicle

oxidizer – A substance, used together with a rocket fuel, to supply oxygen that enables the fuel to burn at high altitudes and in space where atmospheric oxygen is sparse or unavailable

PAM – Payload Assist Module

PARD – Pilotless Aircraft Research Division

parking orbit – A temporary orbit from which a satellite or other spacecraft transfers to another orbit or trajectory in space

PASS – Primary avionics system software

PBAA – Polybutadiene-acrylic acid, a solid-propellant material consisting of an elastomeric (rubberlike) copolymer of butadiene and acrylic acid that permits higher concentrations of solid ingredients and greater fuel content than earlier binders

PBAN – Polybutadiene-acrylic acid-acrylonitrile, a solid-propellant material, successor to PBAA, with greater tear strength

perigee – The nearest point to Earth's center of a satellite orbiting Earth

PFRT – Preliminary flight readiness test

phenolic – A type of resin made from a crystalline acidic compound (phenol) and used for coatings, such as for nozzles and nozzle throats

PIO – Pilot-induced oscillation

pitch – The up-down movement of the nose of a rocket as it flies more or less horizontally

POGS – Polar Orbiting Geomagnetic Survey

polybutadiene – A rubbery solid binder used in propellants such as CTPB, HTPB, PBAA, and PBAN

polymer – A compound consisting of many repeated, linked, simple molecules, with a chemical structure that makes it rubbery, so that it retains its shape while resisting cracking and still permits fuel and oxidizers to be loaded within it before it cures—all useful characteristics for a solid-propellant binder

polyurethane – A class of polymers containing urethane links

PPB – Planning, programming, and budgeting

propellants – Fuels and oxidizers (plus additives) that burn in a combustion chamber to produce expanding gases and thus supply thrust

psi – Pounds per square inch

Q-guidance – A system that permitted much of the computation for guidance (the Q-matrix) to be performed long before launch, leaving little calculation to be done by the computer on the missile

ramjet – A simple kind of jet engine in which the air for combustion is compressed in a tube by the forward motion of the vehicle through the atmosphere instead of by the complex compressor devices used in turbojets

R&DO – Research and Development Operations, function at MSFC

rate gyro – A gyroscope used to detect angular deviations

reaction controls – Controls using thrust to steer or otherwise adjust the attitude of an aircraft, rocket, or spacecraft at altitudes above those where aerodynamic controls are effective

regenerative cooling – Cooling of a liquid-propellant combustion chamber by circulating a propellant around the outside wall

RFNA – Red fuming nitric acid, a liquid oxidizer

RFP – Request for proposals

RLG – Ring laser gyro

RMI – Reaction Motors, Inc.

rocket engine – A propulsion device of a rocket, usually one burning liquid propellants

rocket motor – A propulsion device of a rocket, usually one burning solid propellants

roll – The rotation of a rocket about its longitudinal axis

RP-1 – [Kerosene] rocket fuel

RPL – [Air Force] Rocket Propulsion Laboratory, one of many official names for the "rocket site" at Edwards AFB, California

rpm – Revolutions per minute

SAMSO – [Air Force] Space and Missile Systems Organization

S&ID – [NAA] Space and Information Systems Division

scaling up – Increasing size and performance

SCE – Selective Communications Experiment

servo – An automatic mechanism used for control, as in a servomechanism to actuate flight-control surfaces

SLV – Space-launch vehicle

SP – Special Project, a type of NASA publication

specific impulse – A measure of performance for a propellant combination or propulsion system, expressed as a measurement of thrust per amount of propellant burned per unit of time (in this book, pounds of thrust per pound of propellant burned per second, expressed as lbf-sec/lbm); I_{sp} is the symbol

SRB – Solid rocket booster

SRM – Solid rocket motor

SRMU – Solid rocket motor upgrade

SSD – [Air Force] Space Systems Division, one of two divisions that split from BMD on April 1, 1961

SSME – Space shuttle main engine

ST – (1) flight number of Scout test, e.g., ST-9; (2) model number of a stabilized platform, e.g., ST-124

stabilized platform – A platform used in guidance and control systems that is free to rotate with respect to the rocket but kept stable in space by gyroscopes and gimbals

stage – A component of a rocket or missile containing its own propulsion system and structure. Typically, when one stage has expended its propellants, the next higher stage ignites and the lower stage drops away, reducing the weight to be accelerated to the design speed for the final stage of the vehicle.

Stage-and-a-half – A concept for a missile or launch vehicle in which there is one or more than one booster stage(s) and a sustainer engine, with the booster(s) dropping off part of the way through the launch; different from a two-stage concept in that the boosters are attached to the sides of the sustainer engine rather than below it and burn simultaneously with the sustainer engine for part of its burn time rather than in sequence.

Star – Spherical Thiokol apogee rocket

static testing – Testing of a rocket or rocket system on the ground instead of in flight or on a rocket sled

steel balloon – A propellant tank of the Atlas missile and space-launch vehicle, which had a very thin skin and provided structural support through being inflated with helium, much like a balloon; also used on Centaur

STL – Space Technology Laboratories, successor to the Guided Missile Research Division in the Ramo-Wooldridge Corporation, which merged with Thompson Products to become TRW

strapdown guidance – Guidance provided by a system fixed to the missile or launch-vehicle structure, rather than rotating to maintain a fixed orientation in space like a stabilize platform. Strapdown guidance systems require additional computer power to replace the physical reference provided by a stable platform.

stringer – A longitudinal element to reinforce the skin of an aircraft or rocket structure

submerged nozzle – A nozzle that, instead of extending from the rear of the combustion chamber, is partly or wholly embedded in it. While such a nozzle displaces a small amount of propellant, it also shortens the rocket

and thus reduces its weight and the height necessary for a launch platform.

sustainer – An engine that, after one or more booster engines drop off, stays
with a missile or launch vehicle to carry it and the payload to the designed
speed or to drop off itself if there are upper stages.

SVS – Stage vehicle system

systems engineering – In designing and developing a product or procedure,
the integration of all component systems and other elements, including
the people developing or operating it, so as to achieve the desired goal
most efficiently and effectively

TAT – Thrust-augmented Thor

TEX – Transceiver EXperiment

theoretical specific impulse – A measure of performance for a propellant or
propulsion unit that provides a basis of comparison with other propellants without regard to the particular conditions of employment, such as
altitude

thrust – The force imparted by a rocket engine or motor that impels the
rocket in the desired direction

thrust-to-weight ratio – The amount of thrust per unit of weight of a given
rocket or stage

thrust vector control – Control of the direction of a rocket's or stage's thrust
for purposes of steering

Tiros – Television infrared observation satellite

trajectory – The path of a rocket's or missile's flight

TRW – Thompson Ramo Wooldridge

TV – Test Vehicle

UDMH – Unsymmetrical dimethyl hydrazine, a liquid fuel

USAF – United States Air Force

USN – United States Navy

UTC – United Technology Corporation; United Technology Center

vector – A quantity that has both magnitude and direction; as used here,
primarily the direction

WDD – [Air Force] Western Development Division, which became BMD
on June 1, 1957

WFNA – White fuming nitric acid, a liquid oxidizer

XLR – Experimental liquid rocket (engine)

yaw – The left-right or side-to-side directional motion of the nose of a rocket
flying more or less horizontally

YLR – Operational liquid rocket (engine)

Sources

Archival and Private Sources

Abbreviations

AFFTC/HO	Air Force Flight Test Center History Office, Edwards AFB, Calif.
AFHRA	Air Force Historical Research Agency, Maxwell AFB, Ala.
CSD	Chemical Systems Division, United Technologies Corporation, San Jose, Calif.
JPL	Jet Propulsion Laboratory Archives, Pasadena, Calif.
LC/MD	Library of Congress, Manuscripts Division, Washington, D.C.
MSFC/HO	Marshall Space Flight Center History Office, Huntsville, Ala.
NA	National Archives, Washington, D.C.; College Park, Md.; Laguna Niguel, Calif.
NASM	Smithsonian Institution, National Air and Space Museum, Washington, D.C., and Silver Hill, Md.
NAWCWD	Naval Air Warfare Center, Weapons Division, China Lake, Calif., History Office
NHRC	NASA Historical Reference Collection, NASA Headquarters, Washington, D.C.
OHI	Oral history interview
PI	Private interviews by author or others (to be donated to NHRC)
RATL	Redstone Arsenal Technical Library, Huntsville, Ala.
SMC/HO	[U.S. Air Force] Space and Missiles Systems Center History Office, Los Angeles AFB, Calif.
UP	Unpublished papers (to be donated to NHRC)

Sources

6555th Aerospace Test Group. "Launch Evaluation Report, Titan IIIC—Vehicle No. C-25 (23C-7)." June 20, 1975. File K243.012–132. AFHRA.

———. "Preliminary Report, Launch of Titan IIIC 23C-12 (C-30), 15 Mar 76." April 7, 1976. File K243.012–134. AFHRA.

Adelman, Barnet. OHI by J. D. Hunley. 1996. PI.

Aerojet. "Model Specification." Rocket Propulsion System, Liquid Propellant, AGC Model No. AJ10–104. January 25, 1961. SMC/HO.

"Aerojet AJ10 Series." General Collection, folder B7–020100–01. NASM.

AFBMD. *See* Ballistic Missile Division.

"The Air Force Ballistic Missile Division and the Able Programs." November 27, 1959. "The Able Programs and AFBMD/STL" folder. SMC/HO.

Air Research and Development Command (ARDC). "Space System Development Plan, Able 3 and Able 4 (Earth Satellite, Lunar Satellite, Deep Space Probe) Program, NASA HS-6." 1959. File K243.8636–17. AFHRA.

Alliant Techsystems. "Graphite Epoxy Motor (GEM)." N.d. General Collection, folder B7–031800–01. NASM.

Altman, David. OHI by J. D. Hunley. 1999. PI.

"Apollo Program—All-Up Concept." April 1969. George E. Mueller Collection, box 50, folder 7. LC/MD.

ARDC. *See* Air Research and Development Command.

Arms, W. M. "Thor: The Workhorse of Space—A Narrative History." McDonnell Douglas Astronautics Company, Huntington Beach, Calif. 1972. "Standard Launch Vehicles" files. SMC/HO.

Ballistic Missile Division, U.S. Air Force (AFBMD). "Able Program Progress Report." November 16, 1960. "Program Progress Report—Able 1960" folder. SMC/HO.

———. "ABLE-STAR (AJ10–104)." HQ AFBMD document. August 18, 1961. In "Development of AJ10–104," SMC/HO.

———. "HQ AFBMD Chronology, Space Probes Division, AF Space Boosters." January 1961. "Research & Development—Project Able-5 (Atlas)" folder. SMC/HO.

———. "Progress Report, Able-4 Thor–Pioneer V." 1960. SMC/HO.

———. "Proposed AJ10–104 Final Report." Memo, WDLPM-4 to WDZJP, March 30, 1960. In "Development of AJ10–104," SMC/HO.

Ballistic Missile Organization, History Office Staff. "Chronology of the Ballistic Missile Organization, 1945–1990." August 1993. From private collection of Raymond L. Puffer of Air Force Flight Test Center History Office, Edwards AFB, Calif. UP.

Ballistic Systems Division, U.S. Air Force. "Titan III Launch System Independent Panel Reviews." 1986. Samuel C. Phillips Collection, box 20, folder 8. LC/MD.

Bartley, Charles. Letter to *American Heritage of Invention and Technology*, December 8, 1992. UP.

———. Letter to J. D. Hunley, August 25, 1994. UP.

———. OHI by John Bluth. 1995. JPL.

"Bell Rendezvous." Folder 010680. NHRC.

Bleymaier, J. S. "Factors Relating to the Approval of Titan III Program." 1962. Titan III files. SMC/HO.

"Blue Scout Chronology." Two sections: narrative "Blue Scout History" and dated entries. "History—Blue Scout" folder. SMC/HO.

"Blue Scout History." In "Blue Scout Chronology," SMC/HO.

"Briefing Charts—Able Projects." Folder. SMC/HO.

Bullock, Gilbert D. "The Applications Technology Satellite Program Summary." ATS

Project Office, Goddard Space Flight Center. April 1968. General Collection, "Applications Technology Satellites," folder OA-140000–01. NASM.

Boeing. "Burner II and Burner IIA." News release S-0416. January 29, 1970. General Collection, "Burner II Upper Stage," folder OB-840000–01. NASM.

———. "Burner II for Synchronous Mission Applications." Boeing Aerospace Group, Space Division. June 1967. General Collection, "Burner II Upper Stage," folder OB-840000–02. NASM.

Canney, Howard E., Jr. "Rockets Used by NASA." January 20, 1960. NHRC.

Carlin, Herbert P., et al. "History of the Air Force Materiel Command." 1 Oct. 1995–30 Sept. 1996. Vol. 5. File K226.01. AFHRA.

"Centaur General (1959–89)." Folder 010203. NHRC.

"Centaur Management & Development, Jan. 1961–Mar. 1962." Binder. John L. Sloop Papers, box 22. NHRC.

"Component Research: Guidance." Folder 008768. NHRC.

Convair. "MX-774 Flight Test Report: Jet Controlled Missile." General Collection, folder OM-990774–01, "MX-774 (RTV-A-2, Reports)." NASM.

Convair Division, General Dynamics. "Atlas IIB/Centaur Technical Description: A High-Performance Launch Vehicle for Intelsat VI Class Spacecraft." Report A2 TD-2. April 1983. Microfiche roll E84–10656. SMC/HO.

———. "Atlas Fact Sheet." January 1967. General Collection, folder OA-401060–01, "Atlas Launch Vehicles (SLV-3)." NASM.

———. "Atlas Family of Space Launch Vehicles: Configuration and Performance Summary." Report GDC BNZ-011. Revised May 1972. "Atlas Launch Vehicles—Technical Reports," folder 010191. NHRC.

———. "Summary of Expendable and Reusable Booster Performance and Cost Data," Report GD/C-DCB-65–026. May 26, 1965. File K243.0473–6. AFHRA.

Corliss, William R. "History of the Delta Launch Vehicle." Draft, with comments by L. C. Bruno, September 4, 1973. "Delta Documentation (1959–72)," folder 010246. NHRC.

Delco Electronics Division. "Program 624A—Titan III Inertial Guidance System Preliminary Flight Report, Vehicle C20." May 26, 1971. File K243.012.115. AFHRA.

Delco Systems. "Inertial Guidance System Components." Samuel C. Phillips Collection, box 20, folder 9. LC/MD.

"Delco Systems Operations 1992." George E. Mueller Collection, box 283, folder 4. LC/MD.

"Delta II Becomes New Medium Launch Vehicle." *Astro News* 29, no. 2 (January 23, 1987). Microfilm roll K168.03–2849. AFHRA.

Dembrow, Daniel W., OHI by J. D. Hunley. 1995. PI.

"Development of AJ10–104 (ABLE STAR)." Folder. SMC/HO.

Douglas Aircraft Company. "Saturn IB Payload Planner's Guide." N.d. Bellcom Collection, box 14, folder 6, "Saturn IB." NASM.

Douglas Missile & Space Systems Division. "The Thor History." Douglas report SM-41860. February 1964. Filed under "Standard Launch Vehicles II." SMC/HO.

Ehricke, Krafft A., OHI by [John L. Sloop]. 1974. Folder 010976. NHRC.

"500 Thor Launches." Commemorative brochure. Los Angeles, October 21, 1978. George E. Mueller Collection, box 205, folder 6, "National Aeronautics and Space Administration—Thor." LC/MD.

Felix, Bernard Ross. E-mail comments, January 8, 1997; telephone conversation, February 10, 1997; comments on manuscript, August 22 and 23, 2000. UP and PI.

"Final Report of the Survey Team Established to Investigate the Use of Agena for the National Aeronautics and Space Administration." 1960. History Collection, box JA-12, item 2–954A. JPL.

"Final Report S-IV All-Systems Stage Incident January 24, 1964." Summary, May 11, 1964. George E. Mueller Collection, box 91, folder 14. LC/MD.

Freeman, P. "The Vanguard Control System Development Program." May 1956. General Collection, folder OV-106110–01. NASM.

Geiger, Jeffrey, and Kirk W. Clear. "History of Space and Missile Test Organization and Western Space and Missile Center." November 1, 1987–September 30, 1988. File K241.011. AFHRA.

General Dynamics/Astronautics. "Atlas ICBM Fact Sheet." [January 1962]. General Collection, folder OA-401001–01, "Atlas ICBM (SM-65, HGM-16), Articles, 1957." NASM.

———. "Centaur Primer: An Introduction to Hydrogen-Powered Space Flight." June 1962. In folder 010203, "Centaur General (1959–89)." NHRC.

———. "A Primer of the National Aeronautics and Space Administration's Centaur." N.d. In folder 010203, "Centaur General (1959–89)" NHRC.

General Dynamics Commercial Launch Services. "Mission Planner's Guide for the Atlas Launch Vehicle Family." March 1989. General Collection, folder OA-401095–01, "Atlas I Launch Vehicle." NASM.

General Dynamics/Convair. See Convair Division, General Dynamics.

General Dynamics, Space Systems Division. "Atlas/Centaur: Reliable, Versatile, Available." [1985/6?]. General Collection, folder OA-401070–01, "Atlas Centaur Launch Vehicle." NASM.

Gillam, Isaac. OHI by J. D. Hunley. 1996. PI.

Glasser, Otto J., OHI by John J. Allen. 1984. File K239.0512– 1566. AFHRA.

Goddard Space Flight Center (GSFC). "The Delta Expendable Launch Vehicle." NASA Facts. [1992?]. Folder 010240, "Delta." NHRC.

———. "NASA's Scout Launch Vehicle." NASA Facts. 1992. NHRC.

GSFC. See Goddard Space Flight Center.

Gunn, Charles R. "The Delta and Thor/Agena Launch Vehicles for Scientific and Applications Satellites." GSFC preprint X-470–70–342. 1970. General Collection, folder OD-240014–02, "Delta Launch Vehicle, NASA Report." NASM.

Herzberg, Louis F. "History of the Air Force Rocket Propulsion Laboratory." 1 January–30 June 1964. In "Units, RPL," four unnumbered boxes. AFFTC/HO.

"Highlights from the History of United Technologies." 1988. CSD.

"Hyper Environment Test System (TS 609A)(HETS)." Folder. SMC/HO.

Johnson, V. L. "Delta." In NASA, "Program Review." NHRC.

Kadish, Ronald T. Memo, Col. Kadish, Military Assistant, SAF/AQ, to SAF/OS et al.

on first launch of Pegasus. April 5, 1990. Official Correspondence of USAF Chief of Staff Gen. Larry D. Welch, January 10–June 28, 1990. File K168.03–82. AFHRA.

Lockheed Martin. "Titan IV." March 2000. Titan IV folder, photo storage area. SMC/HO.

LTV Astronautics Division. "The Scout." February 1965. Filed under "Standard Launch Vehicle I." SMC/HO.

Luce, S. "Introduction to Composite Technology." Prepared for Cerritos College Composite Technician Course. N.d. Filed as a book in library. NASM.

Marshall Space Flight Center (MSFC). "Apollo Program Management." Volume 3. December 1967. George E. Mueller Collection, box 64, folder 9. LC/MD.

———, Astrionics Laboratory. "Astrionics System Handbook, Saturn Launch Vehicle." January 2, 1964. Bellcom Collection, box 13, folders 8 and 9. NASM.

———, Saturn Systems Office. "Saturn Illustrated Chronology (April 1957–April 1962)." [1962]. Bellcom Collection, box 13, folder 3. NASM.

Martin Company. "Design Summary, RTV-N-12 Viking, Rockets 1 to 7." January 1954. General Collection, folder OV-550500, "Viking Sounding Rocket." NASM.

———. "The Vanguard Satellite Launching Vehicle: An Engineering Summary." Engineering report 11022. April 1960. Book collection. NHRC.

Martin Marietta Corporation. "Flight Test Objectives/Performance Analysis Titan 23C-4 Postflight Report." File K243.012–126. AFHRA.

———, Astronautics Group. "Titan II Space Launch Vehicle." N.d. General Collection, folder OT-490028–01, "Titan II." NASM.

———. "[Titan IV] System Critical Design Review." 1986. Filed with Titan IV materials. SMC/HO.

———. "Titan IV User's Handbook." December 1988. Microfilm file E87–00454. SMC/HO.

"Minutes of the Saturn V Semi-DCR/AS-502 Evaluation." April 21, 1968. George E. Mueller Collection, box 92, folder 8. LC/MD.

Missile Test Center, U.S. Air Force. "Atlas Fact Sheet." June 20, 1961. General Collection, folder OA-401001–01, "Atlas ICBM (SM-65, HGM-16), Articles, 1957." NASM.

"A Most Reliable Booster System." Booklet. 1977. CSD.

MSFC. See Marshall Space Flight Center.

Mueller, George E. Letter to J. L. Atwood, December 19, 1965. George E. Mueller Collection, box 84, folder 3. LC/MD.

———. Memo with enclosed press release, AAD-2/Dr. Mueller to A/Mr. Webb through AA/Dr. Seamans, October 26, 1963, subject: Reorientation of Apollo Plans. George E. Mueller Collection, box 91, folder 9. LC/MD.

NASA. Agena Program Presentation. [By MSFC at NASA HQ?]. October 1, 1962. History Collection, box JA-36, item 2–2269. JPL.

———. "Apollo 14 Saturn Modified." News release 70–207. December 4, 1970. Folder 010449, "Saturn V, General (1970–)." NHRC.

———. "Centaur Test Demonstrates Precise Guidance." News release 65–271. August 13, 1965. General Collection, folder OA-401070–01, "Atlas Centaur Launch Vehicle." NASM.

———. "First Launch of Centaur Vehicle Scheduled." News release 62–66. April 3, 1962. General Collection, folder OA-401070–01, "Atlas Centaur Launch Vehicle." NASM.

———, Office of Programs. "A History of Vehicles Launched by the National Aeronautics and Space Administration between October 1, 1958 and August 31, 1962." NHRC.

———. "Program Review, Launch Vehicles and Propulsion." June 23, 1962. NHRC.

———. "Saturn V Tests Completed." News release 68–128. July 18, 1968. Folder 010450, "Saturn V, General (1965–1969)." NHRC.

———. Vanguard Division records. Box 1, Project Vanguard Case Files, 1955–1959; box 2, Project Vanguard Case Files, Arnold Engineering and Development Center; box 5, Reports on Project Vanguard, 1955–1959; box 6, Project Vanguard Case Files, Hercules Powder—Allegany Ballistics Laboratory Progress Reports. Record Group (RG) 255, Entry 35. NA (College Park).

"The NASA/Grumman Lunar Module." Pamphlet. General Collection, folder OA-69200–01, "Apollo Lunar Module." NASM.

"NASA Projects in Brief—Agena B." Folder 010194, "Atlas-Agena B & D." NHRC.

Naval Research Laboratory. "Transfer of Project Vanguard to NASA." NRL notice 5400. December 4, 1958. Folder 006633, "Vanguard Documents, Early History." NHRC.

Newell, Homer E., Jr. "Launching of Vanguard III, 18 September 1959." Folder 006643, "Vanguard III (September 18, 1959)." NHRC.

———. Letter, Newell, Office of Space Sciences, NASA, to Gen. Bernard A. Schriever, AFSC, September 27, 1962. History Collection, box JA-150, item 5–1179. JPL.

North American Rockwell, Space Division. "Two Saturn Second Stage Modifications to Be Tested in Space." News release. N.d. Folder 010450, "Saturn V, General (1965–1969)." NHRC.

NRL. *See* Naval Research Laboratory.

Pershing System Description. Volume 1. May 1964. RATL.

Phillips, Samuel C. Draft article for *New York Times*. 1969. Samuel C. Phillips Collection, box 118, folder 4, "Manned Lunar Landing Program, Writings by Phillips." LC/MD.

———. Memo to Dr. Mueller, October 3, 1964. George E. Mueller Collection, box 43, folder 4. LC/MD.

———. Memo, [NASA] MA/S.C. Phillips to M/G.E. Mueller, December 18, 1965, subject: CSM and S-II Review. George E. Mueller Collection, box 84, folder 3. LC/MD.

———. OHI by Frederick I. Ordway. 1988. Samuel C. Phillips Collection, box 138, folder 10. LC/MD.

———. OHIs by Tom Ray. 1970. Folder 001701, "Phillips, Samuel C., Interviews." NHRC.

———. OHI by Robert Sherrod. 1971. Folder 013219, "Phillips, Samuel, Lt. Gen., Apollo Program Director." NHRC.

Piper, Robert F. "History of Titan III, 1961–1963." Air Force monograph. June 1964. SMC/HO.

Pratt & Whitney Division, United Technologies. "RL10 Engine." Brochure. January 1992. Folder 010192, "Atlas II." NHRC.

"Program Test Objectives, Atlas/Able 5." Form STL/OR-60–0000–02011. May 17, 1960. General Collection, folder OA-401062–01. NASM.

"Project Vanguard Report No. 1, Plans, Procedures, and Progress." NRL report 4700. January 13, 1956. Folder 006626, "Vanguard Project Report No. 1." NHRC.

"Project Vanguard Report No. 9, Progress through September 15, 1956." NRL Report 4850. October 4, 1956. Folder 006601, "Vanguard Project: Origins and Progress Reports." NHRC.

"Project Vanguard Report of Progress, Status, and Plans, 1 June 1957." NRL report 4969. Folder 006601, "Vanguard Project: Origins and Progress Reports." NHRC.

"Propulsion J-2." Folder 013782. NHRC.

Ritchey, H. W. "Technical Memoir." c. 1980. UP.

Ritland, O. J. "Able Program Final Progress Report." May 25, 1961. "Program Progress Report—Able 1961" folder. SMC/HO.

Roach, Robert D., Jr. ". . . The Agena Rocket Engine . . . Six Generation of Reliability in Space Propulsion." *Rendezvous* (Bell Aerosystems) 6, no. 6 (1967). In "Bell Rendezvous," NHRC.

Robbins, J. M., and R. W. Feist. "The China Lake Propulsion Laboratories." China Lake, Propulsion System Division, Ridgecrest, Calif. July 1992. NAWCWD.

Rockefeller, Alfred, Jr. "History of Project Able." June 1958. "Project Able History" folder. SMC/HO.

Rosen, Milton W[illiam]. "A Brief History of Delta and Its Relation to Vanguard." Enclosure in Rosen to Constance McL. Green, March 15, 1968. In "Rosen, Milton W.," NHRC.

———. Comments, written (May 8, 2002) and by telephone (May 16–17, 2002), on a draft of this volume's chapter 1. UP.

———. OHI by David DeVorkin. 1983. Space History Division, NASM.

———. "Viking and Vanguard." In "Rocketry in the 1950's, Transcript of AIAA Panel Discussions by Leading Participants." NASA Historical Report 36. October 28, 1971. NHRC.

"Rosen, Milton W. (Misc. Bio)." Folder 001835. NHRC.

Rosen, Milton W., James M. Bridger, and Alton E. Jones. "Rocket Research Report No. XII: The Viking 8 Firing." Naval Research Laboratory, Washington, D.C. 1953. General Collection, folder OM-550500–22. NASM.

Rosen, Milton W., James M. Bridger, and Richard B. Snodgrass. "Rocket Research Report No. XIV: The Viking 9 Firings." Naval Research Laboratory, Washington, D.C. 1954. General Collection, folder OM-550500–22. NASM.

Ross, Chandler C. "Life at Aerojet-General University: A Memoir." Aerojet History Group, 1981. UP.

"SAMSO Chronology." *See* Space Division.

Schubert, W. "Centaur." In NASA, "Program Review," NHRC.

Seltzer, S. M. "Saturn IB/V Astrionics System." N.d. Bellcom Collection, box 13, folder 10. NASM.

Simmons, Ronald L. Biography, résumé, and e-mails. 2002. UP.

"Solid Propulsion Systems." TN 457–72. Naval Weapons Center, China Lake, Calif. "KOB, Solid Propellant Rocket Motors" folder. NAWCWD.

Space and Missile Systems Organization, U.S. Air Force. "Centaur D-1 Payload Users Guide." June 30, 1970. Bellcom Collection, box 12, folder 5. NASM.

Space Division, U.S. Air Force. "Space and Missile Systems Organization: A Chronology, 1954–1979." SMC/HO. Cited as "SAMSO Chronology."

[Space Systems Division, U.S. Air Force]. "Review of the Management Procedures and Decision Making Process of the Department of Defense as Applied to the SLV-5 Program." 1962. Titan III files. SMC/HO.

Space Technology Laboratories (STL). "1958 NASA/USAF Space Probes (ABLE-1), Final Report." Vol. 1 (summary). SMC/HO.

———. "Hard Impact Lunar Flight Experiment." Exhibit 1 to Proposal 26–10, Project Baker. January 27, 1958. Untitled folder. SMC/HO.

———. *STL Space Log.* June 1961. General Collection, folder OS-017000–01, "Samos." NASM.

Stiffler, Ronald-Bel, and Charles V. Eppley. "History of the Air Force Flight Test Center." 1 Jan.–30 June 1960. AFFTC/HO.

———. "History of the Air Force Flight Test Center." 1 Jan.–30 June 1961. AFFTC/HO.

STL. *See* Space Technology Laboratories.

Stuhlinger, Ernst. OHI by J. D. Hunley. 1994. PI.

"Test Stand 1B Final Report." 1961. Folder, "Program Progress Report—Test Stand 1B." AFFTC/HO.

Thackwell, H. L., Jr. "The Application of Solid Propellants to Space Flight Vehicles." 1959. UP.

Thiel, Walter. "On the Practical Possibilities of Further Development of the Liquid Rockets and a Survey of the Tasks to Be Assigned to Research." Memo, Weapons Test Organization, Kummersdorf—Target Range. March 13, 1937. Translated by D. K. Huzel. General Collection, folder CT-168000–01, "Thiel, Walter." NASM.

Thiokol Chemical Corporation. *Aerospace Facts.* House organ, quarterly, Spring 1983. General Collection, folder B7–820030–02, "Thiokol General, Publications, 'Aerospace Facts.'" NASM.

———. "Off-the-Shelf Motor Catalog." N.d. General Collection, folder B7–820060–01, "Thiokol, General, Rocket Specs." NASM.

"Thor Able and Thor Ablestar." Historical Reports, Programs Managed by the Directorate of TRANSIT/ANNA, [AFBMD?]. March 7, 1962. "Research & Development—Transit Project (Able-Star except Transit 1A)" folder. SMC/HO.

"Titan 34D Flight Readiness Meeting." Vandenberg Air Force Base. November 5, 1986. Samuel C. Phillips Collection, box 21, folder 5. LC/MD.

"Titan 34D Recovery Program." 1986. Samuel C. Phillips Collection, box 21, folder 1. LC/MD.

"Twentieth Birthday of Space 'Workhorse.'" *NASA Activities*, May 1979, 16. Folder 010195, "Launch Vehicles: Agena." NHRC.

United Technology Center (UTC). "The 120–Inch-Diameter, Segmented, Solid-Propellant Rocket." Pamphlet. N.d. History Collection, box JA-233, folder 37. JPL.

U.S. Air Force. "Atlas Space Boosters." Fact Sheet 86–40. November 1986. General Collection, folder OA-401060–01, "Atlas Launch Vehicles (SLV-3)." NASM.

[U.S. Army Ordnance Corps/General Electric Company]. "Ordnance Guided Missile & Rocket Programs." Vol. 10, "Hermes Guided Missile Systems, Inception Through 30 June 1955." Technical Report. Filed as a book in library. NASM.

U.S. Navy. Briefing on Polaris management. April 29, 1964. George E. Mueller Collection, box 47, folder 12. LC/MD.

UTC. *See* United Technology Center.

"Vanguard II (Feb. 17, 1959)." Folder 006640. NHRC.

"Vanguard III Satellite Launch Vehicle 7." General Collection, folder OV-106130–01. NASM.

"Vanguard Project, History." General Collection, folder OV-106015–01. NASM.

"Vanguard Project: Origins and Progress Reports." General Collection, folder 006601. NHRC.

"Vanguard Satellite Launching Vehicle 3." General Collection, folder OV-106113–01. NASM.

"Vanguard Test Vehicle 3." Folder 006630. NHRC.

Vought Corporation. "Scout User's Manual." June 1, 1977. General Collection, folder OS-050000–70, "Scout Launch Vehicle Users Manuals, 1977." NASM.

Waldron, Harry N. "A History of the Development of the Inertial Upper Stage for the Space Shuttle, 1969–1985." Draft monograph coordinated but not officially released. 1988. SMC/HO.

Weil, R. "Final Report, Contract No. AF 04(647)-361." Space Technology Laboratories. October 12, 1961. "ABLE 4 & 5" folder. SMC/HO.

Weyland, Herman P., Paul G. Willoughby, Stan Backlund, and J. G. Hill. OHI by J. D. Hunley. 1996. PI.

Wilman, J. Catherine. "Space Division: A Chronology, 1980–1984." SMC/HO.

Published Sources

Besides books and articles, this section includes items "published" on the Internet and papers presented at meetings, which the engineering community counts as publications.

"AbleStar Is Newest Aerojet Triumph: Restartable Space Engine Orbits Two-in-One Payload." *Aerojet Booster* 5, no. 1 (July 1960): 1–2. Copy in folder B7–020030–01, "Aerojet General Publications. . . ." General Collection, NASM.

ACDelco. "About ACDelco." <www.acdelcocanada.com/cmg/pages/history.shtml>.

"Aerojet to Build Able Star, Bigger, Restartable Able." *Missiles and Rockets*, February 15, 1960, 34.

Aeronautics and Space Engineering Board, Assembly of Engineering, National Research Council, Report of Ad Hoc Committee on Liquid Rocket Propulsion Technologies. *Liquid Rocket Propulsion Technology: An Evaluation of NASA's Program.* Washington, D.C.: National Academy Press, 1981. Copy in "NASA, 1981." George E. Mueller Collection, box 198, folder 9. LC/MD.

"Air Force Presses for Early Repeat of Samos Shot." *Missiles and Rockets*, October 17, 1960, 16.

Aitken, Hugh G. J. *The Continuous Wave: Technology and American Radio, 1900–1932.* Princeton, N.J.: Princeton University Press, 1985.

Andrepont, Wilbur C., and Rafael M. Felix. "The History of Large Solid Rocket Motor Development in the United States." Paper AIAA-94–3057, presented at the 30th AIAA/ASME/SAE/ASEE Joint Propulsion Conference and Exhibit, Indianapolis, June 27–29, 1994.

Anselmo, Joseph C. "Air Force Readies Pick of Two EELV Finalists." *Aviation Week & Space Technology*, December 9, 1996, 82–83.

"Aramid Fiber." <www.fibersource.com/f-tutor/aramid.htm>.

Armacost, Michael H. *The Politics of Weapons Innovation: The Thor-Jupiter Controversy.* New York: Columbia University Press, 1969.

ATK Thiokol Propulsion. "Castor 120." <www.thiokol.com/castor3.htm>, accessed December 17, 2002.

"Atlas Launch Vehicle History; Atlas Launch Vehicle Family" *Astronautics and Aeronautics*, November 1975, n.p., between 77 and 78.

Babcock, Elizabeth. *Magnificent Mavericks: Evolution of the Naval Ordnance Test Station from Rockets to Guided Missiles and Underwater Ordnance.* Vol. 3 of *History of the Navy at China Lake, California.* Washington, D.C.: China Lake Museum Foundation, in press.

Baker, David. *Spaceflight and Rocketry: A Chronology.* New York: Facts on File, 1996.

Bartley, Charles E., and Robert G. Bramscher. "Grand Central Rocket Company." In *History of Rocketry and Astronautics: Proceedings of the Twenty-Eighth and Twenty-Ninth History Symposia of the International Academy of Astronautics,* edited by Donald C. Elder and Christopher Rothmund, 267–77. AAS History Series 23. San Diego: Univelt, 2001.

Bedard, Andre. "Composite Solid Propellants." <www.astronautix.com/articles/comlants.htm>.

"Bidirectional Woven Kevlar." <www.aircraftspruce.com/catalog/cmpages/bikevlar.php>.

Biggs, Robert E. "Space Shuttle Main Engine: The First Ten Years." In Doyle, *Liquid Rocket Engine Development,* 69–122.

Bijker, Wiebe E., Thomas P. Hughes, and Trevor J. Pinch, eds. *The Social Construction of Technological Systems: New Directions in the Sociology and History of Technology.* Cambridge, Mass.: MIT Press, 1987.

Bille, Matt, Pat Johnson, Robyn Kane, and Erika R. Lishock. "History and Development of U.S. Small Launch Vehicles." In Launius and Jenkins, *To Reach the High Frontier,* 186–228.

Bille, Matt, and Erika Lishock. *The First Space Race: Launching the World's First Satellites.* College Station: Texas A&M University Press, 2004.

———. "NOTSNIK: The Secret Satellite." Paper AIAA-A02–13657, presented at the 40th AIAA Aerospace Sciences Meeting and Exhibit, Reno, Nev., January 14–17, 2002.

Bilstein, Roger E. "The Saturn Launch Vehicle Family." In *Apollo: Ten Years since Tranquillity Base*, edited by Richard P. Hallion and Tom D. Crouch, 115–23. Washington, D.C.: Smithsonian Institution Press, 1979.

———. *Stages to Saturn: A Technological History of the Apollo/Saturn Launch Vehicles*. SP-4206. Washington, D.C.: NASA, 1980.

———. *Testing Aircraft, Exploring Space: An Illustrated History of NACA and NASA*. Baltimore: Johns Hopkins University Press, 2003.

Boeing. "Delta III Launch Vehicle." <www.boeing.com/defense-space/space/delta/delta3/delta3.htm>.

———. "Delta IV Launch Vehicles." <www.boeing.com/defense-space/space/delta/delta4/delta4.htm>.

———. "Successful First Launch for Boeing Delta IV." Press release, November 20, 2002. <www.spaceref.ca/news/viewpr.html?pid=9864>.

Bonnett, E. W. "A Cost History of the Thor-Delta Launch Vehicle Family." Paper A74–08, presented at the 25th Congress of the International Astronautical Federation, Amsterdam, September 30–October 5, 1974.

Bowles, Mark D. "Eclipsed by Tragedy: The Fated Mating of the Shuttle and Centaur." In Launius and Jenkins, *To Reach the High Frontier*, 415–42.

"Brainpower First in United's Space Venture." *Business Week*, March 5, 1960, 138–44.

Braslow, Albert L. *A History of Suction-Type Laminar-Flow Control with Emphasis on Flight Research*. Monographs in Aerospace History 13. Washington, D.C.: NASA, 1999.

Brennan, William J. "Milestones in Cryogenic Liquid Propellant Rocket Engines." Paper AIAA-67–978, presented at the AIAA 4th Annual Meeting and Technical Display, Anaheim, Calif., October 23–27, 1967.

Broad, William J. "New Age of Precision Brought to Navigation by Modern Gyroscopes." *New York Times*, January 3, 1984.

Bromberg, Joan Lisa. *NASA and the Space Industry*. Baltimore: Johns Hopkins University Press, 1999.

Brooks, Courtney G., James M. Grimwood, and Loyd S. Swenson Jr. *Chariots for Apollo: A History of Manned Lunar Spacecraft*. SP-4205. Washington, D.C.: NASA, 1979.

Butler, Chris. "Reliable Delta to Get More Work." *Missiles and Rockets*, September 24, 1962, 28–29.

Butrica, Andrew. "The Quest for Reusability." In Launius and Jenkins, *To Reach the High Frontier*, 443–69.

———. *Single Stage to Orbit: Politics, Space Technology, and the Quest for Reusable Rocketry*. Baltimore: Johns Hopkins University Press, 2003.

CAIB. *See* Columbia Accident Investigation Board.

Ceruzzi, Paul E. *A History of Modern Computing*. Cambridge, Mass.: MIT Press, 1998.

Chapman, John L. *Atlas: The Story of a Missile*. New York: Harper, 1960.

Chase, Charles A. "IUS Solid Rocket Motors Overview." Paper presented at the JANNAF Propulsion Conference, Monterey, Calif., February 1983.

Chemical Propulsion Information Agency. *CPIA/M1 Rocket Motor Manual*. Vol. 1. Laurel, Md.: CPIA, 1994. Cited as *CPIA/M1*.

———. *CPIA/M5 Liquid Propellant Engine Manual*. Laurel, Md.: CPIA, 1994. Cited as *CPIA/M5*.

Chien, Philip. "The Reliable Workhorse." *Ad Astra*, February 1991, 38–41.

Clark, Evert. "The Moon Program's Business Brain Trust." *Nation's Business*, May 1970, 32–36.

Clark, John D. *Ignition! An Informal History of Liquid Rocket Propellants*. New Brunswick, N.J.: Rutgers University Press, 1972.

Cleary, Mark C. "The Cape: Military Space Operations, 1971–1992." <http://www.globalsecurity.org/space/library/report/1994/cape/PREFACE.htm >.

Colucci, Frank. "Blue Delta." *Space* 3 (May–June 1987): 42–43.

Columbia Accident Investigation Board (CAIB). *Report*, vol. 1 (August 2003). <caib.nasa.gov/news/report/pdf/vol1/full/caib_report_volume1.pdf>.

Covault, Craig. "Atlas V Soars, Market Slumps." *Aviation Week & Space Technology*, August 26, 2002, 22–24.

———. "Boeing Faces Morale, Leadership Issues as Delta IV Nears Critical First Flight." *Aviation Week & Space Technology*, August 26, 2002, 24.

———. "Commercial Winged Booster to Launch Satellites from B-52." *Aviation Week & Space Technology*, June 6, 1988, 14–16.

CPIA. *See* Chemical Propulsion Information Agency.

Crouch, Tom D. *Aiming for the Stars: The Dreamers and Doers of the Space Age*. Washington, D.C.: Smithsonian Institution Press, 1999.

Curry, Robert E., Michael R. Mendenhall, and Bryan Moulton. "In-Flight Evaluation of Aerodynamic Predictions of an Air-Launched Space Booster." TM-104246. Edwards AFB, Calif.: NASA, 1992.

Darling, David. "GRAB." <www.daviddarling.info/encyclopedia/G/GRAB.html>.

———. "MACSAT (Multiple Access Communications Satellite)." <www.daviddarling.info/encyclopedia/M/MACSAT.html>.

Davis, Deane. Correspondence. *Journal of the British Interplanetary Society* 35 (January 1982): 17, 44.

———. "Seeing Is Believing, Or, How the Atlas Rocket Hit Back." *Spaceflight* 25 (May 1983): 196–98.

Dawson, Virginia P. *Engines and Innovation: Lewis Laboratory and American Propulsion Technology*. SP-4306. Washington, D.C.: NASA, 1991.

Dawson, Virginia P., and Mark D. Bowles. *Taming Liquid Hydrogen: The Centaur Upper Stage Rocket, 1958–2002*. SP-2004–4230. Washington, D.C.: NASA, 2004.

Day, Dwayne A. "Corona: America's First Spy Satellite Program." *Quest: The History of Spaceflight Magazine* 4, no. 2 (1995): 4–21.

Defense Metals Information Center, Battelle Memorial Institute. "Report on the Third Maraging Steel Project Review." Memorandum 181. Columbus, Ohio: DMIC, 1963.

Dethloff, Henry C. *Suddenly, Tomorrow Came: A History of the Johnson Space Center, 1957–1990*. SP-4307. Washington, D.C.: NASA, 1993.

DiFrancesco, A., and F. Boorady. "The Agena Rocket Engine Story." Paper AIAA-89–2390, presented at the AIAA/ASME/SAE 25th Joint Propulsion Conference, Monterey, Calif., July 10–13, 1989.

Dooling, Dave. "A Third Stage for Space Shuttle: What Happened to Space Tug?" *Journal of the British Interplanetary Society* 35 (1982): 553–64.

Doyle, Stephen E., ed. *History of Liquid Rocket Engine Development in the United States, 1955–1980.* AAS History Series 13. San Diego: Univelt, 1992.

Dumoulin, Jim. "Space Transportation System Payloads." <science.ksc.nasa.gov/shuttle/technology/sts-newsref/carriers.html>.

Dunar, Andrew J. "Wernher von Braun: A Visionary as Engineer and Manager." In *Realizing the Dream of Flight: Biographical Essays in Honor of the Centennial of Flight, 1903–2003,* edited by Virginia P. Dawson and Mark D. Bowles, 185–212. SP-2005–4112. Washington, D.C.: NASA, 2005.

Dunar, Andrew J., and Stephen P. Waring. *Power to Explore: A History of Marshall Space Flight Center, 1960–1990.* SP-4313. Washington, D.C.: NASA, 1999.

Dunn, W. Paul. "Evolution of the Inertial Upper Stage." *Crosslink* (Aerospace Corporation), Winter 2003. <www.aero.org/publications/crosslink/winter2003/08.html>.

Dyer, Davis, and David B. Sicilia. *Labors of a Modern Hercules: The Evolution of a Chemical Company.* Boston: Harvard Business School Press, 1990.

"EELV Evolved Expendable Launch Vehicle." <www.globalsecurity.org/space/systems/eelv.htm>.

"The Experimental Engines Group." Interview with [W. F.] Bill Ezell, [C. A.] Cliff Hauenstein, [J. O.] Jim Bates, [G. S.] Stan Bell, and [R.] Dick Schwarz. *Threshold: An Engineering Journal of Power Technology* (Rocketdyne), no. 4 (Spring 1989): 21–27.

Ezell, Linda Neuman. *NASA Historical Data Book.* Vol. 2, *Programs and Projects, 1958–1968.* SP-4012. Washington, D.C.: NASA, 1988.

———. *NASA Historical Data Book.* Vol. 3, *Programs and Projects, 1969–1978.* SP-4012. Washington, D.C.: NASA, 1988.

FAS [Federation of American Scientists] Space Policy Project. "Delta." <www.fas.org/spp/military/program/launch/delta.htm>.

———. "Inertial Upper Stage." <www.fas.org/spp/military/program/launch/ius.htm>.

———. "Skynet 4." <www.fas.org/spp/guide/uk/military/comm/skynet_4.htm>.

———. "Titan-4 Launch History." <www.fas.org/spp/military/program/launch/t4table.htm>.

———. "White Cloud Naval Ocean Surveillance System." <www.fas.org/spp/military/program/surveill/noss.htm>.

Feld, Dave. "Agena Engine." Paper 1412–60, presented at the 15th American Rocket Society annual meeting, Washington, D.C., December 5–8, 1960.

Ferguson, Eugene S. *Engineering and the Mind's Eye.* Cambridge, Mass.: MIT Press, 1992.

Fink, Donald E. "Titan 34D Booster Design Completed." *Aviation Week & Space Technology,* May 15, 1978, 48, 51.

Forsyth, Kevin S. "Delta: The Ultimate Thor." In Launius and Jenkins, *To Reach the High Frontier,* 103–46.

Friedman, Norman. *Seapower and Space: From the Dawn of the Missile Age to Net-Centric Warfare.* Annapolis, Md.: Naval Institute Press, 2000.

Friedman, S. Morgan. "The Inflation Calculator." <www.westegg.com/inflation/infl.cgi>.

Fuller, Paul N., and Henry M. Minami. "History of the Thor/Delta Booster Engines." In Doyle, *Liquid Rocket Engine Development*, 39–51.

Furth, F. R. "Project Vanguard—The IGY Earth Satellite." Paper presented at the 24th Annual Meeting of the Institute of the Aeronautical Sciences, January 23–26, 1956. Copy in folder OV-106110–01, "Vanguard Satellite Launch Vehicle." NASM.

Geddes, Philip. "Centaur: How It Was Put Back on Track." *Aerospace Management* 7 (April 1964): 24–29.

Gibson, Cecil R., and James A. Wood. "Apollo Experience Report—Service Propulsion Subsystem." TN D-7375. Washington, D.C.: NASA, August 1973. <http://72.14.253.104/search?q=cache:2gWzFDpCphUJ:ntrs.nasa.gov/archive/nasa/casi.ntrs.nasa.gov/19730023031_1973023031.pdf+%22Apollo+Experience+Report—Service+Propulsion+Subsystem%22&hl=en&ct=clnk&cd=1&gl=us>.

Gibson, James N. *The Navaho Missile Project: The Story of the "Know-How" Missile of American Rocketry.* Atglen, Pa.: Schiffer, 1996.

Glennan, T. Keith. *The Birth of NASA: The Diary of T. Keith Glennan.* Edited by J. D. Hunley. Introduction by Roger D. Launius. SP-4104. Washington, D.C.: NASA, 1993.

Goodstein, Judith R. *Millikan's School: A History of the California Institute of Technology.* New York: Norton, 1991.

Gordon, Robert, et al. *Aerojet: The Creative Company.* Los Angeles: Stuart F. Cooper, 1995.

Gorn, Michael H. *Expanding the Envelope: Flight Research at NACA and NASA.* Lexington: University Press of Kentucky, 2001.

———. *Harnessing the Genie: Science and Technology Forecasting for the Air Force, 1944–1986.* Washington, D.C.: Office of Air Force History, 1988.

———. *The Universal Man: Theodore von Kármán's Life in Aeronautics.* Washington, D.C.: Smithsonian Institution Press, 1992.

Gray, Mike. *Angle of Attack: Harrison Storms and the Race to the Moon.* New York: Norton, 1992.

Green, Constance McLaughlin, and Milton Lomask. *Vanguard: A History.* Washington, D.C.: Smithsonian Institution Press, 1971.

Green, Joseph, and Fuller C. Jones. "The Bugs That Live at -423°." *Analog: Science Fiction, Science Fact* 80, no. 5 (January 1968): 8–41.

Griffiths, G. H., and A. M. Angus. "Low Cost Guidance." *Space* 5 (January–February 1989): 18–19.

Guill, J. H., W. H. Sterbentz, and L. W. Gordy. "The Agena: Transportation Workhorse of Space." Paper 74–079, presented at the 25th Congress of the International Astronautical Federation, Amsterdam, September 30–October 5, 1974.

Guilmartin, John F., Jr., and John Walker Mauer. *A Shuttle Chronology, 1964–1973: Abstract Concepts to Letter Contracts.* 5 vols. JSC-23309. Houston: JSC, 1988.

Haeussermann, Walter. "Developments in the Field of Automatic Guidance and Control of Rockets." *Journal of Guidance and Control* 4, no. 3 (May–June 1981): 225–39.

Hagen, John P. "The Viking and the Vanguard." In *The History of Rocket Technology: Essays on Research, Development, and Utility*, edited by Eugene M. Emme, 122–41. Detroit: Wayne State University Press, 1964.

Hale, N. Wayne, Jr. "Accepting Risk—Some Thoughts for Safety Day." <atc.nasa.gov/host-edEvents/rmc5/Letter%20from%20Wayne%20Hale%20–%20Accepting%20Risk.pdf.>

Hall, R. Cargill. "Earth Satellites, A First Look by the United States Navy." In R. C. Hall, *Rocketry and Astronautics* (3rd–6th Symposia), 2: 253–77.

———, ed. *Essays on the History of Rocketry and Astronautics: Proceedings of the Third Through the Sixth History Symposia of the International Academy of Astronautics.* 2 vols. CP-2014. Washington, D.C.: NASA, 1977. Reprinted as *History of Rocketry and Astronautics: Proceedings of the Third through Sixth Symposia of the International Academy of Astronautics. AAS History Series.* Parts 1 and 2. San Diego: Univelt, 1986.

———. *Lunar Impact: A History of Project Ranger.* SP-4210. Washington, D.C.: NASA, 1977.

———. "Origins and Development of the Vanguard and Explorer Satellite Programs." *Airpower Historian* 11, no. 1 (January 1964): 100–112.

———. "Project Ranger: A Chronology." JPL/HR-2. Pasadena: Jet Propulsion Laboratory, Caltech, 1971.

Hallion, Richard P., and Michael H. Gorn. *On the Frontier: Experimental Flight at NASA Dryden.* Washington, D.C.: Smithsonian Institution Press, 2003.

Hansen, James R. *Engineer in Charge: A History of the Langley Aeronautical Laboratory, 1917–1958.* SP-4305. Washington, D.C.: NASA, 1987.

———. "Learning through Failure: NASA's Scout Rocket." *National Forum* 81, no. 1 (Winter 2001): 18–23.

———. *Spaceflight Revolution: NASA Langley Research Center from Sputnik to Apollo.* SP-4308. Washington, D.C.: NASA, 1995.

Harwood, William B. *Raise Heaven and Earth: The Story of Martin Marietta People and Their Pioneering Achievements.* New York: Simon and Schuster, 1993.

Hatton, Jason P. "Titan Transtage Rockets." <members.aol.com/_ht_a/hattonjasonp/hasohp/TRANSTG.HTML>.

Hawkes, Russell. "United Technology Builds Larger Booster Capabilities." *Aviation Week,* April 10, 1961, 56–65.

Heald, Dan. "LH$_2$ Technology Was Pioneered on Centaur 30 Years Ago." In *History of Rocketry and Astronautics, Proceedings of the Twenty-Sixth History Symposium of the International Academy of Astronautics,* edited by Philippe Jung, 205–21. AAS History Series 21. San Diego: Univelt, 1997.

Heppenheimer, T. A. *Countdown: A History of Space Flight.* New York: Wiley, 1997.

———. *Development of the Space Shuttle, 1972–1981.* Vol. 2 of *History of the Space Shuttle.* Washington, D.C.: Smithsonian Institution Press, 2002.

———. *The Space Shuttle Decision: NASA's Search for a Reusable Space Vehicle.* SP-4221. Washington, D.C.: NASA, 1999. Republished as vol. 1 of *History of the Space Shuttle* (Washington, D.C.: Smithsonian Institution Press, 2002).

Herring, Mack R. *Way Station to Space: A History of the John C. Stennis Space Center.* SP-4310. Washington, D.C.: NASA, 1997.

Heusinger, B. K. "Saturn Propulsion Improvements." *Astronautics & Aeronautics,* August 1964, 20–25.

Hevly, Bruce. "The Tools of Science: Radio, Rockets, and the Science of Naval Warfare." In *National Military Establishments and the Advancement of Science and Technology*, edited by Paul Forman and José M. Sánchez-Ron, 215–32. Dordrecht, Netherlands: Kluwer, 1996.

Heyman, Jos. "Directory of U.S. Military Rockets and Missiles." Appendix 3: "Space Vehicles, SLV-4/SLV-5/SB-4/SB-5/SB-6." <www.designation-systems.net/dusrm/app3/b-6.html>.

Holmes, Jay. "*Able-Star* Makes Technological First." *Missiles and Rockets*, April 25, 1960, 44.

———. "ABL's Altair Runs Up 13–13 Record." *Missiles and Rockets*, June 6, 1960, 29–30.

Hunley, J. D. "The Evolution of Large Solid Propellant Rocketry in the United States." *Quest: The History of Spaceflight Quarterly* 6, no. 1 (1998): 22–38.

———. "Fifty Years of Flight Research at NASA Dryden." In *The Meaning of Flight in the 20th Century: Conference Proceedings, 1998 National Aerospace Conference*, 196–203. Dayton, Ohio: Wright State University, 1999.

———. "Minuteman and the Development of Solid-Rocket Launch Technology." In Launius and Jenkins, *To Reach the High Frontier*, 229–300.

———. *Preludes to U.S. Space Launch Vehicle Technology: Goddard Rockets to Minuteman III*. Gainesville: University Press of Florida, 2008.

Huzel, Dieter K., and David H. Huang. *Design of Liquid Propellant Rocket Engines*. SP-125. Washington, D.C.: NASA, 1967.

———. *Modern Engineering for Design of Liquid-Propellant Rocket Engines*. Revised and updated by Harry Arbit et al. Washington, D.C.: AIAA, 1992.

Iliff, Kenneth W. "Parameter Estimation for Flight Vehicles." *Journal of Guidance, Control, and Dynamics*, September–October 1989, 609–22.

Iliff, Kenneth W., and Curtis L. Peebles. *From Runway to Orbit: Reflections of a NASA Engineer*. SP-2004–4109. Washington, D.C.: NASA, 2004.

"In the Vanguard: First Inertial Guidance System for Satellite Rocket." *Aviation Research & Development*, October 1956, 25.

Isakowitz, Steven J. *International Reference Guide to Space Launch Systems*. Washington, D.C.: AIAA, 1991.

———. *International Reference Guide to Space Launch Systems*. 2nd ed. Updated by Jeff Samella. Washington, D.C.: AIAA, 1995.

Jacob, Margaret C. "Science Studies after Social Construction: The Turn toward the Comparative and the Global." In *Beyond the Cultural Turn: New Directions in the Study of Society and Culture*, edited by Victoria E. Bonnell and Lynn Hunt, 95–120. Berkeley and Los Angeles: University of California Press, 1999.

Jaffe, Leonard. *Communications in Space*. New York: Holt, Rinehart and Winston, 1967.

Jaqua, Vance, and Allan Ferrenberg. "The Art of Injector Design." *Threshold: An Engineering Journal of Power Technology* (Rocketdyne), no. 4 (Spring 1989): 2–11.

Jenkins, Dennis R. "Broken in Midstride: Space Shuttle as a Launch Vehicle." In Launius and Jenkins, *To Reach the High Frontier*, 357–414.

———. *Space Shuttle: The History of the National Space Transportation System, the First 100 Missions*. 3rd ed. Cape Canaveral, Fla.: D. R. Jenkins, 2001.

Jet Propulsion Laboratory (JPL). *Mariner-Mars 1964, Final Project Report.* SP-139. Washington, D.C.: NASA, 1967.

———. *Mariner-Venus 1962, Final Project Report.* SP-59. Washington, D.C.: NASA, 1965.

Johnson, John, Jr. "NASA Chief Isolates Likely Cause of Shuttle Trouble." *Los Angeles Times*, December 3, 2005.

Johnson, Stephen B. "Samuel Phillips and the Taming of Apollo." *Technology and Culture* 42 (2001): 685–709.

———. *The United States Air Force and the Culture of Innovation, 1945–1965.* Washington, D.C.: Air Force History and Museums Program, 2002.

Joliff, Elizabeth C. "History of the Pershing Weapon System." Historical monograph AMC 76M. Unclassified portions only. Redstone Arsenal, Ala.: Army Missile Command, 1974.

Jones, Bryn. "The History of Astronomical Research in the University of Wales." <bryn-jones.members.beeb.net/wastronhist/univwaleshist.html>.

Jortner, Julius. "Analysis of Transient Thermal Responses in a Carbon-Carbon Composite." In *Thermomechanical Behavior of High-Temperature Composites*, edited by Julius Jortner, 19. New York: American Society of Mechanical Engineers, 1982.

JPL. *See* Jet Propulsion Laboratory.

Judge, John F. "Westinghouse Know-How Speeds Flow of 120–in. Cases for Titan III-C's." *Missiles and Rockets*, September 14, 1964, 32–33.

Kennedy, Gregory P. *Vengeance Weapon 2: The V-2 Guided Missile.* Washington, D.C.: Smithsonian Institution Press, 1983.

Kennedy, W. S., S. M. Kovacic, and E. C. Rea. "Solid Rocket History at TRW Ballistic Missiles Division." Paper presented at the AIAA/SAE/ASME/ASEE 28th Joint Propulsion Conference, Nashville, July 6–8, 1992.

Kit, Boris, and Douglas S. Evered. *Rocket Propellant Handbook.* New York: Macmillan, 1960.

Klager, Karl. "Segmented Rocket Demonstration: Historic Development Prior to Their Use as Space Boosters." In *History of Rocketry and Aeronautics: Proceedings of the Twenty-Fifth History Symposium of the International Academy of Astronautics*, edited by J. D. Hunley, 159–70. AAS History Series 20. San Diego: Univelt, 1997.

Knauf, James M., Linda R. Drake, and Peter L. Portanova. "EELV: Evolving toward Affordability." *Aerospace America*, March 2002, 38, 40–42.

Kork, Jyri, and William R. Schindler. "The Thor-Delta Launch Vehicle: Past and Future." Paper SD 32 presented at the 19th Congress of the International Astronautical Federation, New York, October 13–19, 1968.

Kraemer, Robert S. *Rocketdyne: Powering Humans into Space.* With Vince Wheelock. Reston, Va.: AIAA, 2006.

Krebs, Gunter. "Blue Scout-1." <www.skyrocket.de/space/doc_lau_det/blue-scout-1.htm>.

———. "Blue Scout-2." <www.skyrocket.de/space/doc_lau_det/blue-scout-2.htm>.

———. "Blue Scout Junior (SRM-91, MER-6)." <www.skyrocket.de/space/doc_lau/blue_scout_jr.htm>.

——— . "Commercial Titan-3." <www.skyrocket.de/space/doc_lau_det/titan-3_com-mercial.htm>, accessed July 10, 2003.

———. "Intelsat-6 (601, 602, 603, 604, 605)." <www.skyrocket.de/space/doc_sdat/in-telsat-6.htm>.

———. "Scout." <www.skyrocket.de/space/doc_lau_fam/scout.htm>.

——— . "Scout-E1." <space.skyrocket.de/index_frame.htm?http://www.skyrocket.de/space/doc_lau_fam/scout.htm>.

———. "Skynet 4A, 4B, 4C, 4D, 4E, 4F." <www.skyrocket.de/space/doc_sdat/skynet-4.htm>.

——— . "Solrad 1, 2, 3, 4A, 4B/Grab (Dyno)." <space.skyrocket.de/index_frame.htm?http://www.skyrocket.de/ space/doc_sdat/solrad-1.htm >.

——— . "Titan-34D Transtage." <www.skyrocket.de/space/doc_lau_det/titan-34d_transtage.htm>.

———. "Transtage." <www.skyrocket.de/space/doc_stage/transtage.htm>.

———. "USA Military Satellites." <www.skyrocket.de/space/doc_sat/usa_secr.htm>.

Kurten, Pat M. "Apollo Experience Report—Guidance and Control Systems: Lunar Module Abort Guidance System." TN D-7990. Washington, D.C.: NASA, 1975.

Lambright, W. Henry. *Powering Apollo: James E. Webb of NASA.* Baltimore: Johns Hopkins University Press, 1995.

Latour, Bruno. *Science in Action: How to Follow Scientists and Engineers Through Society.* Cambridge, Mass.: Harvard University Press, 1987.

Launius, Roger D. *NASA: A History of the U.S. Civil Space Program.* Malabar, Fl.: Krieger, 1994.

——— . "NASA and the Decision to Build the Space Shuttle, 1969–72." *Historian* 57 (Autumn 1994): 17–34.

——— . "Titan: Some Heavy Lifting Required." In Launius and Jenkins, *To Reach the High Frontier,* 147–85.

Launius, Roger D., and Dennis R. Jenkins, eds. *To Reach the High Frontier: A History of U.S. Launch Vehicles.* Lexington: University Press of Kentucky, 2002.

Layton, Edwin T. "Mirror-Image Twins: The Communities of Science and Technology in 19th Century America." *Technology and Culture* 12 (1971): 562–80.

——— . "Presidential Address: Through the Looking Glass, or News from Lake Mirror Image." *Technology and Culture* 28 (1987): 594–607.

——— . "Technology as Knowledge." *Technology and Culture* 15 (1974): 31–41.

Leiss, Abraham. "Scout Launch Vehicle Program, Phase 6, Part 1." CR-165950. Langley, Va.: NASA, 1982. Available through NASA libraries on microfiche X82–10346.

Lethbridge, Cliff. "Blue Scout Junior Fact Sheet." <www.spaceline.org/rocketsum/blue-scout-junior.html>.

——— . "Commercial Titan III Fact Sheet." <www.spaceline.org/rocketsum/commer-cial-titan-III.html>.

——— . "Titan IVA Fact Sheet." 2001. <www.spaceline.org/rocketsum/titan-IV.html>.

Levine, Arnold S. *Managing NASA in the Apollo Era.* SP-4102. Washington, D.C.: NASA, 1982.

Library of Congress, Science and Technology Division. *Astronautics and Aeronautics,*

1966: Chronology on Science, Technology, and Policy. SP-4007. Washington, D.C.: NASA, 1967.

———. *Astronautics and Aeronautics, 1968: Chronology on Science, Technology, and Policy.* SP-4010. Washington, D.C.: NASA, 1969.

———. *Astronautics and Aeronautics, 1969: Chronology on Science, Technology, and Policy.* SP-4014. Washington, D.C.: NASA, 1970.

———. *Astronautics and Aeronautics, 1970: Chronology on Science, Technology, and Policy.* SP-4015. Washington, D.C.: NASA, 1972.

Lindsey, Robert. "UTC Chief Sees Tighter Rocket Market." *Missiles and Rockets,* April 18, 1966, 22–23.

———. "UTC Solves Scout Stage Spin Problem." *Missiles and Rockets,* August 30, 1965, 28–31.

Liquid Propellant Information Agency. *Liquid Propellant Safety Manual.* Silver Spring, Md.: Applied Physics Laboratory, Johns Hopkins University, 1958.

Logsdon, John M. *The Decision to Go to the Moon: Project Apollo and the National Interest.* Cambridge, Mass.: MIT Press, 1970.

Logsdon, John M., et al., eds. *Exploring the Unknown: Selected Documents in the History of the U.S. Civil Space Program,* vol. 4. SP-4407. Washington, D.C.: NASA, 1999.

Loory, Stuart H. "Quality Control . . . and Success." *New York Herald Tribune,* April 21, 1963.

Mack, Pamela E., ed. *From Engineering Science to Big Science: The NACA and NASA Collier Trophy Research Project Winners.* SP-4219. Washington, D.C.: NASA, 1998.

Maine, Richard E., and Kenneth W. Iliff. *Applications of Parameter Estimation to Aircraft Stability and Control: The Output-Error Approach.* RF-1168. Washington, D.C.: NASA, 1986.

Malm, Robert. "For Want of a Nail." <www.robertemalm.com/for%20want%20of%20 a%20nail.htm>.

Marcus, Gideon. "The Pioneer Rocket." *Quest: The History of Spaceflight Quarterly* 13, no. 4 (2006): 26–30.

Marshall, Andrew C. *Composite Basics.* 3rd ed.; Walnut Creek, Calif.: Marshall Consulting, 1993.

Martin, Richard E. "The Atlas and Centaur 'Steel Balloon' Tanks: A Legacy of Karel Bossart." Paper IAA-89–738, presented at the 40th Congress of the International Astronautical Federation, Málaga, October 7–13, 1989.

———. "A Brief History of the *Atlas* Rocket Vehicle." 3 parts. *Quest: The History of Spaceflight Quarterly* 8, no. 2 (2000): 54–61; no. 3 (2000): 40–45; no. 4 (2000): 46–51.

Martinez, Al. "Rocket Engine Propulsion Power Cycles." *Threshold: An Engineering Journal of Power Technology* (Rocketdyne), no. 7 (Summer 1991): 14–26.

McCool, A. A., and Keith B. Chandler. "Development Trends of Liquid Propellant Engines." In *From Peenemünde to Outer Space,* edited by Ernst Stuhlinger et al., 289–306. Huntsville, Ala.: NASA Marshall Space Flight Center, 1962.

McDougall, Walter A. *. . . the Heavens and the Earth: A Political History of the Space Age.* New York: Basic Books, 1985.

McDowell, Jonathan. Agena missions table. <www.planet4589.org/space/rockets/liq-uid/US/Agena.new>.

———. "BE-3." <www.planet4589.org/space/book/lv/engines/motorlist/be3.html>.

———. "The Present Day." <www.planet4589.org/space/book/lv/engines/kick/THEP-RESENTDAY.html>.

———. "The Scout Launch Vehicle." *Journal of the British Interplanetary Society* 47 (March 1994): 99–108.

———. "Star 20." <www.planet4589.org/space/book/lv/engines/motorlist/star20.html>.

———. "US Reconnaissance Satellite Programs, Part 1: Photoreconnaissance." *Quest: The History of Spaceflight Magazine* 4, no. 2 (1995): 22–33.

McGraw-Hill Encyclopedia of Science and Technology. 9th ed. 20 vols. New York: Mc-Graw-Hill, 2002.

Meyer, Robert R., Jr., and Jack Barneburg. "In-Flight Rain Damage Tests of the Shuttle Thermal Protection System." NASA TM-100438. May 1988. <www.nasa.gov/centers/dryden/pdf/88136main_H-1484.pdf>.

Meyers, J. F. "Delta II—A New Era Under Way." IAF paper 89–196, presented at the 40th Congress of the International Astronautical Federation, Málaga, October 7–12, 1989.

"Military EELV Launches Could Save $10 Billion." *Aviation Week & Space Technology*, August 26, 2002, 23.

Moore, T. L. "Solid Rocket Development at Allegany Ballistics Laboratory." Paper AIAA-99–2931 presented at the 35th Joint Propulsion Conference, Los Angeles, June 20–24, 1999.

Murray, Charles, and Catherine Bly Cox. *Apollo: The Race to the Moon.* New York: Simon and Schuster, 1989.

NASA. "Constellation Program." <www.nasa.gov/mission_pages/constellation/main/index.html> and associated links, accessed July 31, 2007.

———. "Defense Meteorological Satellites Program." <heasarc.nasa.gov/docs/heasarc/missions/dmsp.html>.

———. "NASA and Martin Marietta Sign Commercial Launch Agreement." News release 88–137, November 18, 1988. <spacelink.nasa.gov/NASA.News/NASA.News.Releases/Previous.News.Releases. . ./88–10–09>, accessed July 8, 2003.

———. *National Space Transportation System Reference.* Vol. 1, *Systems and Facilities.* Washington, D.C.: NASA, 1988.

———. "NSTS 1988 News Reference Manual." <science.ksc.nasa.gov/shuttle/technology/sts-newsref/>.

———. "Saturn IB News Reference." MSFC and contractors. December 1965 (changed September 1968). <http://www.boggsspace.com/mall/sib.asp>.

———. "Saturn V News Reference." MSFC, KSC, and contractors. August 1967 (portions changed December 1968). <history.msfc.nasa.gov/saturn_apollo/saturnv_press_kit.htm>.

———. "Scout Launch Vehicle to Retire after 34 Years of Service." News release 94–72. <www.nasa.gov/home/hqnews/1994/94–072.txt>.

"NASA Awards Scout Contracts." *Aviation Week*, March 2, 1959, 22.

NASA Education Division. *Rockets: Physical Science Teacher's Guide with Activities.* EP-291. Washington, D.C.: NASA, 1993.

"NASA Weighing 2 Rocketdyne Engines." *Hartford Courant*, March 23, 2006.

Neufeld, Jacob. *The Development of Ballistic Missiles in the United States Air Force, 1945–1960.* Washington, D.C.: Office of Air Force History, 1990.

Neufeld, Michael J. "Orbiter, Overflight, and the First Satellite: New Light on the Vanguard Decision." In *Reconsidering Sputnik: Forty Years Since the Soviet Satellite*, edited by Roger D. Launius, John M. Logsdon, and Robert W. Smith, 231–57. Amsterdam: Harwood, 2000.

Newell, Homer E. *Sounding Rockets.* New York: McGraw-Hill, 1959.

Newlon, Clarke. "Krafft Ehricke." In *Rocket and Missile Technology*, edited by Gene Gurney, 86–91. New York: Franklin Watts, 1964.

"NOSS Triplets: Naval Ocean Surveillance System—2nd Generation Satellites." <satobs.org/noss.html>, accessed July 11, 2003.

Noyes, W. A., Jr., ed. *Chemistry: A History of the Chemistry Components of the National Defense Research Committee, 1940–1946.* Boston: Little, Brown, 1948.

NSSDC (National Space Science Data Center). "Spacecraft AS-202." <nssdc.gsfc.gov/database/MasterCatalog?sc=APST202>.

Orbital Sciences Corporation. "Pegasus Mission History." <www.orbital.com/SpaceLaunch/Pegasus/pegasus_history.htm>.

Orloff, Richard W. *Apollo by the Numbers: A Statistical Reference.* SP-2000–4029. Washington, D.C.: NASA, 2000.

Osborne, George H., Robert Gordon, and Herman L. Coplen, with George S. James. "Liquid-Hydrogen Rocket Engine Development at Aerojet, 1944–1950." In R. C. Hall, *Rocketry and Astronautics* (3rd–6th Symposia), 2: 279–324.

Pae, Peter. "Delta IV's First Launch Provides a Lift to Boeing." *Los Angeles Times*, November 21, 2002.

Peebles, Curtis. *The Corona Project: America's First Spy Satellites.* Annapolis, Md.: Naval Institute Press, 1997.

———. *Guardians: Strategic Reconnaissance Satellites.* Novato, Calif.: Presidio, 1987.

Pike, Iain. "Atlas: Pioneer ICBM and Space-Age Workhorse." *Flight International* 81 (January 1962): 89–96; 82 (February 1962): 175–79.

"Pioneer Project." <http://www.nasa.gov/centers/ames/missions/archive/pioneer.html >.

Porcelli, G., and E. Vogtel. "Modular, Spin-Stabilized, Tandem Solid Rocket Upper Stage." *Journal of Spacecraft and Rockets* 16, no. 5 (September–October 1979): 338–42.

Povinelli, Louis A. "Particulate Damping in Solid-Propellant Combustion Instability." *AIAA Journal* 5, no. 10 (October 1967): 1791–96.

Powell, Joel W. "The NOTS Air-Launched Satellite Programme." *Journal of the British Interplanetary Society* 50 (1997): 433–40.

———. "Thor-Able and Atlas-Able." *Journal of the British Interplanetary Society* 37 (May 1984): 219–25.

Powell, Joel W., and Lee Caldwell. "New Space Careers for Former Military Missiles." *Spaceflight* 32 (April 1990): 124–25.

Powell, Joel W., and G. R. Richards. "The Atlas E/F Launch Vehicle—An Unsung Work-horse." *Journal of the British Interplanetary Society* 44 (1991): 229–40.

———. "The Orbiting Vehicle Series of Satellites." *Journal of the British Interplanetary Society* 40 (1987): 417–21, 426.

Powell, Robert M. "Evolution of Standard Agena: Corona's Spacecraft." In *Corona: Between the Sun and the Earth; The First NRO Reconnaissance Eye in Space*, edited by Robert A. McDonald, 121–32. Bethesda, Md.: American Society for Photogrammetry and Remote Sensing, 1997.

President. *See* U.S. President

Price, E. W. "Solid Rocket Combustion Instability—An American Historical Account." In *Nonsteady Burning and Combustion Stability of Solid Propellants*, edited by Luigi De Luca, Edward W. Price, and Martin Summerfield, 1–16. Washington, D.C.: AIAA, 1992.

Reddy, Francis. "NASA Christens Its New Rockets." *Astronomy*, July 5, 2006. <www.astronomy.com/asy/default.aspx?c=a&id=4379>.

Richards, G. R., and Joel W. Powell. "The Centaur Vehicle." *Journal of the British Interplanetary Society* 42 (March 1989): 99–120.

———. "Titan 3 and Titan 4 Space Launch Vehicles." *Journal of the British Interplanetary Society* 46 (1993): 123–44.

Robertson, Donald F. "Centaur Canters On." *Space* 7 (March–April 1991): 22–25.

"Rocket Performance: Mass." <www.allstar.fiu.edu/aerojava/rocket5.htm>.

Rockwell International. "Press Information: Space Shuttle Transportation System." January 1984. Copy at NHRC.

Rosen, Milton W. *The Viking Rocket Story*. New York: Harper, 1955.

———. "What Have We Learned from Vanguard?" Paper 719–58, presented at the American Rocket Society 13th annual meeting, New York, November 17–21, 1958. Copy in "Vanguard Satellite Launching Vehicle 3," General Collection, folder OV-106113–01. NASM.

Rumerman, Judy A. "Fairchild Aviation Corporation." <www.centennialofflight.gov/essay/Aerospace/Fairchild/Aero25.htm>.

———. *NASA Historical Data Book*. Vol. 5, *NASA Launch Systems, Space Transportation, Human Spaceflight, and Space Science, 1979–1988*. SP-4012. Washington, D.C.: NASA, 1999.

Scala, Keith J. "The Viking Rocket: Filling the Gap." *Quest: The Magazine of Spaceflight* 2, no. 4 (1993): 34–37.

Schwiebert, Ernest G. *A History of the U.S. Air Force Ballistic Missiles*. New York: Frederick A. Praeger, [1965].

Senate. *See* U.S. Congress. Senate.

Shapley, Deborah. *Promise and Power: The Life and Times of Robert McNamara*. Boston: Little, Brown, 1993.

Shortal, Joseph Adams. *A New Dimension: Wallops Island Flight Test Range, the First Fifteen Years*. RF-1028. Washington, D.C.: NASA, 1978.

Siddiqi, Asif A. *Challenge to Apollo: The Soviet Union and the Space Race, 1945–1974*. SP-2000–4408. Washington, D.C.: NASA, 2000. Later published in two volumes as *Sputnik and the Soviet Space Challenge* and *The Soviet Space Race with Apollo* (Gainsville: University Press of Florida, 2003).

———. *Deep Space Chronicle: A Chronology of Deep Space and Planetary Probes, 1958–2000*. SP-2002–4524. Washington, D.C.: NASA, 2002.

Sloop, John L. *Liquid Hydrogen as a Propulsion Fuel, 1945–1959*. SP-4404. Washington, D.C.: NASA, 1978.

SMC. *See* Space and Missile Systems Center.

"Solid-Fueled Rocket Nears Crucial Test." *Business Week*, June 3, 1961, 42–49.

Space and Missile Systems Center (SMC), History Office, U.S. Air Force. "U.S. Air Force Satellite Launches." <www.losangeles.af.mil/SMC/HO/Slfact.htm>, accessed September 6, 2002. (Copy will be donated to NHRC.)

Space and Missile Systems Center (SMC), Office of Public Affairs, U.S. Air Force. "Pegasus Launch Vehicle." <www.te.plk.af.mil/factsheet/pegfact.html>, accessed December 11, 2002. (Copy will be donated to NHRC.)

"Space Systems Summaries: Titan Launch Vehicle History and Family." *Astronautics and Aeronautics*, June 1975, 76–77.

Spires, David N. *Beyond Horizons: A Half Century of Air Force Space Leadership*. Peterson Air Force Base, Colo.: Air Force Space Command, 1997.

"Stacksat (P87–2): POGS, TEX, and SCE." <samadhi.jpl.nasa.gov/msl/QuickLooks/stacksatQL.html>.

Stambler, Irwin. "Centaur." *Space/Aeronautics*, October 1963, 70–75.

———. "Simplicity boosts Able-Star to reliability record." *Space/Aeronautics*, August 1961, 59, 63–64.

Stangeland, M. L. "Joe." "Turbopumps for Liquid Rocket Engines." *Threshold: An Engineering Journal of Power Technology* (Rocketdyne), no. 3 (Summer 1988): 34–42.

Stehling, Kurt R. "Aspects of Vanguard Propulsion." *Astronautics*, January 1958, 44–47, 68.

———. *Project Vanguard*. Garden City, N.Y.: Doubleday, 1961.

Steier, Henry P. "What Guides the Vanguard." *Missiles and Rockets*, November 1959, 70–72.

Stiff, R. C. "Storable Liquid Rockets." Paper 67–977 presented at the AIAA 4th Annual Meeting and Technical Display, Anaheim, Calif., October 23–27, 1967.

Stone, Irving. "U.S. Schedules Additional Samos Launches." *Aviation Week*, October 17, 1960, 28.

Stuhlinger, Ernst, and Frederick I. Ordway III. *Wernher von Braun, Crusader for Space: A Biographical Memoir*. Malabar, Fla.: Krieger, 1994.

Stumpf, David K. *Titan II: A History of a Cold War Missile Program*. Fayetteville, Ark.: University of Arkansas Press, 2000.

Sutton, E[rnie] S. "From Polymers to Propellants to Rockets—A History of Thiokol." Paper AIAA-99-2929, presented at the 35th AIAA/ASME/SAE/ASEE Joint Propulsion Conference and Exhibit, Los Angeles, June 20–24, 1999.

———. *How a Tiny Laboratory in Kansas City Grew into a Giant Corporation: A History of Thiokol and Rockets, 1926–1996*. Chadds Ford, Pa.: privately printed, 1997.

Sutton, George P. *History of Liquid Propellant Rocket Engines*. Reston, Va.: AIAA, 2006.

——. *Rocket Propulsion Elements: An Introduction to the Engineering of Rockets*. 6th ed. New York: Wiley, 1992.

Swanson, Glen E., ed. *"Before This Decade Is Out...": Personal Reflections on the Apollo Program*. SP-4223. Washington, D.C.: NASA, 1999.

Swenson, Loyd S., Jr., James M. Grimwood, and Charles C. Alexander. *This New Ocean: A History of Project Mercury*. SP-4201. Washington, D.C.: NASA, 1998.

Taylor, Hal. "Atlas Launch Vehicle to Be Uprated." *Missiles and Rockets*, May 17, 1965, 14.

Taylor, John W. R., ed. *Jane's All the World's Aircraft, 1977–1978*. London: Jane's Yearbooks, 1978.

Thackwell, H. L., Jr. "Status of United States Large Solid Rocket Programs." Paper AIAA-65–422, presented at AIAA Second Annual Meeting, San Francisco, July 26–29, 1965.

Thiokol Chemical Corporation. *Rocket Propulsion Data*. 3rd ed. Bristol, Pa.: Thiokol, 1961.

——. "Thiokol's Solid Propellant Upper and Lower Stages." <members.aol.com/SLVehicles7/United_States_2/motors.htm>.

Thomson, Allen. "Titan Launch Dispenser—New Factoids." <satobs.org/seesat/Dec-1998/0032.html>.

"Titan IV." <www.au.af.mil/database/projects/ay1996/acsc/96–004/hardware/docs/titaniv.htm>, accessed July 8, 2003.

"Titan Launch History Summary." <www.aero.org/publications/crosslink/winter2003/07_table.html>.

"Titanium Tanks Machined for Titan 3 Transtage." *Aviation Week & Space Technology*, May 11, 1964, 69.

Tomayko, James E. *Computers in Spaceflight: The NASA Experience*. CR-182505. Washington, D.C.: NASA, 1988. <www.hq.nasa.gov/office/pao/History/computers/Compspace.html>.

——. *Computers Take Flight: A History of NASA's Pioneering Digital Fly-By-Wire Project*. SP-2000–4224. Washington, D.C.: NASA, 2000.

Tucker, Joel E. "History of the RL10 Upper-Stage Rocket Engine." In Doyle, *Liquid Rocket Engine Development*, 123–51.

Turco, J. A. "Evolution of Titan Family Guidance and Control Systems." In *Guidance and Control 1982*, edited by Robert D. Culp, Edward J. Bauman, and W. E. Dorroh Jr., 161–62, supplemented by AAS microfiche 38. San Diego: Univelt, 1982.

United States Air Force (USAF). "Defense Meteorological Satellite Program." <www.af.mil/factsheets/factsheet.asp?fsID=94>.

U.S. Congress. House. "Centaur Launch Vehicle Development Program." Report of the Committee on Science and Astronautics. 87th Cong., 2nd sess., July 2, 1962. H. Rep. 1959.

——. "Centaur Program." Hearings before the Subcommittee on Space Sciences of the Committee on Science and Astronautics. 87th Cong., 2nd sess., May 15 and 18, 1962.

——. "Project Ranger." Report of the Subcommittee on NASA Oversight of the Committee on Science and Astronautics. 88th Cong., 2nd sess., 1964. H. Rep. 1487.

———. "Project Vanguard, a Scientific Earth Satellite Program for the International Geophysical Year." Report to the Committee on Appropriations, by Surveys and Investigations staff. 86th Cong., 1st sess., April 14, 1959. Copy in General Collection, folder OV-106120–01, "Vanguard II Launch," NASM.

———. "Review of Recent Launch Failures." Hearings before the Subcommittee on NASA Oversight of the Committee on Science and Astronautics. 92nd Cong., 1st sess., June 15–17, 1971.

U.S. Congress. Senate. "Summary of the Problems Encountered in the Second Flight of the Saturn V Launch Vehicle." Hearing before the Committee on Aeronautical and Space Sciences. 90th Cong., 2nd sess., April 22, 1968.

U.S. President. Executive Office. National Aeronautics and Space Council. *Aeronautics and Space Report of the President* . . . Washington, D.C.: NASC, 1970–. Published annually for years from 1969, except 1989–1990 volume, which bridged a shift from calendar-year to fiscal-year reporting. Thereafter, annually by fiscal year. For the year 1970 and since 1973, prepared by NASA instead of the National Aeronautics and Space Council.

———. *United States Aeronautics and Space Activities* . . .: *Report to the Congress from the President of the United States*. Washington, D.C.: GPO, 1959–69. Published annually for years 1958–68 by the National Aeronautics and Space Council.

"UTC Prepares for Shuttle Booster Role." *Aviation Week & Space Technology*, January 3, 1977, 15.

Van Nimmen, Jane, and Leonard C. Bruno, with Robert L. Rosholt. *NASA Historical Data Book*. Vol. 1, *NASA Resources 1958–1968*. SP-4012. Washington, D.C.: NASA, 1988.

Vartabedian, Ralph. "NASA Still Vexed by Foam Woes." *Los Angeles Times*, August 16, 2003.

Verger, Ferdinand, Isabelle Sourbès-Verger, and Raymond Ghirardi. *The Cambridge Encyclopedia of Space: Missions, Applications and Exploration*. Cambridge: Cambridge University Press, 2003.

Vicenti, Walter G. *What Engineers Know and How They Know It: Analytic Studies from Aeronautical History*. Baltimore: Johns Hopkins University Press, 1990.

Wade, David Ian. "The Delta Family." *Spaceflight* 38 (November 1996): 373–76.

Wade, Mark. "Blue Scout 1." <www.astronautix.com/lvs/blucout1.htm>.

———. "Blue Scout 2." <www.astronautix.com/lvs/blucout2.htm>.

———. "Blue Scout Junior." <www.astronautix.com/lvs/bluunior.htm>.

———. "HS 393." <www.astronautix.com/craft/hs393.htm>.

———. "Star 20A." <www.astronautix.com/engines/star20a.htm>.

———. "Star 37S." <www.astronautix.com/engines/star37s.htm>.

———. "Thor Able-Star." <www.astronautix.com/lvs/theostar.htm>.

Walker, Chuck. *Atlas: The Ultimate Weapon*. With Joel Powell. Burlington, Ont.: Apogee, 2005.

Wallace, Harold D., Jr. *Wallops Station and the Creation of an American Space Program*. SP-4311. Washington, D.C.: NASA, 1997.

Wallace, Lane E. *Dreams, Hopes, Realities: NASA's Goddard Space Flight Center; The First Forty Years*. SP-4312. Washington, D.C.: NASA, 1999.

———. *Flights of Discovery: 50 Years at the NASA Dryden Flight Research Center.* SP-4309. Washington, D.C.: NASA, 1996.

Waring, Stephen P. "The 'Challenger' Accident and Anachronism: The Rogers Commission and NASA's Marshall Space Flight Center." Paper presented at the National Council on Public History Fifteenth Annual Conference, April 22, 1993.

Westrum, Ron. *Sidewinder: Creative Missile Development at China Lake.* Annapolis, Md.: Naval Institute Press, 1999.

Whalen, David J. *The Origins of Satellite Communications, 1945–1965.* Washington, D.C.: Smithsonian Institution Press, 2002.

"William R. Schindler, NASA Rocket Engineer." *Washington Post,* January 29, 1992.

Williamson, Ray A. "The Biggest of Them All: Reconsidering the Saturn V." In Launius and Jenkins, *To Reach the High Frontier,* 301–33.

Wilson, Andrew. "Burner 29–Boeing's Small Upper Stage." *Spaceflight* 22 (May 1980): 210–13.

———. "Scout—NASA's Small Satellite Launcher." *Spaceflight* 21 (November 1979): 446–59.

Wilson, Brian A. "The History of Composite Motor Case Design." Paper AIAA-93–1782 (a series of slides rather than an actual paper), presented at the 29th Joint Propulsion Conference and Exhibit, Monterey, Calif., June 28–30, 1993.

Winter, Frank H., and George S. James. "Highlights of 50 Years of Aerojet, A Pioneering American Rocket Company, 1942–1992." *Acta Astronautica* 35, nos. 9–11 (1995): 677–98.

Winter, Frank H., and Frederick I. Ordway III. "Pioneering Commercial Rocketry in the United States of America, Reaction Motors, Inc., 1941–1958, Part 2: Projects." *Journal of the British Interplanetary Society* 38 (1985): 155–68.

Wiswell, Robert L., and Michael T. Huggins. "Launch Vehicle & Upper Stage Liquid Propulsion at the Astronautics Laboratory (AFSC)—A Historical Summary." Paper AIAA-1990–1839, presented at the AIAA/SAE/ASME/ASEE 26th Joint Propulsion Conference, Orlando, July 16–18, 1990.

Yaffee, Michael. "Bell Adapts Hustler Rocket Engine for Varied Missions." *Aviation Week,* November 21, 1960, 51, 53, 57, 59, 61.

Yang, Vigor, and William E. Anderson, eds. *Liquid Rocket Engine Combustion Instability.* Washington, D.C.: AIAA, 1995.

Zea, Luis. "Delta's Dawn: The Making of a Rocket." *Final Frontier,* February–March 1995, 46–50.

Zimmerman, C. A., J. Linsk, and G. J. Grunwald. "Solid Rocket Technology for the Eighties." IAF paper 81–353, presented at the 32nd Congress of the International Astronautical Federation, Rome, September 6–12, 1981.

Index

Page numbers for illustrations and their captions are italicized. The letter t *following a page number denotes a table.*

ANS 1. *See* Netherlands Astronomical Satellite

Antares I/IA, 130, 132, 137, 139, 142, 144

Antares II/IIA, 137, 153

Antares III/IIIA, 77, 152–53, 250–51

Apollo missions. *See* Apollo program: flights under

Apollo program, 2, 4, 70, 92, 112, 136, 166, 170–77, 203–6; achievement of goal, 215, 265, 322; Apollo fire, 190, 206; command and service module (CSM), 191–92, 202, 204, 208–10, 213, 215, 374n103; flights under, 188–92, 209–17; lunar module (LM), 209, 213–15; naming of, 367n3; savings from all-up testing, 174, 369n31. *See also* lunar module adapter panels; Mueller, George E.; Phillips, Samuel C.; Saturn launch vehicle

Applications Technology Satellites (ATS), 93

Aquacade, 94

ARC. *See* Ames Research Center; Atlantic Research Corporation

Ares I/V, 2, 323–24

Ariane launch vehicle: as competition for U.S. vehicles, 119, 120, 256

Arma Corporation, 234, 329

Army Air Corps/Forces, 14

Army, U.S.: Army Ballistic Missile Agency (ABMA) of, 72, 159–63, 166; Ordnance Corps/Department of, 129; and propellant tanks for the Saturn program, 160; Signal Corps of, 84. *See also* Army Air Corps/Forces

Arnold Engineering Development Center (AEDC), 43, 56, 167, 251, 292; altitude testing at, 28, 55, 109, 132, 227, 232–33, 252

ARPA. *See* Advanced Research Projects Agency

Athena-H reentry test vehicle: and Castor IV, 353n80

Atlantic Missile Range (AMR), 144–145. *See also* Cape Canaveral

Atlantic Research Corporation (ARC): and propellant development, 141, 327, 331, 334; rockets/motors of, 18, 32

Atlantis, 298

Atlas-Able launch vehicle, 84–86, 354nn7–8

Atlas-Agena launch vehicle, 52, 83, 87–94, *90*, 151; payload weights, 54, 56; size compared with Saturn V, 159; success rates, 91–94

Atlas-Centaur launch vehicle, 83, 94–121, *114*, *117*; characteristics, 118; as commercial vehicle, 119–20; flight tests of, 102–4; number of components in, 121; payload capability of, 112, 120; size compared with Saturn V, 159; success rate of, 118, 124

Atlas launch vehicle, 83–126; characteristics, 91, 93, 359n72; contributions of, 83; engines for, 121–22, 355n15; models, 91–93, 118, 120, 121–24, 264, 322, 359n72, 359n76; modifications, 91, 93; multiple configurations of, 83; payloads, 93, 112, 120, 257; success rates, 91, 118; and Vega stage, 97. *See also* Atlas-Able launch vehicle; Atlas-Agena launch vehicle; Atlas-Centaur launch vehicle; LV-3C; Mercury-Atlas; Project Score; SLV-3A (space launch vehicle); SLV-3C; SLV-3D

Atlas missile, 16, 158; converted to space-launch vehicle, 83; failures, 324; numbers of contractors for, 321. *See also* Mercury-Atlas

ATS. *See* Applications Technology Satellites

Atwood, John Leland, 204–6

Automatic Determination and Dissemination of Just Updated Steering Terms (AD-DJUST), 115–16, 244

AVCO, 44

Baffles: anti-slosh, 50, 52, 112, 167, 199, 203, 286–87, 295, 325; anti-vortex, 274, 286–87; on injectors, 74, 89, 164, 196–97, 233–34, 270, 333; in a turbopump, 274

Ball Aerospace Division, 252

Ballistic Missile Division (BMD), 42, 48, 84, 97, 139. *See also* Ballistic Systems Division; Space Systems Division; Western Development Division

Ballistic Missile Office (BMO), 253

Ballistic Systems Division (BSD), 61, 122, 175, 243; replaced part of Ballistic Missile Division, 394. *See also* Space and Missile Systems Organization

Bartley, Charles, 318, 320, 326; and Grand Central Rocket Company, 25, 320

Battin, Richard H., 329

Bell Aircraft Corporaton, 53–56. *See also* Bell Aerospace Division of Textron

Bell Aerospace Division of Textron, 53, 58–59. *See also* Bell Aircraft Corporation

NACA. *See* National Advisory Committee for Aeronautics
NASA. *See* National Aeronautics and Space Administration
NASA-Air Force Scout Coordinating Committee, 139
National Advisory Committee for Aeronautics (NACA), 23, 95, 129, 171; awards H-1 contract, 352n73
National Aeronautics and Space Administration (NASA), 86, 91, 93, *205* (map of facilities), 252, 277; and Ares I/V, 2, 323–24; and Blue Scout, 138–39; and Centaur, 97, 99, 105, 107, 244; and commercial satellites, 254; and Delta launch vehicle, 62–63, 72, 78; and H-1 contract date, 352n73; and Large Segmented Solid Rocket Motor Program, 278, 280; management/organizational changes in, 171–77; Office of Manned Space Flight (OMSF) in, 171–73; and pogo task force, 211–12; and Program Evaluation and Review Technique, 176; and Project Mona, 43; project planning approach of, 266; and Saturn launch vehicles, 162, 167, 180, 184, 193–94, 202–6, 322; and Scout, 130–31, 137, 145–47, 152, 198; and space shuttle, 266, 268, 270, 284, 286, 289, 291, 294; and Titan-Centaur, 244; and Vanguard, 37, 344n70; and various satellites, 64, 120. *See also* Apollo program; National Advisory Committee for Aeronautics; Project Gemini; Project Mercury; *names of field organizations and officials*
National Oceanic and Atmospheric Administration (NOAA), 61
National Parachute Test Range, 284
National Space Technology Laboratories, 271, 273. *See also* Mississippi Test Facility; Stennis Space Center
Navigation Development System (NDS) spacecraft, 123
Navaho missile, 22, 268; legacies from, 7–8, 327
Naval Air Rocket Test Station, 23–24
Naval Ocean Surveillance System (NOSS) satellites, 124, 261–62
Naval Ordnance Test Station (NOTS), 139; sled track of, 227; and spherical motors,

141; technical contributions of, 23, 136–37, 139. *See also* Notsnik
Naval Research Laboratory (NRL), 10–11, 61, 146; and Vanguard launch vehicle, 17–18, 19–21
Navigation: defined, 208
Navigation Development System (NDS) satellites, 123
Navigation Technology Satellites (NTS) 1, 122–23
Navstar Global Positioning System (GPS) navigational satellites, 78, 80–81, 122–23, 155, 353n83
Navy, U.S., 137, 142; Bureau of Aeronautics (BuAer) in, 24, 140; Bureau of Ordnance in, 130, 140; Bureau of Weapons in, 66; furtherance of rocketry by, 24; Office of Naval Research in, 20; Special Projects Office, 176; use of Blue Scouts, 144. *See also* Polaris; Vanguard launch vehicle; Viking sounding rocket
NDS. *See* Navigation Development System satellites
Negative feedback, 16, 339n18
Netherlands Astronomical Satellite (ANS 1), 152
Neu, Edward A., Jr., 14; "spaghetti" construction developed by, 326–27
Neumann, John von. *See* von Neumann, John
Nielson Engineering and Research, 156
Nimbus meteorological satellites, 55–57
Nitric acid, 12, 23, 321; inhibited red and white fuming (IWFNA and IRFNA), 23–25, 48, 53–54, 59, 63–64; red fuming (RFNA), 23, 341n39; white fuming (WFNA), 23
Nitrocellulose. *See under* propellants: double-base; composite double-base; composite modified double-base; double base
Nitrogen tetroxide: as oxidizer, 261, 295; for thrust vector control, 23, 59, 76, 78, 185, 207, 226, 231, 235, 257
Nitroglycerin. *See under* propellants: double-base; composite double-base; composite modified double-base; and double-base
Nixon, Richard M., 265; on shuttle's reduced costs, 267
NOAA. *See* National Oceanic and Atmospheric Administration

J. D. Hunley is a retired NASA historian. He has written *The Development of Propulsion Technology for U.S. Space-Launch Vehicles, 1926–1991* (2007) and numerous articles in the field of aerospace history. He has also edited more than a dozen books and monographs in that field, including *The Birth of NASA: The Diary of T. Keith Glennan* (1993).